Advance in Digital Signal, Image and Video Processing

Advance in Digital Signal, Image and Video Processing

Editors

Przemysław Falkowski-Gilski
Tadeus Uhl
Zbigniew Łubniewski

MDPI

Basel • Beijing • Wuhan • Barcelona • Belgrade • Novi Sad • Cluj • Manchester

Editors
Przemysław Falkowski-Gilski
Gdansk University
of Technology
Gdansk
Poland

Tadeus Uhl
Maritime University
of Szczecin
Szczecin
Poland

Zbigniew Łubniewski
Gdansk University
of Technology
Gdansk
Poland

Editorial Office
MDPI
St. Alban-Anlage 66
4052 Basel, Switzerland

This is a reprint of articles from the Special Issue published online in the open access journal *Applied Sciences* (ISSN 2076-3417) (available at: https://www.mdpi.com/journal/applsci/special_issues/Digital_Signal_Image_Video_Processing).

For citation purposes, cite each article independently as indicated on the article page online and as indicated below:

Lastname, A.A.; Lastname, B.B. Article Title. *Journal Name* **Year**, *Volume Number*, Page Range.

ISBN 978-3-0365-8454-6 (Hbk)
ISBN 978-3-0365-8455-3 (PDF)
doi.org/10.3390/books978-3-0365-8455-3

Contents

About the Editors

Przemysław Falkowski-Gilski

Przemysław Falkowski-Gilski, Ph.D. is currently working as an Assistant Professor at the Faculty of Electronics, Telecommunications and Informatics, Gdansk University of Technology. In 2018, he received the title of Doctor of Technical Sciences with distinction, discipline Telecommunications. His research interests include: broadcasting and electronic media, coding and compression, mobile and multimedia services, positioning technologies, audio-video signal processing, wireless communication, as well as network quality evaluation.

Tadeus Uhl

Tadeus Uhl, Ph.D., D.Sc., Prof., is currently working as a Full Professor at the Faculty of Economics and Transport Engineering, Maritime University of Szczecin. In 1982 he received the title Ph.D. from Gdansk University of Technology, Poland, and in 1990 the title D.Sc. from Dortmund University, Germany. His research interests include: content compression and feature selection, Internet technologies and packet data transmission in IP networks, quality evaluation, as well as video coding and streaming services.

Zbigniew Łubniewski

Zbigniew Łubniewski, Ph.D., D.Sc., is currently working as an Associate Professor and Head of the Department of Geoinformatics at the Faculty of Electronics, Telecommunications and Informatics, Gdansk University of Technology. His research interests include: underwater acoustics, geographic information systems (GIS), underwater object reconstruction and seabed classification, as well as remote sensing and satellite communication technologies. He also acts as a coordinator of an intercollegiate study on Space and Satellite Technologies.

Preface

It is assumed that high performance quality in signal transmissions will lead to great acceptance and usability. However, with the proliferation of numerous digital systems and services, including audio and video processing, low-quality or best-effort services have gained enormous popularity. This is particularly the case for Internet and mobile technologies, for example, content storing and management, multimedia consumption, and digital maps and images. It is also clearly visible in the growth of the number of users and available applications. Moreover, it shows that the relationship between performance, quality and acceptance is not fully recognized.

The term "quality" can be understood in many different ways. Engineers perceive the term as referring to quality of service (QoS), which is a synonym for network performance and reliability. However, quality can also be defined subjectively. Quality of experience (QoE), which involves the process of comparing perceptual events with a known reference, is focused on defining the characteristics of media transmission systems or services and their acceptance by customers. As has been observed with the proliferation of desktop and mobile platforms, the number of multimedia services and highly capable consumer devices continues to grow. This forces a broader analysis of numerous practical aspects. This issue aims to describe interdisciplinary advances in digital signal, image, and video processing.

In this Special Issue, we invite the international scientific community to publish works highlighting recent advances in digital signal, image, and video processing. Of the total 24 submissions, the acceptance rate was 58 percent. We hope that the 14 published articles will make for pleasant and inspiring reading.

Finally, we would like to take this opportunity to express our most profound appreciation to the MDPI Book staff, the editorial team of the Applied Sciences journal, especially Ms. Christine Zhang, the Assistant Editor of this Special Issue, the talented authors who submitted their work, as well as all hardworking and professional reviewers.

Przemysław Falkowski-Gilski, Tadeus Uhl, and Zbigniew Łubniewski
Editors

applied sciences

MDPI

Editorial

Special Issue on Advance in Digital Signal, Image and Video Processing

Przemysław Falkowski-Gilski [1,*], Tadeus Uhl [2] and Zbigniew Łubniewski [1]

[1] Faculty of Electronics, Telecommunications and Informatics, Gdansk University of Technology, Narutowicza St. 11/12, 80-233 Gdansk, Poland; lubniew@eti.pg.edu.pl
[2] Faculty of Economics and Transport Engineering, Maritime University of Szczecin, Waly Chrobrego St. 1-2, 70-500 Szczecin, Poland; t.uhl@pm.szczecin.pl
* Correspondence: przemyslaw.falkowski@eti.pg.edu.pl

It is assumed that high performance and quality, especially in the case of signal transmission, will lead to great acceptance and usability. However, with the outbreak of numerous digital signal, image and video processing systems and services, low-quality or best-effort services have gained enormous popularity. Moreover, it shows that the relationship between performance, quality and acceptance is not fully recognized. A method for identifying the scrambling types of linear block codes (LBC) by combining correlation features and convolution long short-term memory (LSTM) neural networks (NNs) is proposed in [1]. In order to improve the bit error performance while maintaining low decoding steps, the authors of [2] introduce a neural network subcode that can achieve optimal decoding performance and combine it with the traditional fast successive-cancellation (SC) decoding algorithm.

During imaging, each infrared focal plane linear array scan detection unit determines a row of pixels in the image output. Correcting nonuniformity in high-resolution images without destroying delicate details is challenging. In this paper [3], a single-frame-based nonuniformity correction algorithm is proposed. In another study [4], the requirements for image dehazing methods have been put forward in order to solve the problems associated with nighttime and daytime dehazing, whereas in [5], authors introduce an integrity preservation analysis scheme of video contents working on the blockchain systems.

As we know, content streaming represents a significant part of Internet traffic. The Next-Generation Network (NGN) architecture was proposed for delivering various multimedia services with guaranteed quality. For this reason, the elements of the IP Multimedia Subsystem (IMS) concept, an important part of 4G/5G/6G mobile networks, are used in its service stratum [6]. During the playback, a video player monitors network throughput and dynamically selects the best video quality in the given network conditions. In this work [7], authors improved the throughput estimation using prediction produced by LSTM artificial neural networks (ANNs). Preserving visual quality is a major constraint for any algorithm in image and video processing applications. Advanced Video Coding (AVC) and High-Efficiency Video Coding (HEVC) are the extensively used video coding standards for various video processing applications nowadays [8]. The results of both subjective and objective quality assessments of H.264-, H.265-, and VP9-encoded video are presented in [9].

Even with the aid of visual information, maintaining stable and reliable working conditions is a matter of vital importance for various companies, especially those involving heavy machinery. As an example, oil sheen on the water surface can indicate a source of hydrocarbon in underlying subaquatic sediments [10]. In the machining process, it is important to predict both tool cost and life, and to reduce the equipment downtime. In [11], a new model is proposed for the effective classification of both serviceable and worn cutting edges. Still, due to human exhaustion, as well as unpredicted hazards and dangerous

Citation: Falkowski-Gilski, P.; Uhl, T.; Łubniewski, Z. Special Issue on Advance in Digital Signal, Image and Video Processing. *Appl. Sci.* **2023**, *13*, 7642. https://doi.org/10.3390/app13137642

Received: 22 June 2023
Accepted: 26 June 2023
Published: 28 June 2023

situations, the personnel have to take actions and wisely plan each move. This paper [12] presents a human–computer interaction (HCI)-based system that uses a concentration level measurement function to increase the safety of machine and equipment operators. The goal of another study [13] was to analyze and present an in-depth account of the implementation of mixed agile-oriented tools and practices.

As has been shown, quality evaluation is focused on defining the characteristics of media transmission systems or services and their acceptance by customers, and it can be understood in many different ways. With the outbreak of desktop and mobile platforms, the number of multimedia services and highly capable consumer devices continues to grow. However, in order to safeguard one's privacy while accessing the Internet, it is crucial to have an antivirus program installed on the device [14]. Still, the topic of digital signal, image and video processing is an important and practical subject area. Future studies will surely require a broader analysis of numerous scientific and industrial aspects.

Acknowledgments: Thanks to all the authors and peer reviewers for their valuable contributions to the Special Issue 'Advance in Digital Signal, Image and Video Processing'.

Conflicts of Interest: The authors declare no conflict of interest.

References

1. Tan, J.; Zhang, L.; Zhong, Z. Distinction of Scrambled Linear Block Codes Based on Extraction of Correlation Features. *Appl. Sci.* **2022**, *12*, 11305. [CrossRef]
2. Liu, H.; Zhang, L.; Yan, W.; Ling, Q. Neural-Network-Assisted Polar Code Decoding Schemes. *Appl. Sci.* **2022**, *12*, 12700. [CrossRef]
3. Li, B.; Chen, W.; Zhang, Y. A Nonuniformity Correction Method Based on 1D Guided Filtering and Linear Fitting for High-Resolution Infrared Scan Images. *Appl. Sci.* **2023**, *13*, 3890. [CrossRef]
4. Tang, Y.; Xiang, Y.; Chen, G. A Nighttime and Daytime Single-Image Dehazing Method. *Appl. Sci.* **2023**, *13*, 255. [CrossRef]
5. Lee, W.Y.; Choi, Y.-S. Reliable Integrity Preservation Analysis of Video Contents with Support of Blockchain Systems. *Appl. Sci.* **2022**, *12*, 10280. [CrossRef]
6. Kaczmarek, S.; Sac, M. Performance Evaluation of a Multidomain IMS/NGN Network Including Service and Transport Stratum. *Appl. Sci.* **2022**, *12*, 11643. [CrossRef]
7. Biernacki, A. Improving Streaming Video with Deep Learning-Based Network Throughput Prediction. *Appl. Sci.* **2022**, *12*, 10274. [CrossRef]
8. Christopher, P.R.; Sathasivam, S. Quality Assessment of Dual-Parallel Edge Deblocking Filter Architecture for HEVC/H.265. *Appl. Sci.* **2022**, *12*, 12952. [CrossRef]
9. Klink, J.; Brachmański, S.; Łuczyński, M. Assessment of the Quality of Video Sequences Performed by Viewers at Home and in the Laboratory. *Appl. Sci.* **2023**, *13*, 5025. [CrossRef]
10. Dong, J.; Sitler, K.; Scalia, J.; Ge, Y.; Bireta, P.; Sihota, N.; Hoelen, T.P.; Lowry, G.V. Application of Transfer Learning and Convolutional Neural Networks for Autonomous Oil Sheen Monitoring. *Appl. Sci.* **2022**, *12*, 8865. [CrossRef]
11. David, L.G.; Patra, R.K.; Falkowski-Gilski, P.; Divakarachari, P.B.; Antony Marcilin, L.J. Tool Wear Monitoring Using Improved Dragonfly Optimization Algorithm and Deep Belief Network. *Appl. Sci.* **2022**, *12*, 8130. [CrossRef]
12. Jagoda, J.; Woszczyński, M.; Polnik, B.; Falkowski-Gilski, P. HCI-Based Wireless System for Measuring the Concentration of Mining Machinery and Equipment Operators. *Appl. Sci.* **2023**, *13*, 5396. [CrossRef]
13. Weichbroth, P. A Case Study on Implementing Agile Techniques and Practices: Rationale, Benefits, Barriers and Business Implications for Hardware Development. *Appl. Sci.* **2022**, *12*, 8457. [CrossRef]
14. Samociuk, D. Antivirus Evasion Methods in Modern Operating Systems. *Appl. Sci.* **2023**, *13*, 5083. [CrossRef]

applied sciences

Article

Distinction of Scrambled Linear Block Codes Based on Extraction of Correlation Features

Jiyuan Tan, Limin Zhang and Zhaogen Zhong *

Department of Information Fusion, Naval Aviation University, Yantai 264001, China
* Correspondence: zhongzhaogen@163.com

Abstract: Aiming to solve the problem of the distinction of scrambled linear block codes, a method for identifying the scrambling types of linear block codes by combining correlation features and convolution long short-term memory neural networks is proposed in this paper. First, the cross-correlation characteristics of the scrambling sequence symbols are deduced, the partial autocorrelation function is constructed, the superiority of the partial autocorrelation function is determined by derivation, and the two are combined as the input correlation characteristics. A shallow network combining a convolutional neural network and LSTM is constructed; finally, the linear block code scrambled dataset is input into the network model, and the training and recognition test of the network is completed. The simulation results show that, compared with the traditional algorithm based on a multi-fractal spectrum, the proposed method can identify a synchronous scrambler, and the recognition accuracy is higher under a high bit error rate. Moreover, the method is suitable for classification under noise. The proposed method lays a foundation for future improvements in scrambler parameter identification.

Keywords: scrambled linear block codes; cross-correlation of symbols; partial autocorrelation function; convolutional long short-term memory neural networks

Citation: Tan, J.; Zhang, L.; Zhong, Z. Distinction of Scrambled Linear Block Codes Based on Extraction of Correlation Features. *Appl. Sci.* **2022**, *12*, 11305. https://doi.org/10.3390/app122111305

Academic Editors: Przemysław Falkowski-Gilski, Tadeus Uhl and Zbigniew Lubniewski

Received: 7 September 2022
Accepted: 7 November 2022
Published: 7 November 2022

1. Introduction

In communication systems, a long 0-run or 1-run often appears in transmitted binary sequences [1], and this affects the extraction of timing information in signals. To improve the balance of 0 and 1 symbols in transmitted signals, transmitted signals are scrambled to make it nearly completely random digital sequences [2]. Scrambling processing enhances the reliability and security of signal transmission, so it is widely used in satellite communications, spread spectrum communications, and cryptography [3–5]. In non-cooperative communication, the extraction of scrambled information must be descrambled first, so the blind identification technology of scrambling code parameters is of great significance to the extraction of non-cooperative information and the deciphering of ciphers [6,7].

A scrambling code is divided into a synchronous scrambling code and a self-synchronous scrambling code [6,7]. Among these, the parameter identification of the synchronous scrambling code is the identification of a generator polynomial and the initial state, and the identification of the self-synchronous scrambling code is the identification of a generator polynomial. In recent years, scrambling code parameter identification technology has been widely studied [6,7]. The methods of scrambling parameter identification mainly focus on three types of situations: source unbalance [8–16], source balance (coding scrambling) [17–21], and the estimation of scrambled sequences in a direct-sequence spread spectrum [22–25].

In an actual communication system, signal transmission is often channel-coded, so intercepted scrambled data are usually a source-unbalanced sequence that has been channel-coded and then scrambled. A method for reconstructing the synchronous scrambling code generator polynomial based on a dual code and double search was proposed by Liu [17], in which one must know the scrambling type; in addition, the detection performance index always meeting the requirements cannot be maintained. In [18], a more robust sparse double-search algorithm, which improved the recognition performance of the algorithm under a low signal-to-noise ratio, was put forward. However, knowledge of the constraint length of the encoding is required. In [19], a polynomial identification based on run-length statistics was proposed to solve the self-synchronous scrambling of linear block codes. This method does not require prior information, and the influence of bit errors and channel noise is also not considered. The cost function of a generator polynomial based on the known check vector of the dual code was constructed in [20], and the reconstruction of the generator polynomial under the noisy condition of the convolutional code was completed. The third-order correlation of a synchronously scrambled m-sequence was used in [21] to complete the reconstruction of a generator polynomial under error conditions. This method does not require the information of an a priori vector, nor does it consider the influence of channel noise, but it requires the known coding of the constraint length.

From the above analyses, all the known scrambling code parameter identification methods are based on the known channel coding and scrambling types, but the scrambling code data intercepted by non-cooperative communication parties are unknown, so it is necessary to analyze whether the outgoing signal is scrambled and what scrambling method is adopted that can further identify the parameters. Based on this, a multitype spectrum classification method, which solved the binary classification of linear block codes and linear block code self-synchronous scrambling under error conditions, was proposed in [26]. However, since the biased difference between linear self-synchronous scrambling and linear synchronous scrambling is very small, the distinction between the two leads to poor classification performance. At the same time, the influence of channel noise and recognition performance under high error conditions must be further improved. Reference [21] constructed a rank difference matrix, and when different scrambling methods were used, there was a periodic loss of rank at the integer multiples of the code length, which was used as a basis to judge the scrambling method. Reference [27] proposed to apply neural networks in order to extract rank features for the purpose of classification. However, the structure and parameters of the neural network were not given. Meanwhile, when the BER is greater than 0.01, the matrix tends to be a random matrix, the rank feature is not obvious, and the recognition rate tends to be 0. To summarize, due to the need for accurate scrambling-type information, the existing algorithms cannot achieve the full blind identification of linear block code scrambling. In recent years, deep learning methods have achieved good results in signal recognition [28,29], modulation recognition [30,31], malicious file detection [32,33], and other fields [34,35]. At the same time, deep-learning-based methods avoid manual feature extraction and have high practical value.

To solve the above-mentioned problems and the difficulty of manual feature extraction, under the conditions of a high bit error rate (BER) and a low signal-to-noise ratio (SNR), in this paper, we propose a linear block code scrambling-type identification based on correlation feature extraction. The contributions of this paper are summarized as follows:

(1) We deduce the symbol cross-correlation of the scrambling sequence and construct the partial autocorrelation function, both of which can reflect the correlation characteristics of different scrambling methods.

(2) We construct a shallow network consisting of convolutional neural network (CNN) and long short-term memory (LSTM) neural network models, which can accomplish the identification of scrambled linear block codes under a high BER and a low SNR.

The remainder of this paper is organized as follows: In Section 2, we summarize the principle and mathematical model of the scrambler. In Section 3, we analyze the cross-correlation characteristics of the symbols and construct a partial autocorrelation function. A correlation feature extraction network model is built in Section 4, and we analyze the training process of the network. In Section 5, the proposed method is evaluated using Monte Carlo simulations, and we compare its performance with that of the multi-fractal spectrum method. Finally, a discussion is provided in Section 6.

2. Scrambled Linear Block Code Model

The basis of the scrambling code is the linear-feedback shift register (LFSR), in which the self-synchronous scrambling code introduces an exclusive OR logic between the feedback logic output and the first-stage register, and the obtained result is used as the input of the register. The independent m-sequence generated by the LFSR is added to the information sequence, thereby generating the scrambling sequence. The self-synchronous scrambler and synchronous scrambler are shown in Figure 1a,b, respectively.

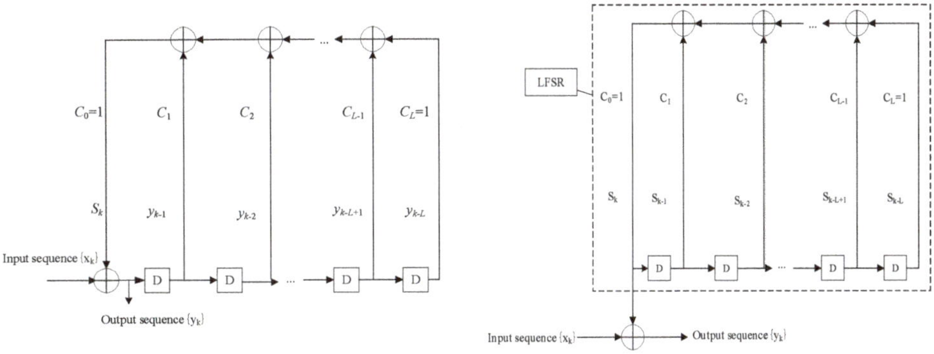

(a) (b)

Figure 1. Scrambler schematic. (**a**) Self-synchronous scrambler; (**b**) Synchronous scrambler.

An L-stage LFSR's feedback network is made up of L registers, such as the first-stage register and the second-stage register, and an L-th stage register from left to right. Each register can have two states, 0 and 1, and the operation $\oplus$ means the addition of modulo-2. When a shift pulse is added, the output content of the previous moment is fed back to the first-level register, the content of each level register is shifted to the next level, and the current output s_t is generated by operation. Therefore, after giving the initial values, $s_0, s_1, \ldots, s_{L-1}$, under the action of the shift pulse, the L-stage LFSR outputs a sequence $\{\mathbf{s}_t\}$, which satisfies the feedback logic

$$s_t = c_1 s_{t-1} \oplus c_2 s_{t-2} \oplus \cdots \oplus c_L s_{t-L}, \tag{1}$$

where $c_i \in GF(2), 1 \leq i \leq L$. This sequence is called the LFSR sequence, and its linearity is mainly reflected in its feedback logic being linear. Any consecutive L term in a sequence of the L-stage LFSR is called the state of the sequence. $(s_0, s_1, \ldots, s_L)$ is called the initial state [16].

The self-synchronous scrambling process can be expressed as

$$y_k = x_k \oplus \sum_{i=1}^{L} \oplus c_i y_{k-i}. \tag{2}$$

Additionally, the synchronous scrambling process can be expressed as

$$y_k = x_k \oplus \sum_{i=1}^{L} c_i s_{k-i}, \tag{3}$$

where $\sum \oplus$ means modulo-2 accumulation. $c_i \in GF(2)$, $i = 1, 2, \ldots, L$ is the feedback coefficient of the LFSR, $GF(2)$ represents the 2-element domain, and the value of c_i is 0 or 1, $c_0 = c_L = 1$. The generator polynomial of the scrambling code can be expressed as

$$f(x) = 1 + c_1 x + c_2 x^2 + \cdots + c_L x^L. \tag{4}$$

In actual communication system information transmission, most of the sources become unbiased sources through channel coding [17], so the signal model studied in this paper is shown in Figure 2 [17,18].

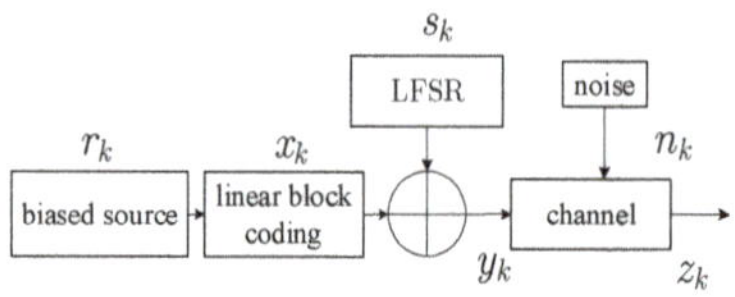

Figure 2. Signal model studied in the present paper.

The noise is Gaussian white noise with mean 0, variance σ^2, and signal amplitude A. The signal-to-noise ratio (SNR) is defined as

$$\mathrm{SNR} = 10 \cdot \log\left(A^2/2\sigma^2\right). \tag{5}$$

For the receiver, first, we intercept a sequence of code elements of a linear block code, and we only need to determine whether the sequence of the code elements is scrambled using LFSR. When scrambling is performed, it is further determined whether the code element sequence employs a self-synchronous scrambler as shown in Figure 1a or a synchronous scrambler as shown in Figure 1b. Further identification of the scrambling parameters is only possible when the above three types of code element sequences are distinguished. In the next section, we focus on how to effectively identify whether the intercepted information sequence is scrambled and what kind of scrambling method is used.

3. Analysis of Correlation Characteristics of Scrambled Linear Block Code

3.1. Cross-Correlation Characteristics of Symbols

It can be seen from Equation (2) that the k-th symbol (y_k) of the self-synchronous scrambling sequence correlates with the previous L symbol (y_{k-i}, $i = 1, 2, \ldots, L$) regardless of the state of the LFSR.

It can be seen from Equation (3) that the k-th symbol (y_k) of the synchronous scrambling sequence correlates with the first L states (s_{k-i}, $i = 1, 2, \ldots, L$) of the LFSR. The relationship between the kth symbol and the state of the LFSR is

$$y_{k-1} = x_{k-1} \oplus \sum_{i=1}^{L} \oplus c_i s_{k-i-1}. \tag{6}$$

Furthermore,

$$s_{k-1} = \sum_{i=1}^{L} \oplus c_i s_{k-i-1} = y_{k-1} \oplus x_{k-1}. \tag{7}$$

The state of the LFSR at each moment satisfies the following relationship [14]:

$$s_k = \sum_{i=1}^{L} c_i s_{k-i}. \tag{8}$$

Combining Equations (2) and (6)–(8), it can be deduced that the following correlation exists between the k-th symbol and the first L symbol:

$$\begin{aligned} y_k &= x_k \oplus s_k = x_k \oplus c_1 s_{k-1} \oplus c_2 s_{k-2} \oplus \cdots \oplus c_L s_{k-L} \\ &= x_k \oplus c_1 \cdot (x_{k-1} \oplus y_{k-1}) \oplus c_2 (x_{k-2} \oplus y_{k-2}) \oplus \cdots \oplus c_L (x_{k-L} \oplus y_{k-L}) \end{aligned} \tag{9}$$

However, when the linear block code is not scrambled, the preceding and following symbols have no correlation with the LFSR, and they are only related to the generator matrix of the linear block code. In summary, the linear block code scrambling correlation distribution is shown in Figure 3.

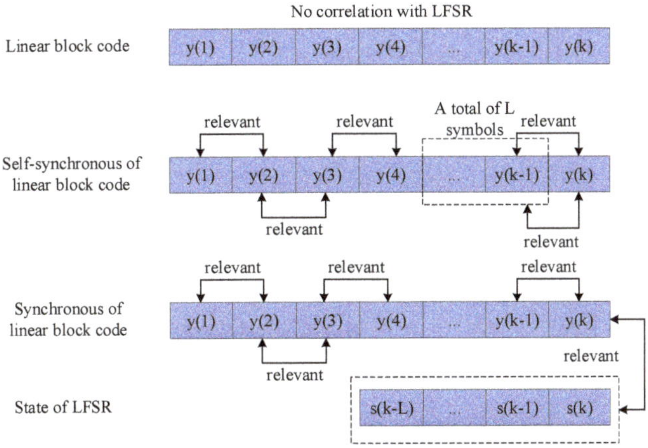

Figure 3. Symbol correlation analysis of linear block code scrambling.

It can be seen in Figure 3 that there is no correlation between the symbols of the linear block code and the LSFR; there is a correlation between the two symbols of the linear block code self-synchronous scrambling, and the symbols have nothing to do with the state of the LFSR; there is a correlation between the symbol and states of the first L LFSRs at the moment of the synchronous scrambling of the symbols of the linear block code, and, at the same time, this leads to a correlation between the symbols. It can be seen from the above analysis that scrambling causes a correlation between the preceding and following symbols. This correlation is related to the state of the LFSR (synchronous scrambling) or is independent (self-synchronous scrambling), so the intercepted symbol sequence can be used as a correlation feature that is used for linear block code scrambling classification.

3.2. Biased Autocorrelation Characteristics of Symbols

From the analysis in Section 3.1, there are correlation characteristics between the symbols scrambled by the linear block code, and they have relatively good correlation characteristics. Next, the biased autocorrelation function [36] of the symbol sequence is constructed as follows:

$$r'_{yy}(\tau) = \frac{1}{N}\sum_{k=0}^{N-\tau-1} Y_k Y_{k+\tau}, \tau \in [0, N-1]. \tag{10}$$

where $Y_k = 1 - 2y_k$.

The unbiased autocorrelation function is defined as

$$r_{yy}(\tau) = \frac{1}{N-\tau}\sum_{k=0}^{N-\tau-1} Y_k Y_{k+\tau}. \tag{11}$$

The relationship between the unbiased and biased autocorrelation functions is expressed as follows:

$$r'_{yy}(\tau) = \frac{N-\tau}{N} r_{yy}(\tau), \tag{12}$$

where $r_{yy}(\tau)$ is an unbiased estimate, and the following formula can be obtained:

$$E\left[r'_{yy}(\tau)\right] = \frac{N-\tau}{N} r_{yy}(\tau). \tag{13}$$

It can be seen from Equation (13) that $r'_{yy}(\tau)$ is a biased estimate but is asymptotically unbiased, and its offset is

$$B = \frac{\tau}{N} r_{yy}(\tau). \tag{14}$$

The variance of the estimator is

$$var\left[r'_{yy}(\tau)\right] = \left(\frac{N-\tau}{N}\right)^2 \cdot var\left[r_{yy}(\tau)\right] \tag{15}$$

It can be deduced that

$$var\left[r'_{yy}(\tau)\right] \le \frac{1}{N} \sum_{m=1+\tau-N}^{N-\tau-1} \left[r^2_{yy}(m) + r_{yy}(m+\tau) r_{yy}(m-\tau)\right] \tag{16}$$

When $N \to \infty$, $var\left[r'_{yy}(\tau)\right] \to 0$, and we have

$$var\left[r'_{yy}(\tau)\right] \le var\left[r_{yy}(\tau)\right]. \tag{17}$$

It can be seen from the above analysis that, although the biased autocorrelation function is biased, it is asymptotically consistent, and the variance in the estimator is smaller than that in the unbiased estimate.

To facilitate comparison and analyses, the biased autocorrelation function is normalized. Under the conditions that the BER is 0.1 and SNR = 6 dB, the normalized partial autocorrelation functions are shown in Figures 4 and 5, respectively.

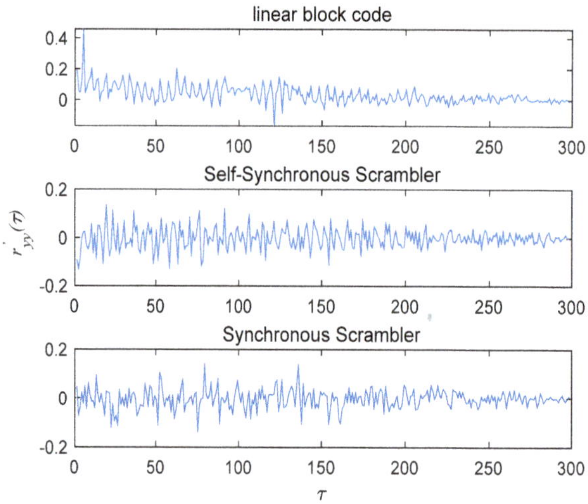

Figure 4. Normalized partial autocorrelation function diagram when BER = 0.1.

It can be seen in Figures 4 and 5 that the normalized partial autocorrelation functions of the different scrambling types have different peaks and periodic changes at each moment under the conditions of bit error and noise, so they can be used as another classification feature.

Since it is difficult to manually extract the above features, a method based on a convolutional LSTM neural network is established for feature extraction and training learning.

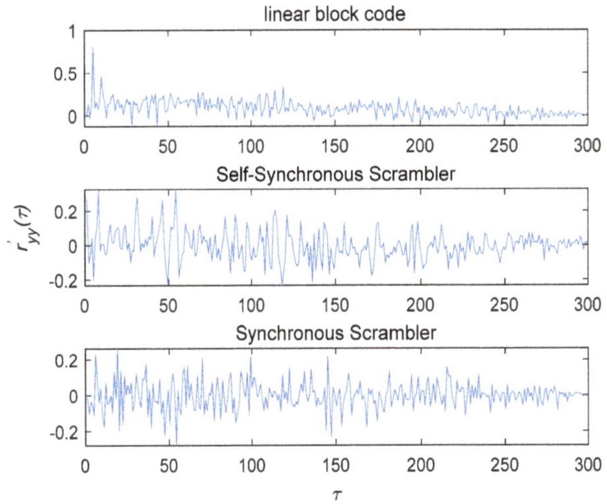

Figure 5. Normalized partial autocorrelation function diagram when SNR = 6 dB.

4. Scrambled Linear Block Code Identification Based on Correlation Features Extraction Network

4.1. Correlation Feature Extraction Network Model

In this paper, the received scrambling sequence features are divided into two lines: one line contains the original symbol sequence of the cross-correlation feature, and the other is the normalized partial autocorrelation function of the sequence; and the two form the $2 \times \beta$-dimensional matrix feature as the input of the network.

It can be seen from the analysis in Section 2 that the scrambling sequence has timing-related characteristics. As a special recurrent neural network, LSTM has achieved good results in processing timing-related information [37,38]. In the present study, CNN and LSTM are combined. The structure of the network, which is mainly composed of two convolutional layers, i.e., one LSTM layer and one fully connected layer, is shown in Figure 6. The activation function of the convolutional layer adopts RELU, the activation function of the LSTM layer adopts tanh, the final fully connected layer uses the normalized exponential function (SoftMax) for classification, and the output is a dimensional one-hot encoding form, corresponding to linear block codes, linear block code self-synchronous scrambling, and linear block code synchronous scrambling.

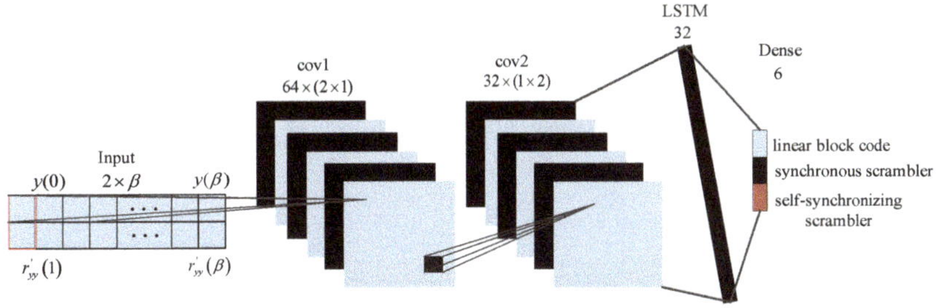

Figure 6. Correlation feature extraction network model diagram.

To prevent overfitting, we first introduce a Dropout layer after each layer of the network to randomly drop some neurons in order to reduce the overfitting phenomenon of the network. Moreover, the EarlyStopping function is used to detect the loss value and to stop training when it is not decreasing.

4.2. Network Training Process

The convolution kernel of convolutional layer 1 is 2×1. Since the input sample is a sequence of the code elements of the receiver signal and autocorrelation features, which form the real and imaginary parts of the input sample, respectively, the input sample is not the received sequence $z = [z(0), z(1), \ldots, z(k-1)]$. Therefore, it is necessary to first combine the real and imaginary parts of the scrambled sequence and to then extract the features based on the correlation.

As can be seen in Figure 3, when the 1×2-dimensional convolution kernel is used to extract the signal features of the scrambled sequences, the deep features obtained from their convolution must show different patterns due to the different correlations of the linear block code, the self-synchronization of the linear block code, and the synchronization of the linear block code. In fact, the process of the convolution of the received sequence using a 1×2-dimensional convolution kernel is similar to the calculation of its correlation features under time delay. However, for the self-synchronization of linear block code and the synchronization of linear block code, the correlation features of code elements separated by more than two code elements are not as obvious as those of two adjacent code elements, so convolutional kernels with lengths longer than 1×2 dimensions do not have this differentiation advantage, and they increase the parameters of network training and increase network complexity. Therefore, the 1×2-dimensional convolution kernel is used to convolve the layer 1 output features in order to obtain correlation features that are more consistent with the nature of scrambled signals.

The weight and bias training update of the correlation feature extraction network is mainly divided into two processes: forward propagation and back propagation. Among them, the forward propagation process is to use the input training samples in order to calculate the neuron activation value of the network, and the back propagation process is to perform a reverse calculation in order to obtain the gradient corresponding to the weight and bias of each error; finally, the gradient-descent algorithm is used to calculate the weight and to offset update adjustment.

Consider the training data $\left\{\left(y^{(1)}, q^{(1)}\right), \left(y^{(2)}, q^{(2)}\right), \cdots, \left(y^{(m)}, q^{(m)}\right)\right\}$ with a sample size of m, where $y^{(i)} (i = 1, 2, \ldots, m)$ is the input scrambled data, and $q^{(i)} (i = 1, 2, \cdots, m)$ is the type label corresponding to the scramble code. The W and b update formulas are, respectively, expressed as follows:

$$W^{t+1} = W^t - \alpha \frac{\partial J(W, b; y, q)}{\partial W^t}, \tag{18}$$

$$b^{t+1} = b^t - \alpha \frac{\partial J(W, b; y, q)}{\partial b^t}. \tag{19}$$

where α is the learning rate, and $J(\cdot)$ is the loss function. The loss function used is categorical_crossentropy, which is a loss function for mul-ticlass classification tasks and is suitable for the multiclass problem.

The expression of $J(\cdot)$ is

$$J(W, b; y, q) = -\frac{1}{m} \sum_{i=1}^{m} q^{(i)} \cdot \log\left(f\left(W, b; y^{(i)}\right)\right) + \lambda \sum \|W\|^2. \tag{20}$$

where λ is the regularization coefficient used to prevent the network from overfitting.

The input $x_m^{(l)}(j)$ and output $y_m^{(l)}(j)$ of each layer of neurons in the network have the following relationship:

$$y_m^{(l)}(j) = f\left(x_m^{(l)}(j) + b_m^{(l)}(j)\right). \tag{21}$$

where l is the layer number of the network, j is the j-th neuron of the feature map, m is the m-th feature map in the network layer number, $f(\cdot)$ is the activation function used by this layer, and b is the bias.

4.3. Input–Output Relationship of Each Network Layer

The input $2 \times \beta$ training data are represented as $I_{N,T}$, where N represents the dimension of the two features, and T is the number of sampling points. The input–output relationship of each network layer is detailed as follows:

(1) Input layer:

$$Input = I_{N,T}. \tag{22}$$

(2) Convolutional layer:

$$y_m^{(2)}(j,k) = f(\sum_{i=1}^{3} I_{j,(k-1)\times 1+i} \times \mathrm{ker}_m^{(2)} + b_m^{(2)}(j,k)). \tag{23}$$

(3) LSTM layer:

$$h_t = f(W_{hx} y_{m,t}^{(3)}(j) + W_{hh} h_{t-1} + b_h), \tag{24}$$

$$h_t^n = f(W_{h^{n-1}h^n} h_t^{n-1} + W_{h^n h^n} h_{t-1}^n + b_h^n), \tag{25}$$

$$y_m^{(4)}(j) = W_{h^N y} h_t^N + b_y, \tag{26}$$

where W_{hx} represents the weight matrix between the input layer and the hidden layer, W_{hh} represents the weight matrix between the hidden layers, h_t represents the hidden activation at time t, $n(n = 1, 2, \ldots, N)$ represents the current network layer, and h_t^n represents the n layer and output at time t.

(4) SoftMax layer:

$$y^5(j) = f(\sum_{i=1}^{n_4} y^{(4)}(i) \times w^{(5)}(i) + b^{(5)}(j)). \tag{27}$$

where $w_i^{(5)}(l)$ represents the connection weight between the fifth and sixth layers, and n_4 is the number of neurons in the fifth layer.

The SoftMax layer calculates the probability estimate of the corresponding category in the sample, and the learned hypothesis function $h_\theta(x)$ is expressed as follows:

$$h_\theta(y^i) = \begin{bmatrix} p(q^{(i)} = 1 | y^{(i)}; \theta) \\ p(q^{(i)} = 2 | y^{(i)}; \theta) \\ \cdots \\ p(q^{(i)} = k | y^{(i)}; \theta) \end{bmatrix} = \frac{1}{\sum_{j=1}^{k} e^{\theta_j^T y^{(i)}}}. \tag{28}$$

where $\theta_1, \theta_2, \ldots, \theta_k \in R^{n+1}$ is the model parameter. To simplify the model, we define the function

$$l\{\alpha\} = \begin{cases} 1, & \alpha = true \\ 0, & \alpha = false \end{cases}. \tag{29}$$

The cost function of the SoftMax layer can be further expressed as

$$J(\theta) = -\frac{1}{m}[\sum_{i}^{m} \sum_{j=1}^{k} l\{q^{(i)} = j\} \log \frac{e^{\theta_j^T y^{(i)}}}{\sum_{l=1}^{k} e^{\theta_j^T y^{(i)}}}. \tag{30}$$

By superimposing the k values, the probability that the network classifies the scrambled data into the j-th category is

$$p(q^{(i)} = j | y^{(i)}; \theta) = \frac{e_j^T y^{(i)}}{\sum_{l=1}^{k} e_j^T y^{(i)}}. \tag{31}$$

where the maximum probability corresponds to the classification category of the scrambling code.

We first extract the autocorrelation and the intercorrelation features of the code element data used as input data; compared with the method in [27], the training and learning of the neural network are carried out, and the biggest advantage of a neural network over manually extracting features is that it can learn the training autonomously, so the recognition effect is better than simply manually extracting features. Compared with the method in Reference [26], a detailed neural network model is constructed, but the autocorrelation features and the intercorrelation features extracted in this paper have better anti-BER and noise performance than the matrix rank features in Reference [26]. When the BER is greater than 0.1, the matrix rank features become extremely insignificant because of erroneous code elements, and this leads to weaker features extracted by the neural network. Therefore, the current method in this paper achieves better results.

5. Results

5.1. Experimental Dataset

Two datasets were constructed, i.e., a scrambled dataset of 2000 frames under the condition of bit error (a bit error rate of 2000), and a scrambled dataset of 2000 frames under the condition of white Gaussian noise (with an SNR in the range of 10–15 dB).

During the training process, 80% of the total samples were randomly selected as training data, with the rest used as test data. The datasets and labels are shown in Tables 1 and 2.

Table 1. Linear block code scrambling dataset under bit error.

Dataset	Label	Number of Signals	Number of Samples (2×400)
linear block code	0	2000×400	2000
self-synchronous scrambler	1	2000×400	2000
synchronous scrambler	2	2000×400	2000

Table 2. Linear block code scrambling dataset with noise.

Dataset	Label	Number of Signals	Number of Samples (2×800)
linear block code	0	2000×800	2000
self-synchronous scrambler	1	2000×800	2000
synchronous scrambler	2	2000×800	2000

The simulation experiment was conducted on an Inteli7-9700K CPU with 16 Gb of memory and an RTX3080 graphics card running Windows 10; the model was built and trained based on TensorFlow.

5.2. Network Model Parameters

Because the Adam optimizer was chosen in this manuscript, the optimizer adaptively adjusts the learning rate, and the automatic learning rate is not a function of epoch. So, there was no need to consider learning rate schedule as a function of epochs. The network model parameters are shown in Table 3.

5.2.1. Effect of LSTM Layers and Parameters on Recognition Rate

We analyzed the effects of the number of LSTM layers and cells on the recognition rate of the algorithm, as shown in Figures 7 and 8, respectively.

Table 3. Network model parameters.

Network Parameters	Numerical Value
batch size	20
learning rate initial value	1×10^{-4}
final number of epochs	100
validation part of training set	25% of the training set
initial value of the weight constraint coefficient	0.01
number of LSTM units with shared weights	6

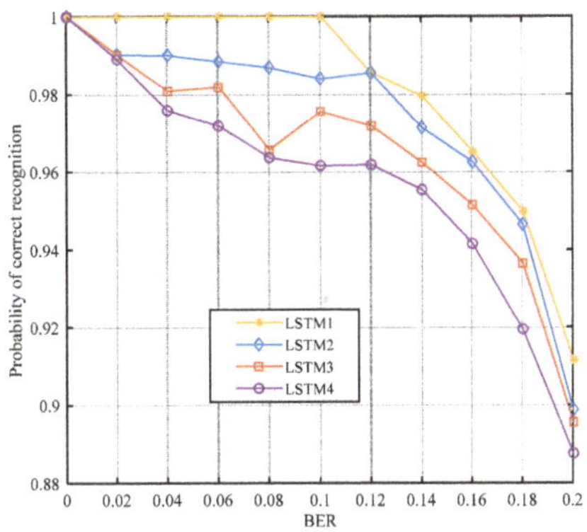

Figure 7. Effect of the number of LSTM layers on the performance of the algorithm.

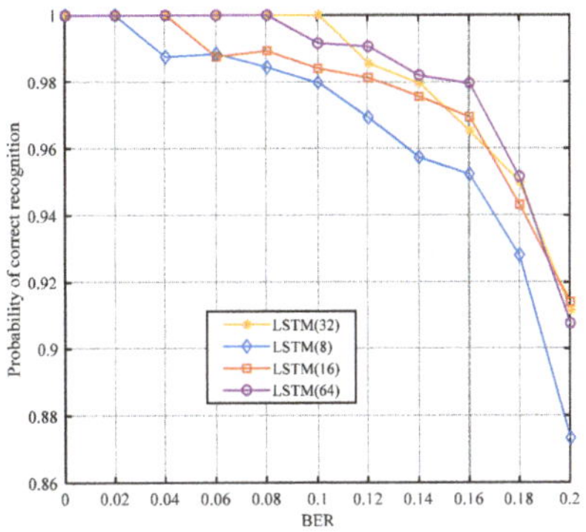

Figure 8. The effect of the number of cells of LSTM on the performance of the algorithm.

In Figures 7 and 8, we can see that the recognition rate is the highest when the number of LSTM layers is 1, and the recognition results are better when the number of cells is 32 or 68. However, the training time of the network increases a lot when the number of cells is 68, so for the number of cells, 32 was chosen, as it has a shorter training time.

5.2.2. Effect of CNN Layers and Parameters on Recognition Rate

We analyzed the effects of the number of CNN layers and cells on the recognition rate of the algorithm, as shown in Figures 9 and 10, respectively.

As can be seen in Figures 9 and 10, the highest recognition rate is achieved when the number of CNN layers is 2; the recognition effect is better when the number of convolutional layer cells in the first layer is 32, and the number of convolutional layer cells in the second layer is 68.

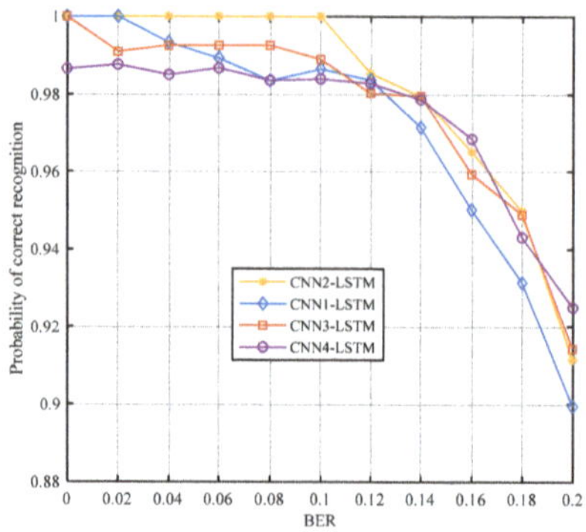

Figure 9. Effect of the number of CNN layers on the performance of the algorithm.

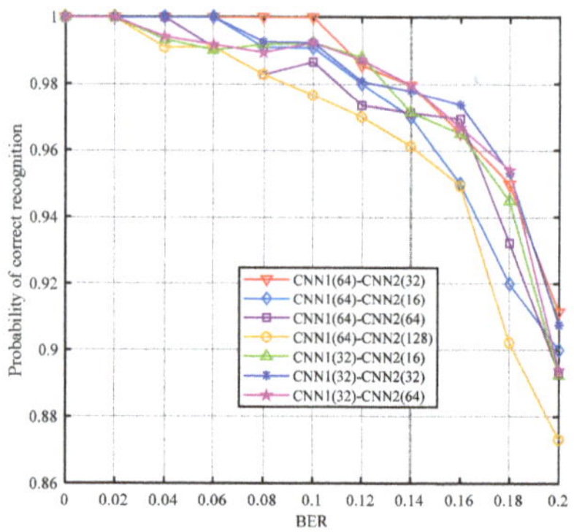

Figure 10. The effect of the number of cells of CNN on the performance of the algorithm.

5.3. Experimental Results

5.3.1. Recognition Rate of The Proposed Algorithm under Bit Error Condition

To verify the validity of the model proposed in this paper, we analyzed the influence of the linear block code parameters, the number of intercepted samples, and the order and terms of the generator polynomial on the network performance under different error conditions.

The relationship between the parameters of the different linear block codes and the recognition rate of the algorithm when the generator polynomial is $f(x) = 1 + x^{18} + x^{23}$ and the number of samples is 400 is shown in Figure 11.

It can be seen in the figure that, under the same simulation conditions, the (15,5) linear block code has the highest recognition rate. At that time, the recognition rate under the four coding parameters can reach more than 85%, indicating good anti-error performance.

The (15,5) linear block code scrambling recognition rate under different sample numbers when the generator polynomial is $f(x) = 1 + x^{18} + x^{23}$ is shown in Figure 12.

It can be seen in the figure that, under the same error conditions, more samples mean a higher recognition rate of the algorithm. When the number of samples is 100 and the error rate is 0.2, the recognition rate is greater than 80%. When the error rate is 0.1, the recognition rate can reach 98%, which shows that the proposed algorithm has better performance under the conditions of a small number of samples and a high BER.

The (15,5) linear block code scrambling recognition rate under different generator polynomials for 400 samples is shown in Figure 13.

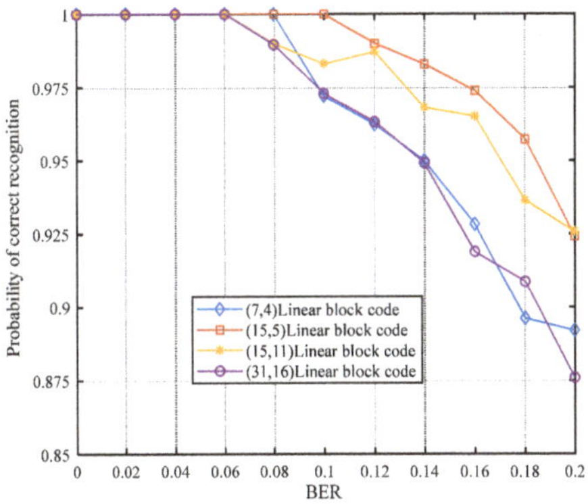

Figure 11. Influence of linear block code parameters on the recognition rate.

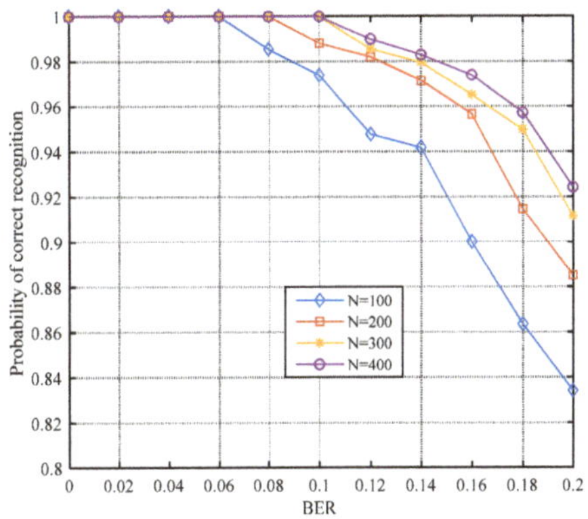

Figure 12. Influence of number of samples on the recognition rate.

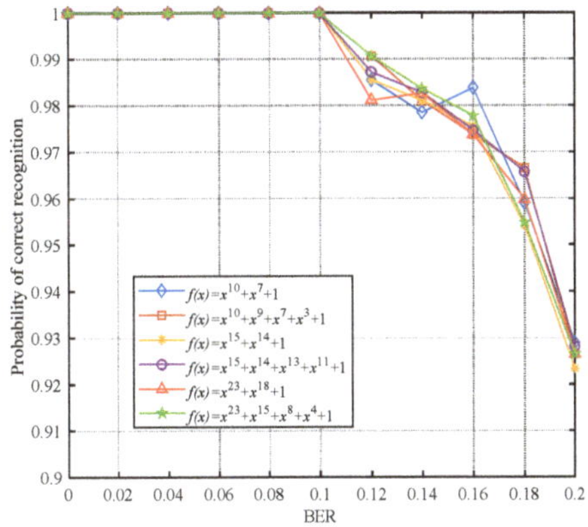

Figure 13. Influence of scrambling code generator polynomial on the recognition rate.

It can be seen in Figure 13 that, with an increase in the bit error rate, the recognition rate gradually decreases. The difference in the recognition rate under a generator polynomial of different orders and terms is not more than 1%. Therefore, the algorithm proposed in this paper is free from the scrambling code generator polynomial order.

The performance of the algorithm proposed in this paper under different linear block code parameters is compared with that of the method proposed in Refs [26,27]. When the generator polynomial is $f(x) = 1 + x^{18} + x^{23}$, the number of samples is 400, and the sequence length is 240,000 bits as shown in Figure 10.

It can be seen in Figure 14 that the recognition rate of the algorithm proposed in this paper when the number of verification samples is only 400 is better than the recognition rate of the algorithm in Reference [26] when the number of samples is 240,000 bits. The recognition rate of the algorithm in Reference [27] decreases rapidly and tends to be 0 when the BER is greater than 0.01. The reason for this is that the rank feature of the matrix is not obvious when the BER is greater than 0.01, which leads to a sharp decrease in the recognition rate. At the same time, the proposed algorithm does not require manual feature extraction and has the characteristics of less data required, simple manual feature extraction, and a high recognition rate.

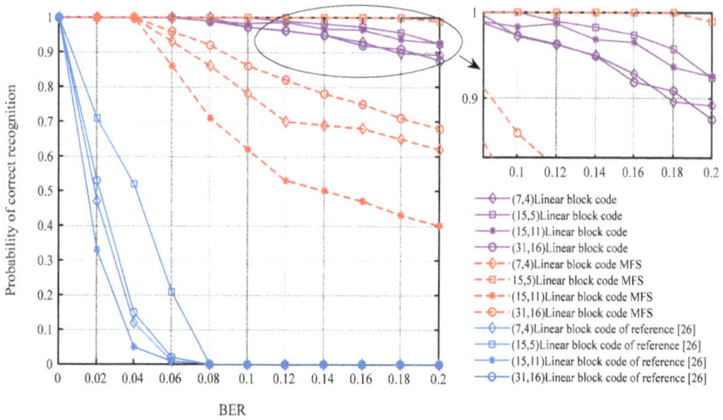

Figure 14. Comparison of algorithms under error conditions. (7,4), (15,5), (15,11), (31,16) Linear block code (in blue color) Li et al. [26].

5.3.2. Recognition Rate of the Proposed Algorithm under the Condition of Gaussian White Noise

We verified the influence of the linear block code parameters, the number of intercepted samples, and the order and number of terms of the generator polynomial on network performance under different types and levels of noise.

The relationship between the parameters of the different linear block codes and the recognition rate of the algorithm when the generator polynomial is $f(x) = 1 + x^{18} + x^{23}$ and the number of samples is 800 is shown in Figure 15.

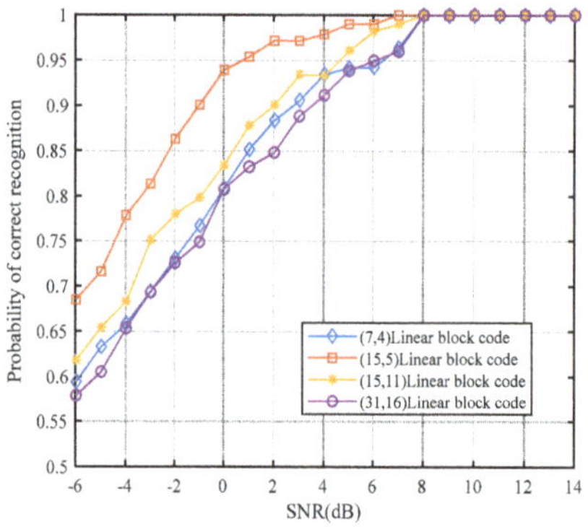

Figure 15. Influence of linear block code parameters on the recognition rate under the Condition of Gaussian White Noise.

With an increase in the SNR, the recognition rate continues to improve. Under the same SNR, the (15,5) linear block code has the highest recognition rate. When SNR = 0 dB, the recognition rate can reach 80%.

The recognition rate of the (15,5) linear block code scrambling under different sample numbers when the generator polynomial is $f(x) = 1 + x^{18} + x^{23}$ is shown in Figure 16.

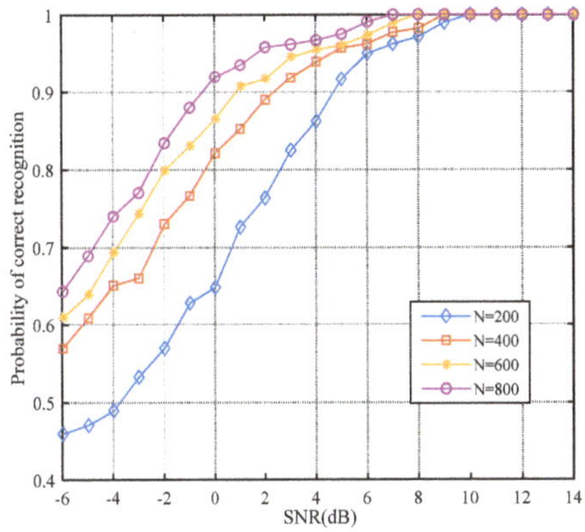

Figure 16. Influence of number of samples on the recognition rate under the Condition of Gaussian White Noise.

It can be seen in the figure that, under the same conditions, the larger the number of samples, the higher the recognition rate. When the SNR = 3 dB and the number of samples is 400, the recognition rate can reach more than 90%.

The (15,5) linear block code scrambling recognition rates under different generator polynomials when the number of samples is 800 are shown in Figure 17.

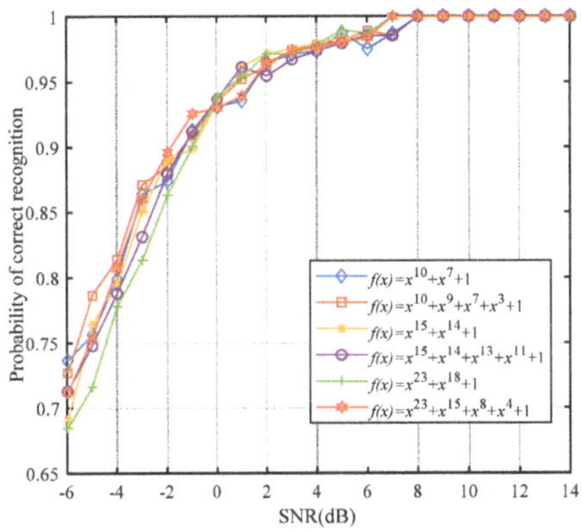

Figure 17. Influence of generator polynomial of scrambling code on the recognition rate.

Under the condition of the same SNR, the parameters of the generator polynomial have very little influence on the recognition rate of the proposed algorithm. Combined with Figure 13, under the conditions of bit error and noise, the algorithm proposed in this paper does not need to consider the influence of different generator polynomials.

6. Discussion

In this study, deep learning is applied to the field of linear block code scrambling-type identification. A linear block code scrambling-type identification method based on correlation feature extraction and a correlation feature extraction network model is proposed.

First, the cross-correlation of the scrambled sequence symbols is deduced according to the scrambling principle, and then a biased autocorrelation function that reflects the correlation is further constructed. To effectively extract the relevant features of the sequence, a correlation feature extraction network model is constructed, and the correlation and autocorrelation features are used as model input for training, which finally achieves the purpose of identifying the type of linear block code scrambling. The simulation results show that, compared with the traditional feature extraction method, the algorithm proposed in this study has a better recognition rate under the BER, and it can recognize the linear block code scrambling-type under the conditions of BER and noise. The identification of scrambling code parameters lays a foundation for future scrambler parameter identification and has significant practical engineering value. Future research will focus on classification recognition with convolutional code scrambling.

Author Contributions: Conceptualization, J.T. and L.Z.; methodology, J.T.; software, Z.Z.; validation, Z.Z., J.T. and L.Z.; formal analysis, Z.Z.; investigation, L.Z.; resources, L.Z.; data curation, Z.Z.; writing—original draft preparation, J.T.; writing—review and editing, Z.Z.; visualization, J.T.; supervision, L.Z.; project administration, L.Z.; funding acquisition, L.Z. All authors have read and agreed to the published version of the manuscript.

Funding: This work was financially supported by the National Natural Science Foundation of China (91538201), the Taishan Scholar Special Foundation (ts201511020), and the Chinese National Key Laboratory of Science and Technology on Information System Security (6142111190404).

Institutional Review Board Statement: Not applicable.

Informed Consent Statement: Informed consent was obtained from all subjects involved in the study.

Data Availability Statement: Data is contained within this article.

Acknowledgments: We thank LetPub (www.letpub.com, accessed on 27 June 2022) for its linguistic assistance during the preparation of this manuscript.

Conflicts of Interest: The authors declare no conflict of interest.

References

1. Chen, Z.L.; Peng, H. Scrambler blind recognition method based on soft information. *J. Commun.* **2017**, *38*, 174–182.
2. Liang, C.; Wang, F.P.; Wang, Z.J. Low complexity method for spread sequence estimation of DSSS signal. *Syst. Eng. Electron.* **2009**, *20*, 41–49.
3. Wu, L.P.; Li, Z.; Chen, L.C. A PN sequence estimation algorithm for DS signal based on average cross-correlation and eigen analysis in lower SNR conditions. *Sci. China Inf. Sci.* **2010**, *53*, 1666–1675. [CrossRef]
4. Scholtz, R. The spread spectrum concept. *IEEE Trans. Commun.* **1997**, *28*, 748–755. [CrossRef]
5. Ahlswede, R.; Csiszar, I. Common randomness in information theory and cryptography. I. Secret sharing. *IEEE Trans. Inf. Theory* **1993**, *39*, 1121–1132. [CrossRef]
6. Jonsson, F.; Johansson, T. Theoretical analysis of a correlation attack based on convolutional codes. In Proceedings of the 2000 IEEE International Symposium on Information Theory, Sorrento, Italy, 6 August 2002. [CrossRef]
7. Massey, J.L. Shift-register synthesis and BCH decoding. *IEEE Trans. Inf. Theory* **1969**, *15*, 122–127. [CrossRef]
8. Liu, X.B.; Koh, S.N.; Wu, X.W. Reconstructing a linear scrambler with improved detection capability and in the presence of noise. *IEEE Trans. Inf. Forensics Secur.* **2012**, *7*, 208–218. [CrossRef]
9. Ma, Y.; Zhang, L.M. Reconstruction of Scrambler with Real-time Test. *J. Electron. Inf. Technol.* **2016**, *38*, 1794–1799.
10. Xie, H.; Wang, F.H.; Huang, Z.T. Blind reconstruction of linear scrambler. *J. Syst. Eng. Electron.* **2014**, *25*, 560–565. [CrossRef]
11. Cluzeau, M. Reconstruction of a Linear Scrambler. *IEEE Trans. Comput.* **2007**, *56*, 1283–1291. [CrossRef]
12. Vijayakumaran, S. LFSR identification using Groebner bases. In Proceedings of the Twenty Second National Conference on Communication, Guwahati, India, 4–6 March 2016. [CrossRef]
13. Xie, H.; Han, Z.Z.; Ding, S. Scrambling Sequence Estimation Method Based on Propagation Operator Algorithm. *J. Syst. Eng. Electron.* **2017**, *39*, 2327–2332.
14. Guo, X.; Su, S.; Qian, H. Scrambling Code Blind Identification in SDH Signal Intelligent Reception. In Proceedings of the 2021 2nd Information Communication Technologies Conference (ICTC), Nanjing, China, 7–9 May 2021. [CrossRef]
15. Zhang, L.M.; Tan, J.Y.; Zhong, Z.G. Blind identification of self-synchronous scrambling codes based on cosine coincidence. *J. Electron. Inf. Technol.* **2022**, *44*, 1412–1420.
16. Tan, J.Y.; Zhang, L.M.; Zhong, Z.G. Reconstruction of a Synchronous Scrambler Based on Average Check Conformity. *Math. Probl. Eng.* **2022**, *2022*, 6318317. [CrossRef]

17. Liu, X.B.; Koh, S.N.; Chui, C.C. A Study on Reconstruction of Linear Scrambler Using Dual Words of Channel Encoder. *IEEE Trans. Inf. Forensics Secur.* **2013**, *8*, 542–552. [CrossRef]
18. Ma, Y.; Zhang, L.M.; Wang, H.T. Reconstructing Synchronous Scrambler With Robust Detection Capability in the Presence of Noise. *J. Commun.* **2015**, *10*, 397–408.
19. Zhang, M.; Lv, Q.T.; Zhu, Y.X. Blind identification of self-synchronous scrambling codes based on linear block codes. *J. Appl. Sci.* **2015**, *33*, 178–186.
20. Shu, N.H.; Min, Z.; Xin, H.L. Reconstruction of Feedback Polynomial of Synchronous Scrambler Based on Triple Correlation Characteristics of M-sequences. *Ieice Trans. Commun.* **2018**, *E101.B*, 1723–1732.
21. Han, S.; Zhang, M. A Method for Blind Identification of a Scrambler Based on Matrix Analysis. *IEEE Commun. Lett.* **2018**, *22*, 2198–2201. [CrossRef]
22. Gu, X.; Zhao, Z.; Shen, L. Blind estimation of pseudo-random codes in periodic long code direct sequence spread spectrum signals. *IET Commun.* **2016**, *10*, 1273–1281. [CrossRef]
23. Kim, D.; Song, J.; Yoon, D. On the Estimation of Synchronous Scramblers in Direct Sequence Spread Spectrum Systems. *IEEE Access* **2020**, *8*, 166450–166459. [CrossRef]
24. Kim, D.; Yoon, D. Blind Estimation of Self-Synchronous Scrambler in DSSS Systems. *IEEE Access* **2021**, *9*, 76976–76982. [CrossRef]
25. Kim, Y.; Kim, J.; Song, J. Blind Estimation of Self-Synchronous Scrambler Using Orthogonal Complement Space in DSSS Systems. *IEEE Access* **2022**, *10*, 66522–66528. [CrossRef]
26. Li, X.H.; Zhang, M.; Han, S.N. Distinction of self-synchronous scrambled linear block codes based on multi-fractal spectrum. *J. Syst. Eng. Electron.* **2016**, *27*, 968–978. [CrossRef]
27. Wang, Z.F.; Zhai, L.Q.; Wei, D. Blind Recognition Algorithm for Scrambled Channel Encoder Based on the Features of Signal Matrix and Layered Neural Network. In Proceedings of the 2021 15th International Symposium on Medical Information and Communication Technology (ISMICT), Xiamen, China, 14–16 April 2021. [CrossRef]
28. Song, Y.Q.; Liu, F.; Shen, T.S. A novel noise reduction technique for underwater acoustic signals based on dual-path recurrent neural network. *IET Commun.* 2022; *in press*. [CrossRef]
29. Gong, W.; Tian, J.; Liu, J. Underwater Object Classification Method Based on Depth wise Separable Convolution Feature Fusion in Sonar Images. *Appl. Sci.* **2022**, *12*, 3268. [CrossRef]
30. O'Shea, T.J.; Corgan, J.; Clancy, T.C. Convolutional Radio Modulation Recognition Networks. In Proceedings of the International Conference on Engineering Applications of Neural Networks, Aberdeen, UK, 2–5 September 2016; pp. 213–226. [CrossRef]
31. Saif, W.S.; Ragheb, A.M. Performance Investigation of Modulation Format Identification in Super-Channel Optical Networks. *IEEE Photonics J.* **2022**, *14*, 8514910. [CrossRef]
32. Acarturk, C.; Sirlanci, M.; Balikcioglu, P.G. Malicious Code Detection: Run Trace Output Analysis by LSTM. *IEEE Access* **2021**, *9*, 9625–9635. [CrossRef]
33. Ahn, G.; Kim, K.; Park, W.; Shin, D. Malicious File Detection Method using Machine Learning and Interworking with MITRE ATT&CK Framework. *Appl. Sci.* **2022**, *12*, 10761. [CrossRef]
34. Sagduyu, Y.E. Adversarial Deep Learning for Over-the-Air Spectrum Poisoning Attacks. *IEEE Trans. Mob. Comput.* **2020**, *20*, 306–319. [CrossRef]
35. Pan, Y.W.; Yang, S.H.; Peng, H. Specific Emitter Identification Based on Deep Residual Networks. *IEEE Access* **2019**, *7*, 54425–54434. [CrossRef]
36. Vaseghi, S.V. *Advanced Digital Signal Processing and Noise Reduction*, 4th ed.; John Wiley & Sons: New York, NY, USA, 2009.
37. Xiao, Q.; Chang, X.; Zhang, X. Multi-Information Spatial–Temporal LSTM Fusion Continuous Sign Language Neural Machine Translation. *IEEE Access* **2020**, *8*, 216718–216728. [CrossRef]
38. Bandara, K.; Bergmeir, C.; Hewamalage, H. LSTM-MSNet: Leveraging Forecasts on Sets of Related Time Series With Multiple Seasonal Patterns. *IEEE Trans. Neural Netw. Learn. Syst.* **2021**, *32*, 1586–1599. [CrossRef] [PubMed]

applied sciences

MDPI

Article

Neural-Network-Assisted Polar Code Decoding Schemes

Hengyan Liu, Limin Zhang *, Wenjun Yan and Qing Ling

Institute of Information Fusion, Naval Aviation University, Yantai 264001, China
* Correspondence: imzlm@163.com

Abstract: The traditional fast successive-cancellation (SC) decoding algorithm can effectively reduce the decoding steps, but the decoding adopts a sub-optimal algorithm, so it cannot improve the bit error performance. In order to improve the bit error performance while maintaining low decoding steps, we introduce a neural network subcode that can achieve optimal decoding performance and combine it with the traditional fast SC decoding algorithm. While exploring how to combine neural network node (NNN) with R1, R0, single-parity checks (SPC), and Rep, we find that the decoding failed sometimes when the NNN was not the last subcode. To solve the problem, we propose two neural network-assisted decoding schemes: a key-bit-based subcode NN-assisted decoding (KSNNAD) scheme and a last subcode NN-assisted decoding (LSNNAD) scheme. The LSNNAD scheme recognizes the last subcode as an NNN, and the NNN with nearly optimal decoding performance gives rise to some performance improvements. To further improve performance, the KSNNAD scheme recognizes the subcode with a key bit as an NNN and changes the training data and label accordingly. Computer simulation results confirm that the two schemes can effectively reduce the decoding steps, and their bit error rates (BERs) are lower than those of the successive-cancellation decoder (SCD).

Keywords: polar code; neural network; decoder

Citation: Liu, H.; Zhang, L.; Yan, W.; Ling, Q. Neural-Network-Assisted Polar Code Decoding Schemes. *Appl. Sci.* **2022**, *12*, 12700. https://doi.org/10.3390/app122412700

Academic Editors: Przemysław Falkowski-Gilski, Tadeus Uhl and Zbigniew Lubniewski

Received: 7 November 2022
Accepted: 8 December 2022
Published: 11 December 2022

Publisher's Note: MDPI stays neutral with regard to jurisdictional claims in published maps and institutional affiliations.

1. Introduction

To reduce interference in the communication process and enable the system to automatically check and correct errors to improve the reliability of data transmission, channel coding technology is widely used in various wireless communication systems. Currently, commonly used coding techniques include Reed–Solomon (RS) codes [1], low-density parity-check (LDPC) codes [2,3], turbo codes [4,5], Bose–Chaudhuri–Hocquenghem codes [6–8], convolutional codes [9], and polar codes [10]. Polar codes have been strictly proven to reach the Shannon limit under the binary discrete memoryless channel and binary erasure channel when the code length tends to infinity [11], and due to their excellent short-code performance, they became the control channel selection coding scheme for 5G enhanced mobile broadband scenarios at the 2016 3GPP RAN1#87 conference [12]. As the only channel coding that has been proven to reach the Shannon limit, polar codes are very important for 5G communication and future 6G communication. In the field of wireless communication, the decoding of polar codes has received extensive attention globally. Examples are the traditional belief propagation (BP) algorithm [13], the successive-cancelation (SC) algorithm [14], the early stop BP algorithm [15], other low complexity BP algorithms [16], the fast SC decoding scheme [17–19], and successive cancellation stack [20]. These traditional algorithms and their improvements can only reduce the decoding complexity or improve the decoding performance but cannot satisfy the requirements of improving communication quality and reducing decoding steps at the same time.

In recent years, the rapid development of deep learning in computer vision and language translation has attracted widespread attention from experts and scholars in the field of communication. Some algorithms directly train deep learning as a polar decoder, Chao, Z.W. et al. explored a polar code residual network decoder with a denoiser and a decoder. The denoiser used a multi-layer perceptron (MLP) [21], a convolutional neural network (CNN) [22], and a recurrent neural network (RNN) [23], while the decoder used a CNN [24]. The decoding performances nearly reached those of a traditional successive cancelation list (SCL) [25] algorithm when the code length was less than 32. In terms of reducing decoding complexity, Xu, W.H. et al. applied a neural network to polar code decoding, but the application was only used for limited code length [26]. Hashemi, S. et al. used the deep learning algorithm to reduce the complexity of the SC algorithm [27]. However, all of them cannot be directly applied to decoding longer polar codes. To apply the machine learning in polar code decoding, researchers combine machine learning with traditional polar decoding algorithms.

There are many combinations of BP algorithm and machine learning. In terms of reducing the complexity, Gruber, T. et al. proposed a neural network BP decoder that can be executed at any code length, which reduced the decoding bit error rate (BER) and saved 50% of the hardware cost [28]. Cammerer, S. et al. proposed a segmentation decoding scheme based on neural networks, which complied with the BP decoding law and divided polar codes into single-parity checks (SPCs) and repetition code (RC) nodes and decoded them using a neural network to achieve the same bit error performance as the SC and BP algorithms with lower complexity [29]. In terms of improving decoding performance, Teng, C.F. et al. used the check relationship between the cyclic redundancy check (CRC) remainder and the decoding result to propose a comprehensive loss function of polar code frozen bits, which realized the unsupervised decoding of the polar codes, and the decoding performance was better than that of the traditional BP decoding algorithm [30]. Gao, J. et al. proposed a BP structure similar to the res-net network, which achieved better decoding performance than the standard BP algorithm [31]. In terms of simultaneously improving decoding performance and reducing complexity, Xu, W. et al. used deep learning to improve the decoding performance of the BP algorithm and used lower computational complexity to achieve a bit error performance close to that of the CRC-aided SCL (CA-SCL) algorithm [32].

The SC algorithm and machine learning are less frequently combined. Doan, D. et al. divided the polar code into equal-length nodes and decoded them using neural networks [33] and obtained a lower complexity than [29]. However, when the code length is longer, the number of neural network models that need to be called for decoding is higher, and the increase in the calculation amount is not negligible. None of the above studies have achieved a low-complexity low-bit-error decoding performance by combining the SC algorithm and machine learning.

To achieve a low-complexity low-bit-error decoding performance by combining SC algorithm and machine learning, this paper proposes a decoding method combining neural network nodes (NNNs) and traditional nodes. This scheme avoids the extra complexity caused by frequent calls of the neural network model and proposes a decoding scheme that can not only reduce the delay, but also improve the decoding performance.

The remainder of this paper is organized as follows: Section 2 introduces the knowledge required for polar code decoding, including the SC decoding scheme, traditional special nodes, and the fast SC decoding scheme. Section 3 introduces the system model and the decoding scheme based on the NNN and CNN structure and then analyzes the performance of the model in decoding. Two NNN recognition strategies are explored. Section 4 presents the simulation results over the additive white Gaussian noise (AWGN) and Rayleigh channel. Section 5 summarizes the paper.

2. Preliminaries

2.1. SC Decoding Algorithm

A polar code $(x_1, x_2, \cdots, x_N)$ with length $N = 2^n$ can be encoded by $(x_1, x_2, \cdots, x_N) = (u_1, u_2, \cdots, u_N)\mathbf{F}^{\otimes n}$, $\mathbf{F} = \begin{bmatrix} 1 & 0 \\ 1 & 1 \end{bmatrix}$, $\otimes$ is Kronecker operation. Consider two independent and identically distributed random variables u_1 and u_2, which are encoded by polar codes:

$$(x_1, x_2) = (u_1, u_2)\mathbf{F} = (u_1, u_2)\begin{bmatrix} 1 & 0 \\ 1 & 1 \end{bmatrix} = (u_1 \oplus u_2, u_2). \tag{1}$$

Obviously, at this time, $N = 2, n = 1$. As shown in Figure 1, (x_1, x_2) passes through a binary discrete memoryless channel (BDMC) with transition probability $W(y|x)$. Then, the received signal (y_1, y_2) is obtained, but we always use its log likelihood ratio (LLR) (L_1, L_2). The process of solving (u_1, u_2) from (L_1, L_2) is the decoding process.

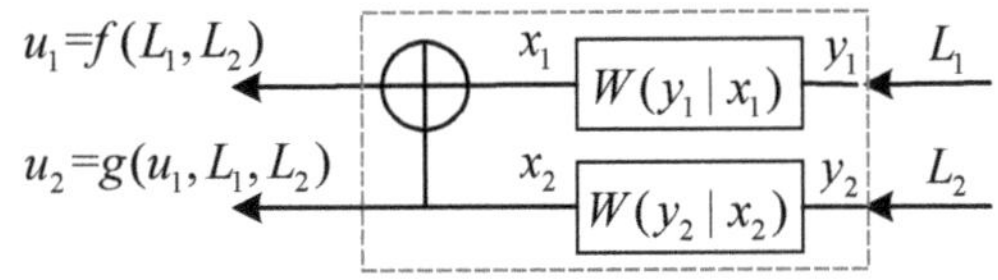

Figure 1. Basic unit model of a polar code.

As shown in Figure 1, we solve u_1 using the $u_1 = f(L_1, L_2)$ operation, and solve u_2 using the $u_2 = g(u_1, L_1, L_2)$ operation; indeed, f operation and g operation are conditional probabilities. Specifically, we solve $u_1 = 0$ when $\frac{\Pr(y_1 y_2 | u_1=0)}{\Pr(y_1 y_2 | u_1=1)} \geq 0$; otherwise, $u_1 = 1$.

$$\begin{aligned}
\Pr(y_1 y_2 | u_1) &= \tfrac{\Pr(y_1 y_2 u_1)}{\Pr(u_1)} = \tfrac{\sum_{u_2 \in \{0,1\}} \Pr(y_1 y_2 u_1 u_2)}{\Pr(u_1)} \\
&= \tfrac{\sum_{u_2 \in \{0,1\}} \Pr(y_1 y_2 | u_1 u_2) \Pr(u_1 u_2)}{\Pr(u_1)} \\
&\overset{(a)}{=} \tfrac{1}{2} \textstyle\sum_{u_2 \in \{0,1\}} \Pr(y_1 y_2 | x_1 x_2) \\
&\overset{(b)}{=} \tfrac{1}{2} \textstyle\sum_{u_2 \in \{0,1\}} \Pr(y_1 | x_1) \Pr(y_2 | x_2) \\
&= \tfrac{1}{2} \textstyle\sum_{u_2 \in \{0,1\}} \Pr(y_1 | u_1 \oplus u_2) \Pr(y_2 | u_2)
\end{aligned} \tag{2}$$

where (a) means $\Pr(u_1) = 0.5$; $\Pr(u_1 u_2) = 0.25$; (b) means W is a BDMC.

$$\begin{aligned}
\ln \tfrac{\Pr(y_1 y_2 | u_1=0)}{\Pr(y_1 y_2 | u_1=1)} &= \ln \tfrac{\sum_{u_2 \in \{0,1\}} \Pr(y_1 | u_2) \Pr(y_2 | u_2)}{\sum_{u_2 \in \{0,1\}} \Pr(y_1 | 1 \oplus u_2) \Pr(y_2 | u_2)} \\
&= \ln \tfrac{\Pr(y_1|0)\Pr(y_2|0) + \Pr(y_1|1)\Pr(y_2|1)}{\Pr(y_1|1)\Pr(y_2|0) + \Pr(y_1|0)\Pr(y_2|1)} \\
&= \ln \tfrac{\frac{\Pr(y_1|0)\Pr(y_2|0)}{\Pr(y_1|1)\Pr(y_2|0)} + 1}{\frac{\Pr(y_2|0)}{\Pr(y_2|1)} + \frac{\Pr(y_1|0)}{\Pr(y_1|1)}} = \ln \tfrac{1 + e^{L_1 + L_2}}{e^{L_1} + e^{L_2}}
\end{aligned} \tag{3}$$

where $L_1 = \ln \frac{\Pr(y_1|0)}{\Pr(y_1|1)}$ and $L_2 = \ln \frac{\Pr(y_2|0)}{\Pr(y_2|1)}$ are the LLRs of the received signal. Equation (3) is called the f operation, which is commonly estimated as follows:

$$\ln \frac{1 + e^{L_1 + L_2}}{e^{L_1} + e^{L_2}} \approx \text{sign}(L_1)\text{sign}(L_2)\min\{|L_1|, |L_2|\}, \tag{4}$$

where $\text{sign}(-a) = -1$; $\text{sign}(a) = 1$; $a > 0$; min indicates taking the minimum value.

$U_2 = 0$ if $\ln \frac{\Pr(y_1 y_2 u_1 | u_2=0)}{\Pr(y_1 y_2 u_1 | u_2=1)} \geq 0$, where

$$\begin{aligned}
\Pr(y_1 y_2 u_1 | u_2) &= \tfrac{\Pr(y_1 y_2 u_1 u_2)}{\Pr(u_2)} = \tfrac{\Pr(y_1 y_2 | u_1 u_2) \Pr(u_1 u_2)}{\Pr(u_1)} \\
&= \tfrac{1}{2} \Pr(y_1 y_2 | u_1 u_2) = \tfrac{1}{2} \Pr(y_1 y_2 | x_1 x_2) \\
&= \tfrac{1}{2} \Pr(y_1 | x_1) \Pr(y_2 | x_2) \\
&= \tfrac{1}{2} \Pr(y_1 | u_1 \oplus u_2) \Pr(y_2 | u_2)
\end{aligned} \tag{5}$$

The same can be obtained as follows:

$$\begin{aligned} \ln \frac{\Pr(y_1 y_2 u_1 | u_2=0)}{\Pr(y_1 y_2 u_1 | u_2=1)} &= \ln \frac{\Pr(y_1|u_1)\Pr(y_2|0)}{\Pr(y_1|1 \oplus u_1)\Pr(y_2|1)} \\ &= \ln \frac{\Pr(y_1|u_1)}{\Pr(y_1|1 \oplus u_1)} + \ln \Pr(y_2|0)\Pr(y_2|1) \\ &= (1-2u_1)L_1 + L_2 \end{aligned} \tag{6}$$

Equation (6) is called the g calculation. We use a tree diagram to represent the decoding process of Figure 1, which is more intuitive.

As shown in Figure 2, the circles represent nodes; the upper node is the root node; and the lower two nodes are leaf nodes. After the root node calculates the LLR of the data, u_1 is obtained using $u_1 = f(L_1, L_2)$ operation. Then, it returns the result to the root node. After that, u_2 is obtained using the $u_2 = g(u_1, L_1, L_2)$ operation, and then, u_2 is returned to the root node to finally obtain the decoding results. When the code length is 4, the decoding diagram is as shown in Figure 3.

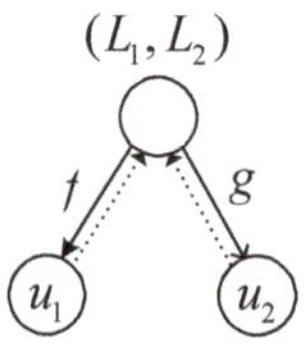

Figure 2. Basic unit model of a polar code.

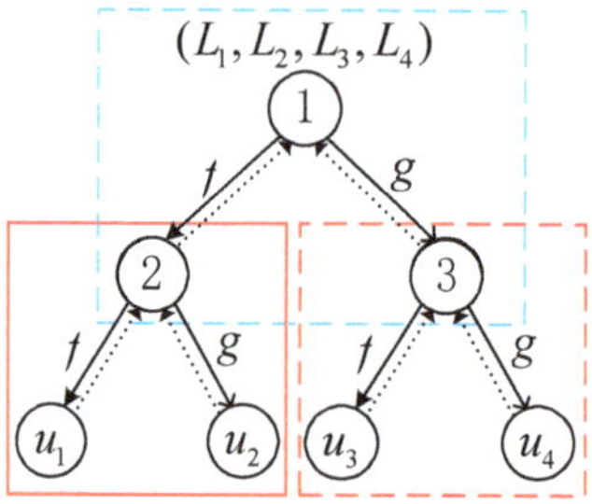

Figure 3. Polar code decoding tree diagram with length 4.

As shown in Figure 3, the polar code of length 4 can be divided into three basic units: the red solid line block diagram in the lower left corner, the red dotted line block diagram in the lower right corner, and the blue dot-dash line block diagram above. According to the algorithm in Figure 2, nodes u_1, u_2, 1, u_3, u_4, 3 and 1 are calculated sequentially, and the calculation function is marked next to the line. Any polar code with a code length of $N = 2^n$ can perform SC decoding in sequence according to this rule, where n is a positive integer. However, the frozen bit transmission information is fixed and does not need to be decoded. Therefore, in order to reduce the amount of calculation in the actual execution process, the frozen bit part is not included in the recursive calculation. Let white circles denote nodes that contain only frozen bits, black circles denote nodes that contain only information bits, and gray circles denote nodes with both frozen bits and information bits. Unless otherwise specified, the meanings of the nodes in this paper are the same as above. A polar code with a code length of $N = 32$ and information bit length $K = 16$ constructed through the Gaussian approximation (GA) method is used, and the signal-to-noise ratio is 2.5 dB. Then, the polar code tree diagram is as shown in Figure 4. We usually measure the decoding time complexity by decoding steps. The decoding steps are calculating by adding the numbers of operation f, operation g and data return. As shown in Figure 4, the downward solid arrow with solid line represents the f operation; the downward non-solid arrow with chain line represents the g operation; the upward solid arrow with dotted line represents the direction of data return, which is calculated through (1). The serial numbers in the figure record the counting process of decoding time step. The total number of decoding step is 81. Note that the R0 does not need a data return. Unless otherwise specified, the meanings of the arrows in this paper are the same as above.

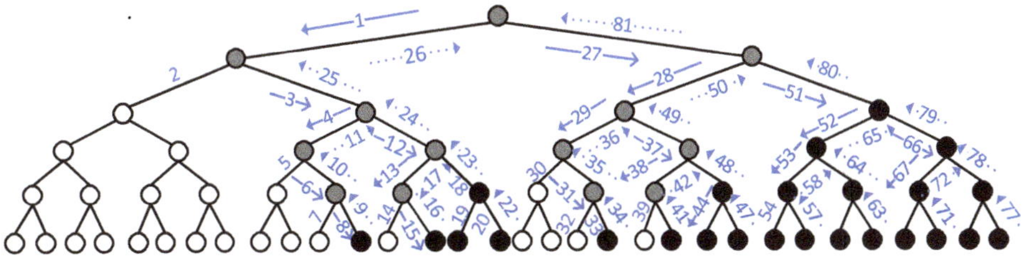

Figure 4. Schematic diagram of the SC decoding time step.

2.2. Decoding of Traditional Special Nodes

This section introduces four special nodes: R0 node, R1 node, Rep node, and SPC node, which are distinguished by the positions of the frozen bits and information bits. A node that has all frozen bits is an R0 node; a node that has all information bits is an R1 node; a node that has only one information bit at the last bit position is a Rep node; a node that has only one frozen bit at the first bit position is an SPC node. Assuming that a polar code $s_1^{N_S} = (s_1, s_2, \cdots, s_{N_S}), s_i \in \{0,1\}$ with length N_S is transmitted, after the BDMC with transition probability $\Pr(y|s), s \in \{0,1\}$, the received signal $y = (y_1, y_2, \cdots, y_{N_S})$ is chosen to obtain the sequence $(\beta_1, \beta_2, \cdots \beta_{N_S})$, where

$$\beta_i = (1 - sign(\ln \frac{\Pr(y_i|0)}{\Pr(y_i|1)}))/2, \tag{7}$$

If the node is R1, $(\beta_1, \beta_2, \cdots \beta_{N_S})$ is its fast decoding result. If the node is R0, there is no need to decode. The decoding method of the Rep node and SPC node are as follows.

2.2.1. Fast Decoding of a Rep Node

The log-likelihood ratio (LLR) sum of a Rep node with a code length of N_S can be calculated as follows:

$$S_q = \sum_{i=1}^{N_S} \ln \frac{\Pr(y_i|0)}{\Pr(y_i|1)}, \tag{8}$$

where S_q is the LLR sum and Pr denotes probability. The decoding result of Rep is 0 when $S_q \geq 0$ and 1 otherwise.

2.2.2. Fast Decoding of an SPC Node

For an SPC node with a code length of N_S, if $\sum_{i=1}^{N_S} \beta_i = 0$, $(\beta_1, \beta_2, \cdots \beta_{N_S})$ is its decoding result. Otherwise, find β_p, which is the smallest absolute LLR of $(\beta_1, \beta_2, \cdots \beta_{N_S})$. Let $\beta_p = 1 - \beta_p$. Then, $(\beta_1, \beta_2, \cdots \beta_{N_S})$ is the decoding result.

2.3. Traditional Fast SC Decoding

When we recognize a polar code with length N to a string of subcodes, we first judge if the whole polar code can be recognized as a subcode; if not, we will judge the former, and the latter $N/2$ polar code can be recognized as a subcode. If so, the recognition stops, and if not, we divide the $N/2$ polar code by the $N/4$ polar code and continue the recognition, until the original is totally recognized as subcodes. When decoding a polar code with a code length of 32, as discussed in Section 2.1, the polar codes are first recognized as special nodes, as shown in Figure 5.

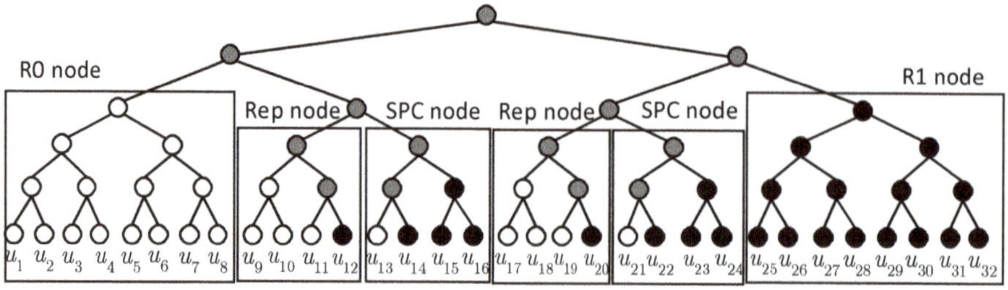

Figure 5. Schematic diagram of subcode classification.

After node recognition, the nodes are decoded as discussed in Section 2.2. The decoding sequence is the same as that in Section 2.1.

The decoding process is shown in Figure 6. Though the traditional fast SC decoding costs 19 decoding steps which is much lower than SC decoding, it cannot improve the bit-error performance. Maintaining the low time step of the fast SC algorithm while improving the bit-error performance is the goal of the algorithm in this paper.

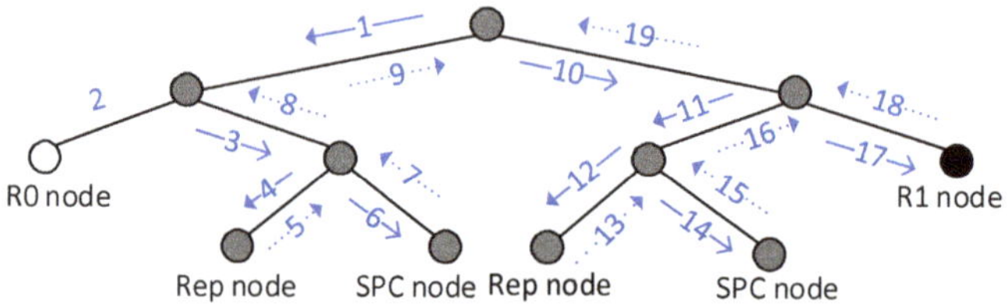

Figure 6. Schematic diagram of the decoding process based on traditional subcodes.

3. Neural Network-Assisted Polar Code Decoding Scheme

Considering that the traditional fast SC decoding algorithm can effectively reduce the decoding time step, in this paper, we will reduce the BER of fast SC decoding structure. Xu Xiang studied the decoding performance of the polar code neural network decoder based on CNNs, RNNs, long short-term memory, and MLP under different network designs and parameter settings [34]. In comparison, it can be seen that the CNN-based decoder can achieve a better performance with fewer parameters. Therefore, we consider using the CNN to improve the decoding performance and call the scheme neural network assisted (NNA) decoding scheme.

3.1. System Model

Figure 7 shows the block diagram of the proposed decoding scheme system, which is composed of training and decoding parts. The "Training Part" is executed first, followed by the "Decoding Part". The Training Part is composed of "Polar Encoder1" and "Neural Network Training"; the Decoding Part comprises "Polar Encoder2", "Mapper", "Noise", and "NNA Decoding Scheme". Note that the Mapper Layer, Noise Layer, and Decoder Layer are trained in the training part, but only the model of the Decoder Layer is saved and called in NNA Decoding. Although the decoder layer decodes $\dot{y}_1^{N_S}$ to obtain $\hat{\dot{u}}_1^{N_S}$, the purpose of the training part is to obtain the trained Decoder Layer model and not the $\hat{\dot{u}}_1^{N_S}$.N_S and K_S denote the code length and information length of the codewords to be trained, respectively, while N and K denote the code length and information length of the codewords to be decoded, respectively. u_1^N denotes a row vector $(u_1, \cdots, u_N)$. The two parts have different inputs, $\dot{u}_1^{N_S}$ and u_1^N, and different diagram structures, but $\dot{u}_1^{N_S} \mapsto \dot{y}_1^{N_S}$ and $u_1^N \mapsto y_1^N$ are mapped in the same way. Taking $u_1^N \mapsto y_1^N$ as an example, consider single-carrier signaling in which N binary streams u_1^N are encoded with a channel encoder that generates a codeword of length N. The codeword x_1^N is then mapped to binary phase-shift keying (BPSK) symbols s_1^N by the mapper. Then, s_1^N is transmitted over noisy channels by the noise block, and finally, the signal y_1^N is received. When the channel noise is additive gaussian white noise (AWGN), y_1^N is given by $y_1^N = s_1^N + n_w$, $n_w \sim N(0, \sigma_w^2)$. Similarly, $\dot{y}_1^{N_S}$ can be obtained from $\dot{u}_1^{N_S}$. Note that $\dot{u}_1^{N_S}$ is determined by the Node Recognizer. We can obtain the LLR values through $2y_1^N/\sigma_w^2$, where σ_w^2 denotes Gaussian noise power. In this paper, the channel-side information is assumed to be known at the receiver. The details of the neural network (NN) model and decoding scheme are introduced in the following section.

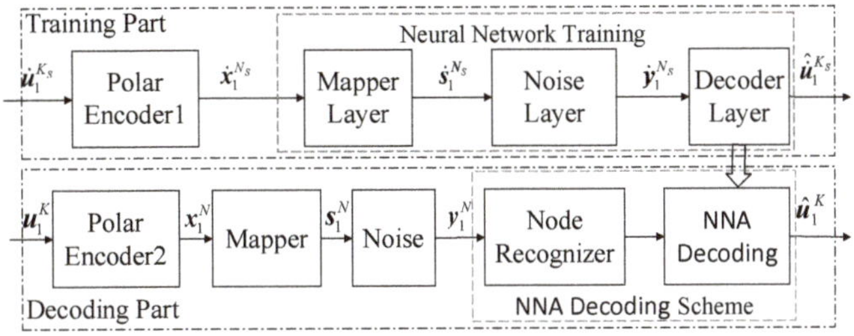

Figure 7. Block diagram of the NNA decoding scheme.

3.2. Neural Network Model and NNN

The NN model is trained by the training part in Figure 7. We write matrix $\dot{U}$ with size $2^{K_S} \times K_S$ to denote all the possible information sequences and matrix $\dot{X}$ with size $2^{K_S} \times N_S$ to denote all the possible codewords, resulting in, $\dot{x}_1^{N_S} \in \dot{X}$. Let A denote the position set of the information bits. The decoder layer is trained to find the $\dot{u}_1^{K_S}$ corresponding to $\dot{y}_1^{N_S}$, with $\dot{u}_1^{K_S} \in \dot{U}$. To fully train the NN, additional layers without trainable parameters are added to perform certain actions, namely mapping the codewords to BPSK symbols and adding noise. Therefore, the NN model approximates $\dot{x}_1^{N_S}$ to $\dot{u}_1^{K_S}$. Therefore, $\dot{X}$ represents the training data in this machine learning problem and $\dot{U}$ is the label. We can generate as much training data as desired by Monte Carlo simulations. The activation functions used in the decoder layer are sigmoid functions and a rectified linear unit (ReLU), which are defined as follows:

$$g_{sigmoid}(z) = \frac{1}{1+e^{-z}}, \tag{9}$$

$$g_{ReLU}(z) = \max\{0, z\}. \tag{10}$$

Mini-batch gradient descent is used to train the NN, and in each batch, different data are fed into the NN. Table 1 summarizes the architecture of the proposed NN model. In this scheme, the batch size, which is equal to the size of the label set and that of the codeword set, is 2^{K_S}. Thus, the training complexity scales exponentially with K_S, which makes training the NN decoder for long polar codes not feasible in actuality. To solve this problem, we define the short codes (N_S is generally less than 32) decoded by the NN decoder as NNNs and combine them with the traditional special nodes mentioned in Section 2.2 to derive an NNA decoding scheme. Note that an NNN could be one of the R1, R0, Rep, or SPC nodes, or none of them. Arıkan's system polar code coding scheme [35] is used in this paper unless otherwise stated.

Table 1. The structure of the NN.

Layer	Output Dimensions	Convolutional Filters	Activation Function
Input	N	/	/
Mapper (Lambda)	N	/	/
Noise (Lambda)	N	/	/
NN Layer 1	$16K_S$	128	ReLU
NN Layer 2	$8K_S$	64	ReLU
NN Layer 3	$4K_S$	32	ReLU
Output	K_S	/	Sigmoid

The neural network structure in this paper consists of a mapper layer, a noise layer, and a decoding layer. All three layers need to be trained. In real-life decoding, only the decoding layer is needed. The structure of the NN is shown in Table 1. The mapper layer and the noise layer are both custom-function layers generated by the Keras tool in Tensorfow2. In this paper, Gaussian white noise is used unless otherwise stated; the Gaussian construction method is used to generate polar codes; the modulation method is BPSK. The input of the network during training is $\dot{X}$. The label is $\dot{U}$. Each round of training adds different noises, i.e., the data of each round of training are different from each other to prevent overfitting.

A trained CNN can classify the received signal, find the corresponding information sequence, and finally decode the polar code. A well-trained neural network can achieve the performance of MAP decoding [26]. The decoding results of the NNN with $N_S = 16$ and $K_S = 4$ are shown in Figure 8.

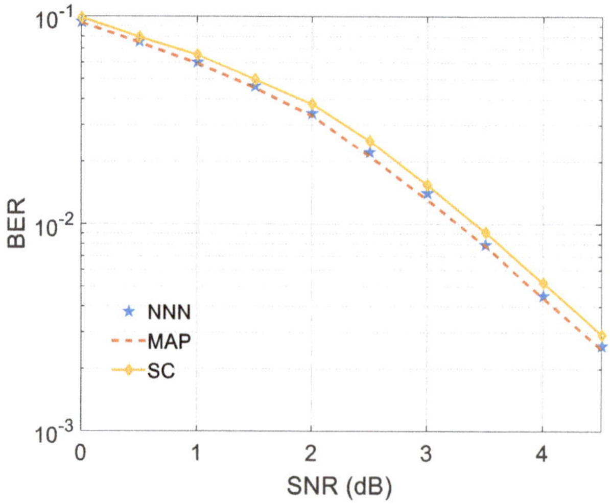

Figure 8. Comparison of the NNN decoding and SC decoding algorithms.

As shown in Figure 8, NNN decoding performance is better than SC algorithm, almost the same as the optimal MAP decoding algorithm, while the traditional fast decoding adopts a suboptimal decoding algorithm, and its decoding performance is similar to SC. Therefore, using well-trained CNN to decode subcodes can obtain better decoding performance than traditional classified subcodes.

3.3. Neural Network-Assisted (NNA) Decoding

The NNA decoding scheme is composed of a node recognizer and a decoder. The former recognizes the code as a concatenation of smaller constituent codes that were simply named nodes, which includes the NNN, and then the latter decodes these nodes. Note that the NNN is decoded by the trained model from the neural network model. Since the decoding scheme proposed in this paper is only for the receiving end, we use BER as the metric to measure the decoding performance. To avoid the computational complexity of the recursive structure of the successive-cancellation decoder (SCD), a vector of size $2 \times N - 1$ is used to store the intermediate calculation results of the SC decoding algorithm. The decoding error rate will decrease as the SNR increases, i.e., when a BER is 6×10^{-7} at a lower signal noise rate (SNR), the BERs at higher SNRs with the same parameters are definitely lower than 6×10^{-7}. Let 6×10^{-7} denote BER threshold. If a BER is lower than the threshold, it will not be displayed in the figures. Therefore, the calculations of higher SNRs with BERs lower than 6×10^{-7} can be skipped to further improve the calculation speed.

To determine how to recognize the NNN, we must determine how it works. The progress of the signal process is shown in the training part of Figure 7. When $N = 16$, K ranges from 1 to 16. As shown in Figure 7 and as explained in Section 3.1, "Polar Encoder1" generates 2^{K_S} possible codewords of N_S bits size in the training part. Considering the multiple possibilities of node recognition, we trained the polar code with $N_S = 16$, $K_S = 1 - 16$, with $N_S = 8$, $K_S = 1 - 8$, and with $N_S = 4$, $K_S = 1 - 4$ in epochs. The gradient of the loss function is calculated in each epoch using the Adam optimizer [36], and the trained model is finally saved to be used in the decoding part.

When only four kinds of traditional special nodes need to be identified, we first identify the polar code with the entire length of N and judge whether the code is one of the above four kinds of nodes. If so, we directly perform maximum likelihood decoding. If not, we continue to identify the first half and the second half, and the above process is repeated until the original code is completely decomposed into nodes. Although the recognition process is a recursive process, this process is only performed once, and the computational complexity increase is limited. This method of transferring half of the original sequence into the recursive operation to identify the subcode is called dichotomy.

Assuming that the length of the NNN is 8, if we execute dichotomy, when the subcode with a code length of 8 does not belong to the four traditional special nodes mentioned before, it will be recognized as an NNN. In practice, the decoding performance will be greatly reduced in some cases. To find the reasons and explore a proper division method of the NNN, we recognized and decoded polar codes with a code length of 16 and information bit length 1–16. In this experiment, the codewords were constructed using the GA method. Taking polar codes with $N = 16$ and $K = 3$ as an example, we write R0(8) to denote an R-0 node with a code length of 8, SPC(4) to denote the SPC node with a code length of 4, and NNN(8/3) to denote an NNN with $N_S = 8$ and $K_S = 3$. When the code is recognized in the traditional way, the node-type structure is R0(8) + R0(4) + SPC(4) and the node can be recognized as R0(8) + NNN(8/3) by the node recognizer as well, that is, the NNN structure is R0(4) + SPC(4). Note that when $N_S = 16$, the code can also be recognized as NNN(16/3).

We simulated multiple times and found that when the NNN is the last node of the original code, the decoding performance is always better than that of the traditional fast SC decoding algorithm. When an NNN is decoded as the non-last subcode, the polarization effect cannot be generated sometimes. This is because the information bits position determined by encoder 1 is different from the actual NNN determined by the recognizer. So, the decoding model is mismatched and errors are propagated in the decoding tree, resulting in decoding failure.

There are two ways to solve the abovementioned problems: one is to use an NNN as the last decoded subcode directly, and the other is to consider adapting the model and improving the correct probability of the first error, thus preventing the decoding error propagation. The two solutions lead to two decoding schemes, which are described below.

3.4. Last Subcode Neural Network-Assisted Decoding (LSNNAD) Scheme

Let N_S denote the code length of the last subcode neural network node (LSNNN). The main idea of this scheme is to identify the last subcode as LSNNN. Note that at this time, LSNNN may be one of the four traditional subcodes mentioned above, or it may be different from the above four subcodes. The type, length, position, and order are determined at the moment when the codeword is constructed. In this experiment, 0 and 1 are used to represent the frozen bits and information bits, i.e., the sequence $information\_pos \in \{0,1\}^N$ of code length N is obtained after the codeword is constructed. When $information\_pos = 1$, the information bits are stored. Otherwise, the frozen bits are stored. In this experiment, all frozen bits are 0. The LSNNN identification flowchart is shown in Figure 9.

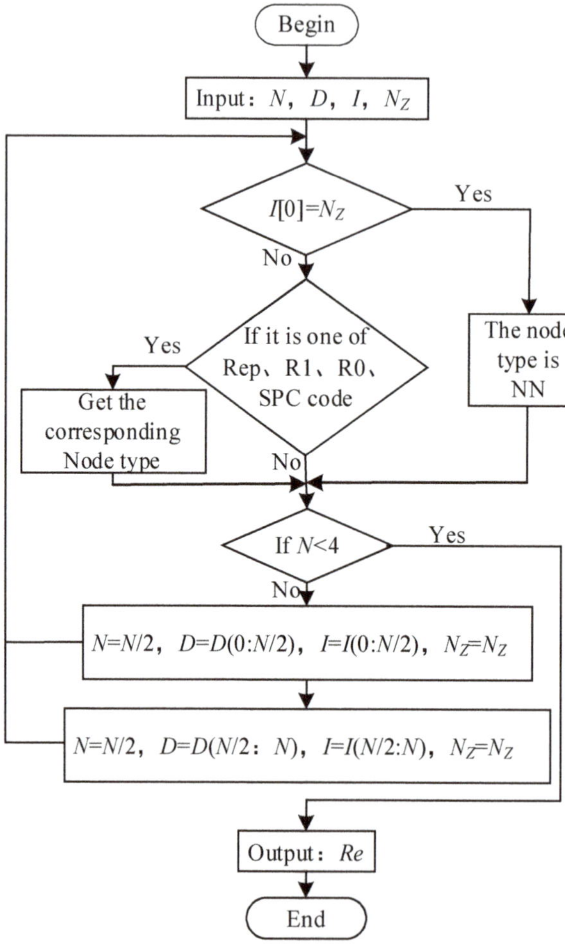

Figure 9. LSNNN identification flowchart.

The input N is the same as in Figure 1. Let D denote the position of bits and A, an arbitrary subset of $\{1, \cdots, N\}$, denote the position of information bits. I denotes the vector of the index, the initial content of which is in the range 0 to $N-1$, and it represents the sequence position currently being processed. $I[0]$ denotes the first element of the vector I and represents the starting position of the code being processed. D, I, and N constantly change during the calculation process depicted in the flowchart (Figure 9). $D(0:N/2)$

denotes the first half of the vector D, $D(N/2:N)$ denotes the second half of the vector D, and $N_Z = N - N_S$ denotes the starting point of the NNN that is already fixed when the total code length and length of the NNN are determined. The output *Re* is a matrix with five rows, with the first to the last rows denoting the node type, the starting point, K, N, and the information position of every node, respectively. In the node identification process, the decoding depth of each node is recorded, i.e., the node position in the SC decoding tree is recorded. The decoding tree after dividing the subcodes according to the above rules is shown in Figure 10.

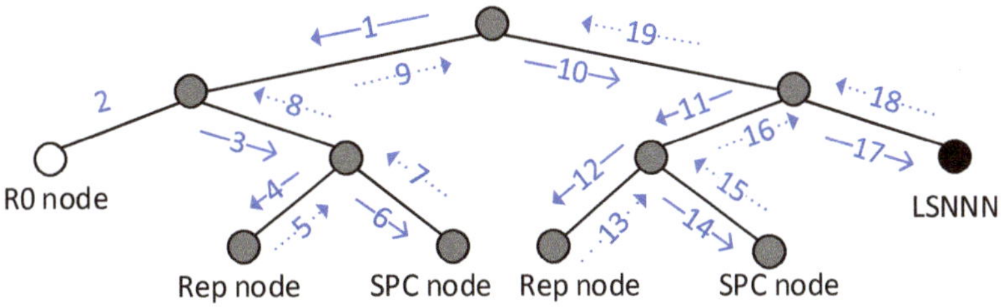

Figure 10. Schematic diagram of the decoding process based on traditional subcodes combined with LSNNN.

3.5. Key-Bit-Based Subcode Neural Network-Assisted Decoding (KSNNAD) Scheme

Although the LSNNAD minimizes the loss of decoding errors caused by model mismatches, the performance improvement is limited. So, we must find another scheme to improve the decoding performance.

Ref. [37] analyzed the SC decoding error of polar codes and found that the first error is usually caused by channel noise, and the frequency of decoding failure caused by this error increases with the increase of code length. Ref. [38] analyzed and proved that when the polar code is divided into several R1 nodes, the first error in SC decoding of the polar code is likely to occur at the first information bit of the R1 nodes. Therefore, the decoding performance of the SC algorithm can be improved as long as the first error is correct. It can be seen from the above analysis that the first error is likely to occur at the first information bit position. This bit is defined as a key bit, and the node containing this bit is recognized as an NNN. In this scheme, all the NNNs are recorded and trained, and the models are saved to be called in the decoding part so that the model mismatched error is solved. Note that the label $\dot{U}$ is the decoding result whose key bit is flipped in the decoding process. So, the training part of the KSNNAD does not need a polar encoder.

The scheme is called a KSNNAD scheme, and the NNN is called the key bit based neural network node (KSNNN). We solve another problem: how to recognize a KSNNN.

The simplest idea of recognizing a KSNNN is to select the key bit as the first bit of the KSNNN. Assuming that the position is *Pos* and the subcode length is N_S, the position of the last bit of the KSNNN is $Pos + N_S$. However, the dichotomy does not ensure that the key bit is the first position of the KSNNN; it may be the middle position. Next, we need to recognize a polar code with a KSNNN and the other four traditional subcodes mentioned above. Considering the complexity, the code length of KSNNN is limited to $N_S \leq 16$. The identification flowchart is shown in Figure 11.

Taking the polar code with $N = 32$, which is mentioned above as an example, the recognized result and decoding process are shown in Figure 12.

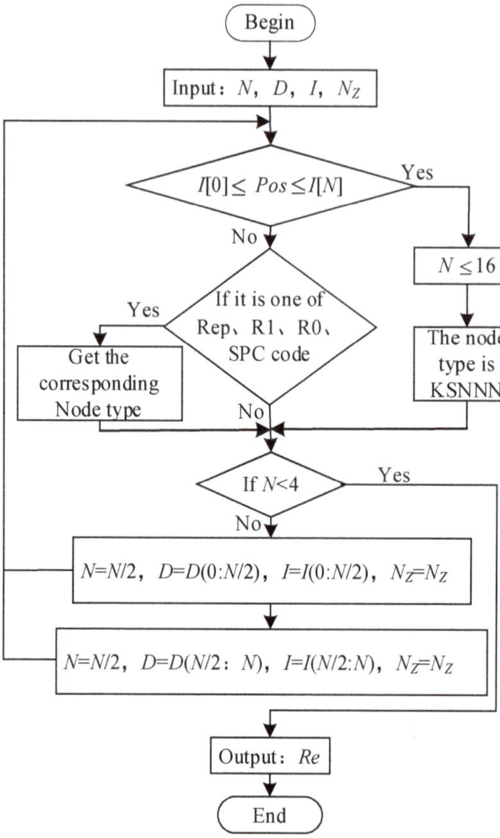

Figure 11. KSNNN identification flowchart.

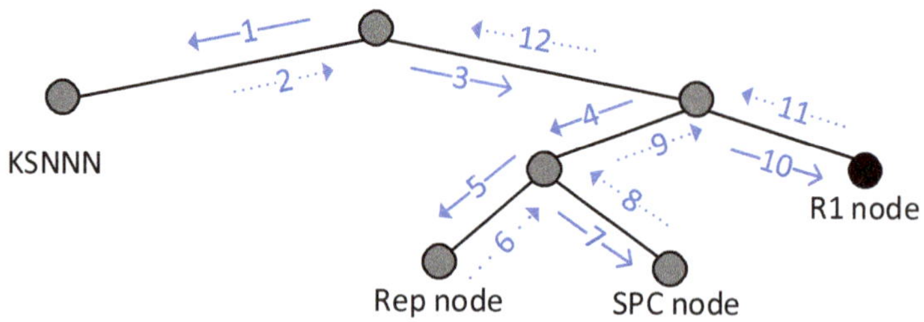

Figure 12. Schematic diagram of the decoding process based on traditional subcodes combined with KSNNN.

As shown in Figure 13, the time step required for decoding at this time is 12. The structure of the KSNNN is R0(8) + Rep(4) + SPC(4), compared with Figure 6. The structure of the original polar code is KSNNN(16) + Rep(4) + SPC(4) + R1(8). In the actual execution process, the subcode classification procedure is first executed. All possible forms of KSNNN are recorded. The key bit flip results corresponding to all possible situations of each specific form are trained as labels. Finally, the models are saved and called in the decoding part separately.

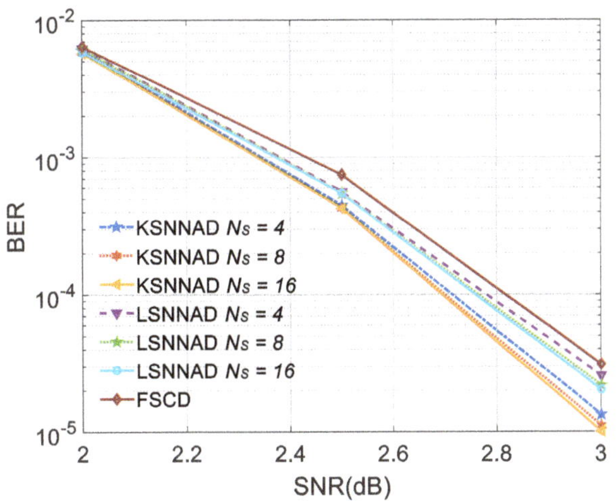

Figure 13. Decoding results of different N_S.

4. Numerical Simulation

This section simulates over the AWGN channel for Sections 4.1–4.4. Section 4.5 is simulated over a Rayleigh channel. The SNR is expressed as E_s/N_0, where E_s is the energy per symbol and N_0 is the single-sided noise power spectral density. We used Keras as the front end of the Tensorflow2 platform with an NVIDIA 3080 graphics card and Intel(R) Core(TM) i9 CPU to run the NN model, with the configuration given in Table 1. NNNs in Sections 4.2–4.5 have the code length N_S= 16. The polar code was constructed at 2.5 dB unless otherwise stated. Since the decoding scheme in this paper is proposed for the receiving end, the decoding performance is focused on the BER.

4.1. Performance with Different N_S

When simulated with a code length of 1024, a code rate of 0.5 and N_S of 4, 8 and 16, the BER performance are shown in Figure 13. The BER curve of KSNNAD with N_S 4 is marked as KSNNAD $N_S = 4$, while LSNNAD with N_S 4 marked as LSNNAD $N_S = 4$, and so on.

As shown in Figure 13, combining NNN with R1, R0, SPC, and Rep can indeed improve decoding performance, but the improvement differs with different N_S and different schemes. Both KSNNAD and LSNNAD have bit error performance improvement as N_S increases and the longer the N_S, the smaller the improvement. KSNNAD outperforms LSNNAD due to performance boost at the key bit, regardless of the value of N_S.

4.2. Performance with Different Code Rates

When simulated with a code length of 1024 and code rate of 0.25, 0.5, and 0.75, the BER performance is shown in Figure 13. The BER curve of KSNNAD with code rate 0.25 is marked as KSNNAD K = 0.25, while LSNNAD with code rate 0.25 marked as LSNNAD K = 0.25, FSC decoding scheme with code rate 0.25 marked as FSCD K = 0.25, and so on.

As shown in Figure 14, both of the KSNNAD and LSNNAD have bit error performance improvement compared with SC and the lower the code rate, the better the improvement. When the code rate is 0.25, compared to SC decoder, LSNNAD needs 0.049 dB less to reach BER 10^{-4}, while KSNNAD needs about 0.078 dB less.

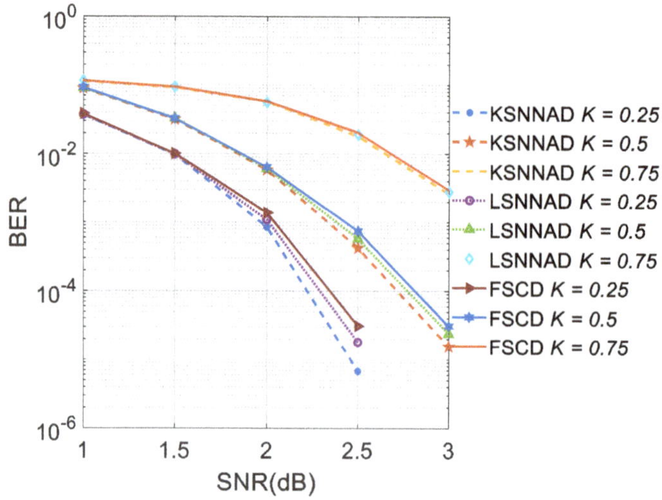

Figure 14. Decoding results of different code rates.

4.3. Performance with Different Code Lengths

This section simulates the KSNNAD, LSNNAD and SC BER performance with a code rate of 0.5 and code lengths of 64, 128, 256, 512, and 1024. The results are shown in Figure 14. The BER curve of KSNNAD with a code length of 64 is marked as KSNNAD $N = 64$, while LSNNAD with a code length of 64 marked as LSNNAD $N = 64$, the FSC decoding scheme with code length 64 marked as FSCD $N = 64$, and so on.

As shown in Figure 15, the decoding performance of the KSNNAD and LSNNAD are better than that of the SC decoder. When the code length is 1024, compared to the SC decoder, LSNNAD needs 0.049 dB less to reach BER 10^{-4}, while KSNNAD needs approximately 0.127 dB less.

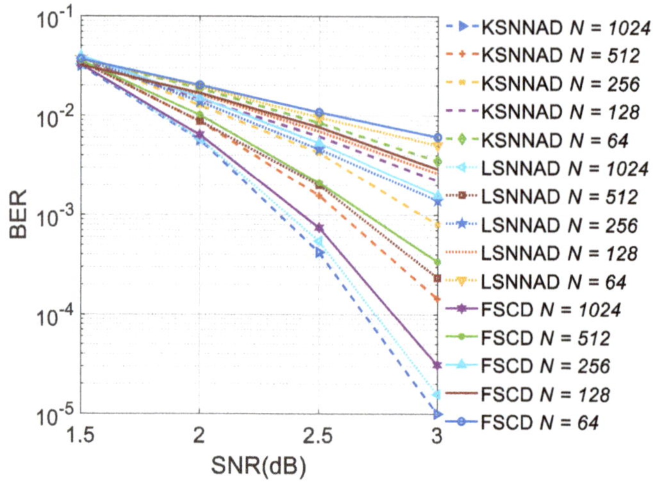

Figure 15. Decoding results of different code lengths.

4.4. Performance Comparison under AWGN Channel

This section simulates the BER performance comparison of different algorithms with a code length of $N= 1024$ and a code rate of $R= 0.5$. The BER performance of the LSNNAD and that of the KSNNAD are compared with those proposed in [17,19,25,33,39]. The simplified SC decoder proposed in [17] is marked as SSCD [17], the SCL decoder proposed in [25] with a list length of 2 is marked as SCLD L = 2 [25], the fast SC flip decoder proposed in [19] with T_{max}= 2 is marked as FSCFD T_{max}= 2 [19], the recursive trellis decoding scheme proposed in [39] with τ= 4 is marked as RTD τ= 4 [39], and so on. The results are shown in Figure 16.

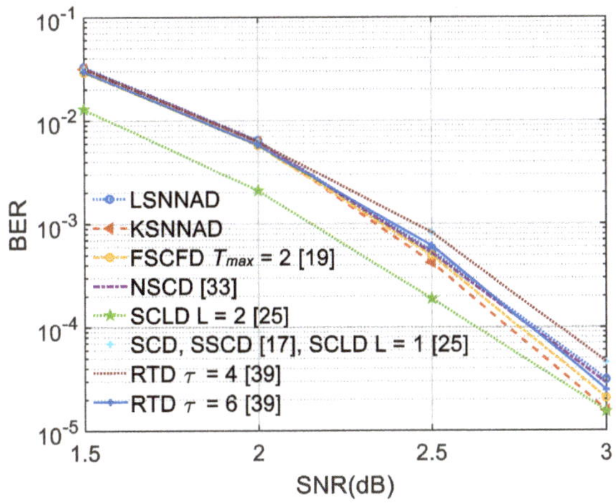

Figure 16. Decoding results of different decoding schemes.

As shown in Figure 16, the BER curves of SCD, SSCD and SCLD with L = 1 are just the same. The BER performance of LSNNAD scheme was slightly worse than NSCD and RTD with τ= 6, whereas the performance of KSNNAD is better than FSCFD with T_{max} = 3 and worse than SCL decoder with L = 2.

4.5. Performance Comparison under Rayleigh Channel

A transmitted signal s passes through the Rayleigh flat-fading channel, and the received signal can be expressed as follows:

$$g_{ReLU}(z) = \max\{0, z\}. \quad (11)$$

where $h(t)$ is the fading factor, which obeys the Rayleigh distribution; $\theta(t)$ is the phase; and $n(t)$ is the complex white Gaussian noise. The comparison between the two decoding schemes and SC algorithm is shown in Figure 17.

As shown in Figure 17, the decoding performance from poor to excellent over Rayleigh channel is as follows: SCD, LSNNAD, and KSNNAD.

4.6. Decoding Complexity

4.6.1. Number of Arithmetic Operations

Compared with FSCD, the difference in the LSNNAD scheme and the KSNNAD scheme lies in node recognition and the training of the NN model. The complexity of NNN recognition and NN training can be ignored because they can be performed offline before the decoding. As to the complexity of node recognition online, LSNNAD needs at most $(2+N)\cdot\log_2 N - 18$ comparisons in the worst situation but only $2N/3 - 16$ comparisons in the best situation, while KSNNAD needs at most $(4+N)\cdot\log_2 N - 18$ comparisons but only $2N/3 - 16$ comparisons in the best situation.

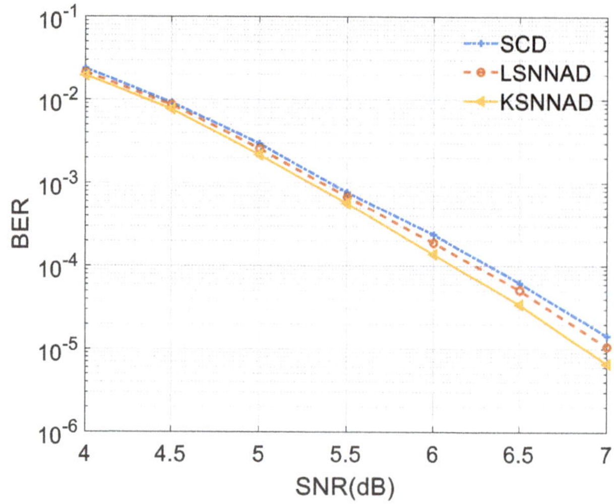

Figure 17. Comparison of decoding results.

Since the subcodes are not of equal length, a specific expression of the arithmetic operation cannot be given for each specific polar code. Let N_S denote code length of different special nodes. When decoding a polar code with a code length of 1024, a code rate of 0.5, and construction SNR 2.5 dB, number of special nodes with different node length for different decoding schemes is shown in Table 2, number of arithmetic operations for different decoding steps is shown in Table 3, and total number of arithmetic operations for different decoding schemes is shown in Table 4. Among them, LSNNAD and KSNNAD were simulated under the NN code length of $N_S = 16$.

Table 2. Number of special nodes for different decoding schemes.

	N_S	R1	R0	Rep	SPC	Type-I	Type-II	Type-III	Type-IV	Type-V	NNN
FSCFD [19]	8	1	0	6	8	0	2	1	1	9	0
	16	1	0	4	4	0	0	0	0	1	0
	32	0	1	0	1	0	1	1	0	0	0
	64	0	0	1	1	0	0	0	1	1	0
	128	0	0	1	1	0	0	0	0	0	0
FSCD	8	3	3	5	5	0	0	0	0	0	0
	16	4	4	9	7	0	0	0	0	0	0
	32	1	2	0	1	0	0	0	0	0	0
	64	0	0	1	1	0	0	0	0	0	0
	128	0	0	1	1	0	0	0	0	0	0
LSNNAD (N_S = 16)	8	3	3	5	5	0	0	0	0	0	0
	16	5	4	9	7	0	0	0	0	0	1
	32	2	2	0	1	0	0	0	0	0	0
	64	0	0	1	1	0	0	0	0	0	0
	128	0	0	1	0	0	0	0	0	0	0
KSNNAD (N_S = 16)	8	3	3	5	5	0	0	0	0	0	0
	16	4	5	9	7	0	0	0	0	0	1
	32	1	3	0	1	0	0	0	0	0	0
	64	0	1	1	1	0	0	0	0	0	0
	128	0	0	0	1	0	0	0	0	0	0

Table 3. Number of arithmetic operations for different decoding steps.

Decoding Step	Arithmetic Operation	FSCFD [19]	FSCD	LSNNAD	KSNNAD
Subcode recognition	Comparison	5984	6128	6264	6315
R1	Addition Multiplication	$9N_S - 7$ $N_S - 1$	$2N_S$ $4N_S$	$2N_S$ $4N_S$	$2N_S$ $4N_S$
Rep	Addition Multiplication	$8N_S - 6$ $N_S - 1$	N_S $3N_S$	N_S $3N_S$	N_S $3N_S$
SPC	Addition Multiplication	$11N_S - 7$ $2N_S - 1$	N_S 0	N_S 0	N_S 0
Type-I	Addition Multiplication	$8N_S - 7$ $N_S - 1$	0 0	0 0	0 0
Type-II	Addition Multiplication	$10N_S - 5$ N_S	0 0	0 0	0 0
Type-III	Addition Multiplication	$10N_S - 3$ $N_S + 1$	0 0	0 0	0 0
Type-IV	Addition Multiplication	$10N_S - 3$ $N_S + 1$	0 0	0 0	0 0
Type-V	Addition Multiplication	$41N_S - 7$ $69N_S - 1$	0 0	0 0	0 0
NNN	Addition Multiplication	0 0	0 0	$164K_S^2$ $164K_S^2$	$164K_S^2$ $164K_S^2$

Table 4. Total number of arithmetic operations for different decoding schemes.

Arithmetic Operation	FSCFD [19]	FSCD	LSNNAD	KSNNAD	RTD [39] (τ=4)	RTD [39] (τ=6)
Comparison	5984	6128	6264	6315	3587	60,270
Addition	15,081 ($T_{max} + 1$)	864	43,152	43,056	5358	69,256
Multiplication	11,653 ($T_{max} + 1$)	1248	43,280	43,208	0	0

One symbolic operation, taking absolute values and subtracting, is recorded as one addition; and one logarithmic and exponential operation is recorded as two multiplications. The operations of each decoding process are shown as Table 3.

An exclusive or operation is recorded as an addition, so FSCFD's CRC check needs N additions at most. Calling the model once will introduce $164K_S^2$ multiplications and $164K_S^2 + 13K_S$ additions in LANNAD and KSNNAD. Now we have the number of arithmetic operations for different special nodes and the subcode recognition, as well as the number of special nodes, so the total number of arithmetic operations for different decoding schemes can be calculated, and the results are shown in Table 4.

As shown in Table 4, when $T_{max} = 2$, FSCFD and KSNNAD are the closest in complexity, and the performance of KSNNAD is better than FSCFD, according to simulation results in Figure 16. RTD with $\tau = 6$ is much more computationally intensive than KSNNAD but does not perform as well as KSNNAD in Figure 16.

4.6.2. Decoding Steps

Since the subcodes are not of equal length, a specific expression of the time step cannot be given for each specific polar code. Table 5 shows the decoding steps and node structure of LSNNAD, KSNNAD, and the algorithms proposed in Refs. [17,33] for a polar code with a code length of 128, a code rate of 0.5, and construction SNR 2.5 dB. Among them, NSCD proposed in Ref. [33], LSNNAD and KSNNAD were simulated under the NN code length of $N_S = 16/N_S = 8$. Let None (64) denote 64 bits that have not been recognized by any nodes. Note that the 64 bits are not 64 consecutive bits but several continuous strings scattered of bits. Further, 21R1 denotes that there are 21 R1 nodes to be recognized in total, but we do not provide every length of the R1 nodes.

Table 5. Comparison of decoding steps.

Decoding Algorithm	Decoding Steps	Node Structure
SCD	317	None(128)
SSCD [17]	146	21R1 + None (64)
NSCD [33]	28/60	8NNN/16NNN
LSNNAD	66/70	2R0 + 6Rep + 4SPC + 5R1 + NNN/ 2R0 + 6Rep + 4SPC + 6R1 + NNN
KSNNAD	70/74	3R0 + 5Rep + 4SPC + 6R1 + NNN/ 4R0 + 5Rep + 4SPC + 6R1 + NNN

It can be seen from Table 5 that the decoding steps of Ref. [33] were greatly affected by the length of the NNN. Moreover, they call the NN model multiple times, and each call will give rise to $164K_S^2$ multiplications and $164K_S^2 + 13K_S$ additions. In contrast, the decoding schemes proposed in this paper are less affected by the length of the NNN because they only call the decoding model once. Compared with the traditional SCD algorithm, they can still effectively reduce the running time step of the decoding algorithm.

5. Conclusions

This paper proposes two NNA decoding schemes named LSNNAD and KSNNAD. There are two differences between the two schemes: one is the recognition method. After recognizing the polar code as a combination of NNN, R1, R0, SPC, and Rep node, decoding is performed and the trained NN model is called. The other is the training label. The LSNNAD scheme's training label is the original information bits, and the KSNNAD scheme's training label is the right decoding result with the flipping key bit derived from a definitely matched model. The simulation results verified the effectiveness of the two schemes. Since KSNNAD can effectively improve the decoding accuracy of the first error location, its decoding BER performance curve is better than that of the LSNNAD.

Author Contributions: Methodology, H.L.; software, W.Y.; validation, H.L.; formal analysis, H.L.; investigation, H.L.; resources, H.L.; data curation, H.L.; writing—original draft preparation, H.L.; writing—review and editing, W.Y.; supervision, Q.L.; project administration, L.Z.; funding acquisition, W.Y. All authors have read and agreed to the published version of the manuscript.

Funding: This research was funded by the NATIONAL NATURAL SCIENCE FOUNDATION OF CHINA, grant number 62271499.

Institutional Review Board Statement: Not applicable.

Informed Consent Statement: Not applicable.

Data Availability Statement: Not applicable.

Acknowledgments: We are very grateful for Trifonov P's help with this paper in the simulation comparison.

Conflicts of Interest: The authors declare no conflict of interest.

References

1. Helleseth, T.; Klve, T. Algebraic coding theory. In *Coding Theory*; John Wiley & Sons: San Francisco, CA, USA, 2007; pp. 13–96.
2. Gallager, R. Low-density parity-check codes. *IRE Trans. Inf. Theory.* **1962**, *8*, 21–28. [CrossRef]
3. Machay, D.J.C. Good codes based on very sparse matrices. *IEEE Trans. Inf. Theory* **1999**, *45*, 399–431. [CrossRef]

4. Berrou, C.; Glavieux, A.; Thitimajshima, P. Near Shannon limit error-correcting coding and decoding: Turbo codes. In Proceedings of the IEEE International Conference on Communications, Geneva, Switzerland, 23–26 May 1993; pp. 1064–1070.
5. Berrou, C.; Glavieux, A. Near optimum error-correcting coding and decoding: Turbo codes. *IEEE Trans. Commun.* **1996**, *44*, 1261–1271. [CrossRef]
6. Bose, R.C.; Chaudhuri, D.K. On a class of error correcting binary group codes. *Inform. Control* **1960**, *3*, 68–79. [CrossRef]
7. Hocquenghem, A. Codes correcteurs d'erreurs. *Chiffres* **1959**, *2*, 147–156.
8. Reed, I.S.; Solomon, G. Polynomial codes over certain finite fields. *J. Soc. Ind. Appl. Math.* **1960**, *8*, 300–304. [CrossRef]
9. Virerbi, A.J. Error bounds for convolutional codes and an asymptotically optimum decoding algorithm. *IEEE Trans. Inf. Theory* **1967**, *13*, 260–269. [CrossRef]
10. Arikan, E. Channel polarization: A method for constructing capacity-achieving codes. In Proceedings of the IEEE International Symposium on Information Theory, Toronto, ON, Canada, 6–11 July 2008; pp. 3051–3073.
11. Shannon, C.D. A mathematical theory of communication. *Bell Syst. Tech. J.* **1948**, *27*, 379–423. [CrossRef]
12. Liu, Y.J. *Channel Codes: Classical and Modern*, 1st ed.; Electronic Industry Press: Beijing, China, 2022; pp. 1–576.
13. Jin, S.; Liu, X. A memory efficient belief propagation decoder for polar codes. *China Commun.* **2015**, *12*, 34–41.
14. Arikan, E. Channel polarization: A method for constructing capacity-achieving codes for symmetric binary-input memoryless channels. *IEEE T. Inform. Theory* **2009**, *55*, 3051–3073. [CrossRef]
15. Choi, S.; Yoo, H. Area-efficient early-termination technique for belief-propagation polar decoders. *Electronics* **2019**, *8*, 1001. [CrossRef]
16. Abbas, S.M.; Fan, Y.; Chen, J. Low complexity belief propagation polar code decoder. In Proceedings of the IEEE Workshop on Signal Processing Systems, Hangzhou, China, 14–16 October 2015; pp. 1–6.
17. Alamdar-yazdi, A.; Kschischang, F.R. A simplified successive-cancellation decoder for polar codes. *IEEE Commun. Lett.* **2011**, *15*, 1378–1380. [CrossRef]
18. Sarkis, G.; Giard, P.; Vardy, A.; Thibeault, C.; Gross, W.J. Fast polar decoders: Algorithm and implementation. *IEEE J. Sel. Areas Commun.* **2014**, *32*, 946–957. [CrossRef]
19. Ardakani, M.H.; Hanif, M.; Ardakani, M.; Tellambura, C. Fast successive-cancellation-based decoders of polar codes. *IEEE Trans. Commun.* **2019**, *67*, 2360–2363. [CrossRef]
20. Chen, K.; Niu, K. Stack decoding of polar codes. *Electron. Lett.* **2012**, *48*, 695–696.
21. Kruse, R.; Mostaghim, S.; Borgelt, C.; Steinbrecher, M. Multi-layer perceptrons. In *Computational Intelligence*; Springer: Berlin, Germany, 2022; pp. 53–124. [CrossRef]
22. Strigl, D.; Kofler, K.; Podlipnig, S. Performance and scalability of GPU-based convolutional neural networks. In Proceedings of the 18th Euromicro Conference on Parallel, Distributed and Network-based Processing, Pisa, Italy, 17–19 February 2010; pp. 317–324.
23. Ben, N.M.; Chtourou, M. On the training of recurrent neural networks. In Proceedings of the 8th International Multi-Conference on Systems, Signals & Devices, Sousse, Tunisia, 22–25 March 2011; pp. 1–5.
24. Cao, Z.; Zhu, H.; Zhao, Y.; Li, D. Learning to denoise and decode: A novel residual neural network decoder for polar codes. In Proceedings of the IEEE 92nd Vehicular Technology Conference, Victoria, BC, Canada, 18 November–16 December 2020; pp. 1–6.
25. Tal, I.; Vardy, A. List decoding of polar codes. *IEEE Trans. Inf. Theory.* **2015**, *61*, 2213–2226. [CrossRef]
26. Xu, W.H.; Wu, Z.Z.; Ueng, Y.L. Improved polar decoder based on deep learning. In Proceedings of the IEEE International Workshop on Signal Processing Systems, Lorient, France, 3–5 October 2017; pp. 1–6.
27. Hashemi, S.; Doan, N.; Tonnellier, T.; Gross, W.J. Deep-learning-aided successive-cancellation decoding of polar codes. In Proceedings of the 2019 53rd Asilomar Conference on Signals, Systems, and Computers, Pacific Grove, CA, USA, 03–06 November 2019; pp. 532–536.
28. Gruber, T.; Cammerer, S.; Hoydis, J.; Ten Brink, S. On deep learning-based channel decoding. In Proceedings of the 51st Annual Conference on Information Sciences and Systems, Baltimore, MD, USA, 22–24 March 2017.
29. Cammerer, S.; Gruber, T.; Hoydis, J.; Ten Brink, S. Deep learning-based decoding of polar codes via partitioning. In Proceedings of the IEEE Global Communications Conference, Singapore, 4–8 December 2017.
30. Teng, C.F.; Wu, A.Y. Unsupervised learning for neural network-based polar decoder via syndrome loss. *Bell Syst. Tech. J.* **1948**, *27*, 379–423.
31. Gao, J.; Zhang, D.X.; Dai, J.C. Resnet-like belief-propagation decoding for polar codes. *IEEE Wirel. Commun. Le.* **2021**, *10*, 934–937. [CrossRef]
32. Xu, W.H.; Tan, X.S.; Be'ery, Y.; Ueng, Y.L.; Huang, Y.M.; You, X.H.; Zhang, C. Deep learning-aided belief propagation decoder for polar codes. *arXiv* **2019**, arXiv:Signal Processing. [CrossRef]
33. Doan, D.; Hashemi, S.H.; Gross, W.J. Neural successive cancellation decoding of polar codes. In Proceedings of the IEEE 19th International Workshop on Signal Processing Advances in Wireless Communications, Kalamata, Greece, 25–28 June 2018; pp. 1–5.
34. Xu, X. *Study on Decoding Algorithms of Polar Codes Based on Deep Learning*; Beijing Jiaotong University: Beijing, China, 2019.
35. Arikan, E. Systematic polar coding. *IEEE Commun. Lett.* **2011**, *15*, 860–862. [CrossRef]
36. Kingma, D.P.; Ba, J.L. Adam: A Method for Stochastic Optimization. Available online: http://arxiv.org/abs/1412.6980 (accessed on 30 January 2017).

37. Li, S.B.; Deng, Y.Q.; Gao, X. Generalized segmented bit-flipping scheme for successive cancellation decoding of polar codes with cyclic redundancy check. *IEEE Access* **2019**, *7*, 83424–83436. [CrossRef]
38. Zhang, Z.Y.; Qin, K.J.; Zhang, L.; Chen, G.T. Progressive bit-flipping decoding of polar codes: A critical-set based tree search approach. *IEEE Access* **2018**, *6*, 57738–57750. [CrossRef]
39. Trifonov, P. Recursive trellis decoding techniques of polar codes. In Proceedings of the IEEE International Symposium on Information Theory, Los Angeles, CA, USA, 21–26 June 2020; pp. 407–412.

applied sciences

MDPI

Article

A Nonuniformity Correction Method Based on 1D Guided Filtering and Linear Fitting for High-Resolution Infrared Scan Images

Bohan Li [1,2], Weicong Chen [1,2] and Yong Zhang [2,*]

1 University of Chinese Academy of Sciences, Beijing 100049, China
2 Shanghai Institute of Technical Physics Chinese Academy of Sciences, Shanghai 200083, China
* Correspondence: zhangyong@mail.sitp.ac.cn

Abstract: During imaging, each infrared focal plane linear array scan detector detection unit determines a row of pixels in the image output. This sensor's nonuniformity appears as horizontal stripes. Correcting nonuniformity in high-resolution images without destroying delicate details is challenging. In this paper, a single-frame-based nonuniformity correction algorithm is proposed. A portion of a single-frame picture is intercepted initially. The 1D column guided filter is applied to smooth the captured image in the vertical direction. Then, the smooth image and high-frequency component with horizontal stripes and texture information are obtained. The subsequent step is to use the smooth portion of the image as the guided image and the high-frequency portion of the image as the input, so that the estimated nonuniformity noise of the image may be extracted using a 1D row guided filter. The segment of the corrected image is then obtained by subtracting the estimated nonuniformity noise from the segment of the raw image. The correction coefficients could be obtained by performing a linear regression fit on the pre- and post-guided filtering image segments. With the correction coefficients, the entire image could be corrected. Based on qualitative and quantitative analysis, the proposed algorithm outperforms other current advanced algorithms in terms of nonuniformity correction and real-time performance.

Keywords: guided filtering; nonuniformity correction; infrared focal plan linear array detector; linear regression fitting

Citation: Li, B.; Chen, W.; Zhang, Y. A Nonuniformity Correction Method Based on 1D Guided Filtering and Linear Fitting for High-Resolution Infrared Scan Images. *Appl. Sci.* **2023**, *13*, 3890. https://doi.org/10.3390/app13063890

Academic Editors: Przemysław Falkowski-Gilski, Tadeus Uhl and Zbigniew Lubniewski

Received: 17 February 2023
Revised: 12 March 2023
Accepted: 15 March 2023
Published: 18 March 2023

1. Introduction

In recent years, infrared imaging systems have been greatly improved in terms of technical conditions and manufacturing processes, and they are now widely used in the fields of thermal imaging, fire detection, and aerial small target reconnaissance. The common infrared detectors are divided into infrared focal plane array detectors and infrared focal plane linear array scan detectors. Nevertheless, when the infrared detector is imaging, the response coefficient of the detector will fluctuate correspondingly under the impact of variations in the ambient temperature and the voltage of the imaging circuit, resulting in fixed pattern noise in the infrared system imaging [1,2]. This type of noise is typically difficult to eliminate directly at the hardware level, and the most prevalent techniques for compensating for hardware faults fall into two categories: reference-based correction methods [3] and scene-based correction methods [4,5].

The reference-based correction methods generally require the acquisition of blackbody temperature images, which are later used for the corrections performed by calibration. Common methods include single-point correction [1,6], two-point correction [7,8], multi-point correction [9,10], and polynomial fit correction [11,12]. Prior to being put into service, infrared detectors are typically adjusted using this procedure. Changes in temperature, voltage, and other environmental conditions will affect the detector response coefficients when the infrared detector is in use. Currently, if the reference-based correction method is continued, it is necessary to cease equipment use and recalibrate infrared detectors, which severely impairs the normal operation of infrared detectors. To remedy this issue, scene-based non-uniform correction methods use the textural qualities of the acquired scenes to correct the images in real time, allowing for the detector to be recalibrated without the need for the blackbody.

In 2012, Tender et al. [13] proposed a nonuniformity correction method for infrared focal plane array detectors based on histogram statistics. This method can correct the nonuniformity of the image by modifying the histogram of the adjacent column of the image of an infrared array. However, this algorithm requires an image histogram forward transform as well as an inverse transform, and it has a significant temporal complexity. He et al. [14] first proposed a guided image filtering method in 2012. Y. Cao et al. [15] applied the guided filtering method to the nonuniformity correction of the infrared focal plane array detector in 2016. This method effectively removed the nonuniformity noise of the array image by constructing a one-dimensional guided filtering window and conducting twice guided filtering operations. Guided filtering needs the calculation of linear coefficients for each window, making it more time-consuming for high-resolution images. In 2018, Y. Cao et al. [16] proposed a denoising method based on multi-level wavelet transforms and guided filtering. This method corrects the nonuniformity of the array image by conducting guided filtering operations on the vertical high-frequency components of the image after the wavelet transform. After the wavelet transforms, the guided filtering only targets the high-frequency image components, which drastically reduces the temporal complexity of this method. However, the approach processes the wavelet transform in the frequency domain and tends to blur the image details. After that, Ende Wang et al. [17,18] combined TV regularization, wavelet transform, and guided filtering to the nonuniformity correction of the focal plane array detector. Due to frequency domain denoising, all of these methods may blur visual features and produce ghost effects. In addition, the multi-frame nonuniformity correction methods [19–23] and nonuniformity correction methods [24–27] based on a deep learning model are also applied to the field of image nonuniformity correction.

All of the aforementioned techniques are nonuniformity correction techniques for infrared focal plane array detectors. In this paper, based on the detector's features, a linear rectification model of an infrared focal plane linear array scan image is proposed using guided filtering. In this paper, we first extract a portion of the infrared linear array scan image and then apply 1D column guided filtering operations to a blurred image and the high-frequency component of the original image. The high-frequency component image is taken as the input image, and the blurred image is taken as the guide image. The 1D row guided filtering is performed to extract the nonuniformity noise of the linear array scan image. By subtracting the estimated nonuniformity noise image from the original image, the partial linear scan array image corrected by the guided filter is obtained. Finally, using the linear regression model, the linear correction coefficients are calculated to complete the nonuniformity correction of the entire frame.

Compared to the other correction algorithms [13–27], the method in this paper has the following advantages. First of all, this approach employs only a portion of a linear array scan image to calculate the linear model's correction coefficients and complete the frame's correction. The image resolution used for guided filtering calculation is significantly lowered as well as the computational complexity, which is well suited to satisfy the real-time demands of high-resolution infrared focal plane linear array scan images. Secondly, since the same row of pixels in the infrared scan image is determined only by a single detection unit, the algorithm in this study combines the principles of infrared column detector calibration and guided filtering to develop a one-dimensional linear regression coefficient model, which further improves the image's correction effect. The proposed approach corrects the nonuniformity in the spatial domain without degrading the image's high-frequency texture features or introducing ghost artifacts. Thirdly, the approach presented in this paper is a single-frame image algorithm that does not require consideration of the influence of camera jitter and scene heat radiation variations on nonuniformity correction and has high robustness. Finally, the approach presented in this research is superior to the correction algorithm based on deep learning that disregards, without considering the creation of the dataset.

2. Proposed Approach

Our proposed infrared image nonuniformity correction approach comprises two processing stages: (1) to intercept a portion of an entire visual frame and apply two one-dimensional guided filtering operations to it; (2) to use the portion of the image before and after guided filtering to calculate the linear correction coefficients and apply them to correct the whole frame image. The result of our proposed algorithm's non-uniform correction on the real infrared scanning image is depicted in Figure 1. The processing pipeline of guided filtering is depicted in Figure 2. The complete processing pipeline is schematically illustrated in Figure 3.

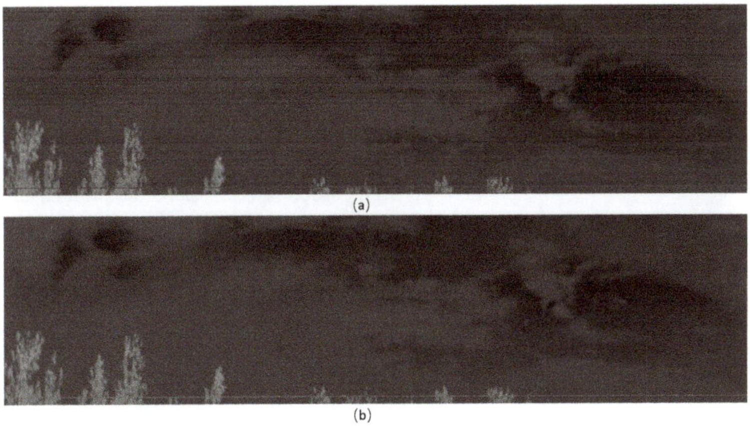

Figure 1. Non-uniform correction effect of real infrared scanning image. (**a**) Raw infrared scanning image. (**b**) Non-uniform correction results of our proposed algorithm.

2.1. Nonuniformity Correction Model Based on 1D Guided Image Filtering

The guided filter denoising model was first proposed by He [14] in 2012. In 2015, Cao [15] applied the guided filter denoising model to the nonuniformity correction of infrared uncooled array detectors. They obtained good results. In this subsection, we propose a novel non-uniform correction model based on two 1-D guided filtering for infrared focal plane linear array scan detectors. The implementation of guided filtering [14] is based on the assumption that the guided picture and the output image captured in a limited window have a linear connection, as follows:

$$q_i = a_k v_i + b_k, \forall i \in w_k \tag{1}$$

where q_i is the pixel value of the output image, (a_k, b_k) is the linear coefficients in the small window, and v_i is the pixel value of the guided image. w_k is the window, and noise characteristics must be taken into account when designing the window form. The guided filtering is completed by constraining the minimum difference between the output image q_i and the input image u_i:

$$E(a_k, b_k) = \sum_{i \in w_k} ((a_k v_i + b_k - u_i)^2 + \varepsilon a_k) \tag{2}$$

where $E(a_k, b_k)$ is the cost function, ε is a regularization parameter penalizing large a_k. The results of coefficients (a_k, b_k) [14] is shown as

$$a_k = \frac{\frac{1}{|w|}\sum_{i\in w_k} v_i u_i - \mu_k \bar{u}_k}{\frac{1}{|w|}\sum_{i\in w_k} v_i^2 - \mu_k^2 + \varepsilon} \tag{3}$$

$$b_k = \bar{u}_k - a_k \mu_k \tag{4}$$

where μ_k is the mean of the guide image v in the window, $\bar{u}_k$ is the mean of the input image u_i in the window w_k, and $|w|$ is the number of pixels in the window w_k. ε is a manually set parameter, which determines the strength of local averaging. Its effect is similar to the range variance parameter σ_r^2 in a bilateral filter [28]. However, a pixel i is included in all overlapping windows w_k that cover i, so the value of q_i in (4) is not the same when computed in various windows. A basic technique is to calculate the mean of all conceivable values of q_i. Thus, after computing (a_k, b_k) for each window w_k in the image, we compute the output of the filtering by

$$q_i = \frac{1}{|w|}\sum_{k|i\in w_k}(a_k v_i + b_k) = \bar{a}_i v_i + \bar{b}_i \tag{5}$$

where $k|i \in w_k$ defines a number of overlapping windows w_k covering the pixed i, $\bar{a}_i = \frac{1}{|w|}\sum_{k\in w_i} a_k$, and $\bar{b}_i = \frac{1}{|w|}\sum_{k\in w_i} b_k$ are the coefficient averages of all windows overlapping pixel i. For the nonuniformity of the infrared focal plane linear array scan detector, this paper designed a guided filtering correction model, as shown in Figure 2.

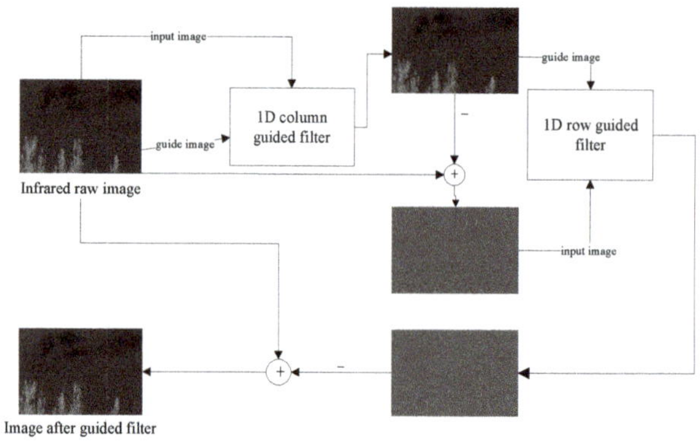

Figure 2. Guided filter processing workflow of the proposed method for horizontal stripe.

Since the nonuniformity of the focal plane linear array scan image appears as horizontal strips, the algorithm first takes the infrared raw image I_p as a guided image and an input image. It smooths the image through a 1D column guided filter window, and at this point, Equations (3) and (4) become Equations (6) and (7).

$$a_k' = \frac{\sigma_k^2}{\sigma_k^2 + \varepsilon} \tag{6}$$

$$b_k = (1 - a_k')\mu_k \tag{7}$$

where μ_k and σ_k^2 are the mean and variance of the image I_p in the window w_k. Equation (1) is used to conduct guided filtering operations, and the output image I_u is the smoothed image in the vertical direction. An image I_n containing nonuniformity noise and part of the image's high-frequency texture is obtained by subtracting I_u from I_p, and the formula is given by

$$I_n = I_p - I_u \tag{8}$$

In order to better extract the nonuniformity noise from the image I_n, a 1D row guided filtering window is designed in this paper, with I_u as the guided image and I_n as the input image. The output image I_s is obtained by guided filtering using Equation (5). We assume that the image I_s contains most of the nonuniformity noise and little horizontal high-frequency texture detail. The image after nonuniformity correction is obtained by subtracting I_s from I_p, as follows:

$$I_q = I_p - I_s \tag{9}$$

2.2. Linear Fitting Correction Model Based on Guided Filtering

Nonuniformity has been effectively corrected by the guided filter model. However, the image's high-frequency information may be unnecessarily blurred by the process of guided filtering. Since the infrared focal plane linear array scan detector imaging speed is fast, there would be tens of thousands of columns of images per second, and the guided filter needs to calculate the linear coefficients (a_k, b_k) of each window, as shown in Equations (3) and (4). The algorithm has a high time complexity and large computational overhead, making it difficult to ensure real-time performance. Traditionally, the nonuniformity correction methods of an infrared array detector involve collecting the temperature of a blackbody and employing a linear model with two or more points to adjust. This is consistent with the assumption that the guided image and the output image are linearly related in the window. Therefore, the small window of guided filtering is discarded. The guided filtering window is a one-dimensional horizontal window, and its length is equal to the width of the guided image. The linearity coefficients are calculated using the image I_q after the guided filtering as the input image and the raw infrared image I_p as the guided image [29]. At this moment, the coefficients of the guided filter only need to be calculated once, and the computation of the linear coefficients is turned into least squares linear regression, which significantly decreases the algorithm's temporal complexity, as follows:

$$a(i) = \frac{\sum\limits_{j=1}^{W} I_q(i,j)(I_p(i,j) - \sum\limits_{k=1}^{W} I_p(i,k)/W)}{\sum\limits_{j=1}^{W} (I_p(i,j) - \sum\limits_{k=1}^{W} I_p(i,k)/W)^2} \tag{10}$$

$$b(i) = \frac{\sum\limits_{j=1}^{W} I_q(i,j) - a(i) \sum\limits_{j=1}^{W} I_p(i,j)}{W} \tag{11}$$

where W is the width of the image after a guided filtering process, $I_p(i,j)$ is the image that was partially extracted from the full-frame raw infrared image with a horizontal strip, and $I_q(i,j)$ is $I_p(i,j)$ after guided filtering, as shown in Figure 3. The obtained linear correction coefficients are used to correct the whole frame of the raw infrared image:

$$\hat{I}(i,j) = a(i) \times \hat{I}_p(i,j) + b(i) \tag{12}$$

where $a(i), b(i)$ are the correction coefficients of the i th row of the image, $\hat{I}_p(i,j)$ is the whole frame raw infrared image, and $\hat{I}(i,j)$ is the whole frame image after nonuniformity correction.

2.3. Proposed Workflow of the Proposed Method for Strip Nonuniformity Correction

The overall flow chart of the proposed algorithm in this paper is shown in Figure 3. There are specific steps in the algorithm.

1. A portion of the raw infrared scan image is intercepted and used to calculate the linear correction coefficients. After the experiments (see Section 3.2), the linear correction coefficients are calculated using the 1500 image columns chosen for this work.

2. The selected image is used to eliminate the horizontal strip caused by nonuniformity with the guided filtering model proposed in Section 2.1.
3. We use the linear fitting model based on guided filtering proposed in Section 2.2 to calculate the linear correction coefficients $a(i), b(i)$.
4. The whole frame of the infrared image is corrected using linear coefficients as shown in the Formula (12).

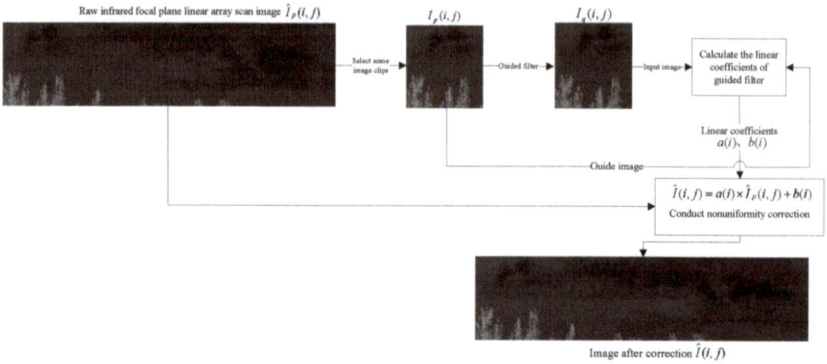

Figure 3. The overall flow chart of the proposed algorithm for nonuniformity correction.

3. Experiments and Comparison

This paper compares this algorithm against the most recent nonuniformity correction algorithms for infrared detectors to validate the algorithm's correction capabilities. The nonuniformity correction algorithm based on multi-level wavelet transform and guided filtering was proposed by Cao [16] in 2018. The nonuniformity correction algorithm based on wavelet decomposition and TV regularization was proposed by Wang [18] in 2020, and the other is the algorithm based on wavelet decomposition and image column equalization proposed by Wang [17] in 2019. The data in this paper are images from the FLIRADAS dataset, which includes 4224 infrared images, each with a resolution of 480 × 640. By analyzing the noise of the focal plane linear array, the simulated horizontal stripe is added to the image of the FLIRADAS dataset in order to quantitatively and qualitatively evaluate the algorithm's correction effect. The infrared long-wave scan detector acquires the real image set of the long-wave infrared scan, with a total of 50 frames and a resolution of 3053 × 55,000. This research uses 20 frames of scan images without nonuniformity to design the algorithm's parameters. For qualitative examination, 30 frames of real scan images with nonuniformity are used.

3.1. Noise Modeling Analysis

The causes of nonuniformity in infrared images are very complicated. From the perspective of signal transmission, the nonuniformity of sensors is mainly caused by two reasons. One is the inconsistent response of the photosensitive element of the infrared detection unit, while the other is the impact of the readout circuit's noise. According to the analysis in Song [3], the detector response model can be approximated as a linear model when the infrared detector operates normally, as indicated in the Formula (13):

$$Y(i) = g(i)R(i) + o(i) + \sigma(i) \tag{13}$$

where $Y(i)$ denotes the response value output by the i-th unit in the direction of the detector column. $R(i)$ indicates the actual amount of infrared radiation received by the detection device. $g(i)$ and $o(i)$ are the gain coefficient and bias coefficient of the infrared detection device, respectively. $\sigma(i)$ represents random noise. Random noise $\sigma(i)$ can be ignored when the signal-to-noise ratio of the infrared detector is high. For infrared scanning detectors, the output values of the same row of images have the same gain and bias coefficients, so the nonuniformity of the detector appears as visible horizontal stripes. In conventional infrared detector nonuniformity correction, the black body is typically employed for image acquisition first, followed by the linear model for two-point correction [3], as illustrated by Equation (14). The objective is to ensure that each detector unit has an identical response curve.

$$\hat{I}(i,j) = k(i) \times v(i,j) + m(i) \tag{14}$$

where $\hat{I}(i,j)$ represents the image after correction, $k(i)$ represents the gain correction coefficient of the linear model, and $m(i)$ represents the bias correction coefficient. Analyzing the linear correction model of nonuniformity noise, this paper proposes an analog nonuniformity model of infrared detectors, as demonstrated by Equation (15).

$$I_{1n}(i,j) = g(i) \times I_1(i,j) + b(i) \tag{15}$$

where $I_1(i,j)$ is the infrared image without nonuniformity, $I_{1n}(i,j)$ is the image with added analog horizontal stripes, $i \in [1, H]$. H is the image's height. $g(i)$ is a sequence of Gaussian distribution with a mean value of 1, and $b(i)$ is a sequence of Gaussian distribution with a mean value of 0. The horizontal stripe of an image is more apparent as the variance of the Gaussian sequence increases.

3.2. Image Correction Effect of Image Size Used in Linear Fitting Model

The two-step approach for infrared scan images is proposed in this research. First, a portion of the high-resolution scan image is intercepted, and then the intercepted image is corrected using a one-dimensional guided filter model. The next step is to calculate the coefficients of the linear fitting model and then use the correction coefficients to correct the full image frame. This section primarily studies the effect of intercepted image size on the correction. The experimental data consists of 20 frames of real infrared scanning images without nonuniformity to which horizontal stripe was added using Formula (15), where parameter $g(i)$ follows a Gaussian distribution with a mean of 1 and a variance of 0.02 and parameter $b(i)$ follows a Gaussian distribution with a mean of 0 and a variance of 0.02. The scan images with various column counts are intercepted, and the PSNR of the images after correction is computed. Table 1 displays the average findings of the full dataset.

Table 1. Image correction effect of image size used in linear fitting model.

Images\Colums	500	1000	1500	2000	2500
PSNR	39.79	40.33	40.76	40.78	40.74

Table 1 demonstrates that when 1500 scan image columns are picked for the calculation of correction coefficients, the image can already reach a good level. PSNR increases somewhat when additional scan image columns are selected to calculate correction coefficients. As the number of image columns increases, the PSNR of the corrected image tends to stabilize and does not grow with the number of image columns. This demonstrates that this approach is practical for the nonuniformity correction of the infrared scan image and can only use a portion of the image to calculate the correction coefficients needed to finish the correction of the entire image frame.

3.3. Image Correction Effect of Noise

The previous section's experiment demonstrated that when the number of intercepted image columns is 1500, the effect of the picture after correction becomes more effective and stable. The purpose of this section's experiment is to investigate the effect of noise on image correction. This research adds Gaussian white noise with various variances to the intercepted image. The diagram of the experimental example is presented in Figure 4.

Figure 4. Infrared scan image with the addition of Gaussian noise. Due to the very high resolution of the infrared column scan image (3053 × 55,000), only a portion of the scanned image (3053 × 8192) is displayed in this figure. The image within the red border is the image utilized for the linear fitting model in Section 2.1, with a resolution of 3053 × 1500 pixels.

We add Gaussian white noise with varying variances to this portion of the image, and then use the Formula (15) to add horizontal stripes to the image. $g(i)$ conforms to a Gaussian distribution with a mean of 1, and $b(i)$ conforms to a Gaussian distribution with a mean of 0. $g(i)$ and $b(i)$ have the same variances which are shown in the first row of Table 2. The experimental data consist of 20 frames of infrared scan images without nonuniformity; the experimental outcomes are presented in Table 2.

Table 2. Influence of Gaussian noise in the intercepted image on correction algorithm.

Variance	25	36	49	64	81	100	121	144	169	196	225
PSNR	43.90	44.02	44.14	44.12	44.17	44.12	43.79	43.93	44.03	43.55	43.41

The image of the infrared column scan is an 8-bit image. The first row of Table 2 is the variance of the added Gaussian noise with a mean of 0 and the variance ranges from 25 to 255. The noise intensity increases as the variance of the Gaussian distribution increases. The second row is the average PSNR of the dataset following correcting. Table 2 demonstrates that when the variance of the added Gaussian noise grows, the PSNR of the image collection swings within a given range, and the connection is not linear. Consequently, the Gaussian noise has little effect on the denoising effect of the algorithm described in this paper.

3.4. *Qualitative Analysis of the Correction Effect of the Algorithms*

On the dataset, the algorithm proposed in this research is compared to the nonuniform correction algorithms Cao [16], Wang [17], and Wang [18] developed in recent years, and their correction effects are evaluated qualitatively. The data collection consists of five infrared images from the FLIRADAS dataset and two frames of real infrared scan images with nonuniformity. Figure 5 depicts the five raw infrared images picked from the FLIRADAS dataset. After adding stripes, the images are depicted in Figure 6. The noise parameter $g(i)$ in Formula (15) follows a Gaussian distribution with a mean value of 1 and a variance of 0.02, while $b(i)$ follows a Gaussian distribution with a mean value of 0 and a variance of 0.02. The resolution of the image in Figure 6 is relatively high, and when the image is magnified, the stripes of nonuniformity can be seen.

Since the comparative algorithms Cao [16], Wang [17], and Wang [18] all used wavelet transform in the process, different wavelet bases have a substantial impact on the image correction effect. By comparing the effects of each algorithm on different wavelet bases, the one with the best effect is selected for nonuniformity correction, and the results are compared to the algorithm presented in this paper. The common wavelet-based haar wavelet, sym8 wavelet, and db4 wavelet are used for comparison. The nonuniformity correction effects of Cao [16], Wang [17], and Wang [18] with respect to various wavelet bases, are shown in Figure 7.

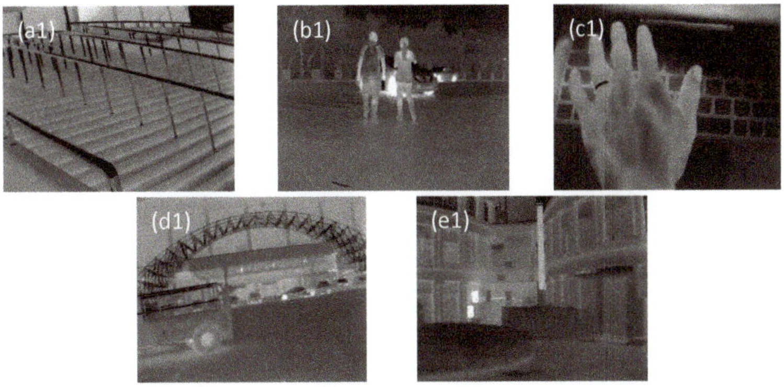

Figure 5. Original image (**a1**) stairs (**b1**) people (**c1**) hands (**d1**) bus (**e1**) buildings.

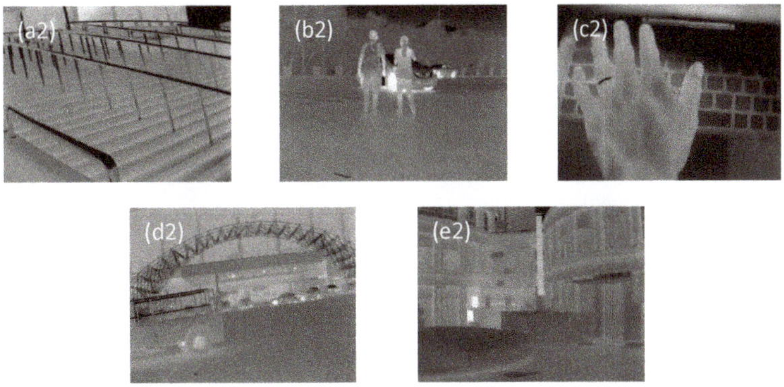

Figure 6. Image after adding analog stripes (**a2**) stairs (**b2**) people (**c2**) hands (**d2**) bus (**e2**) buildings.

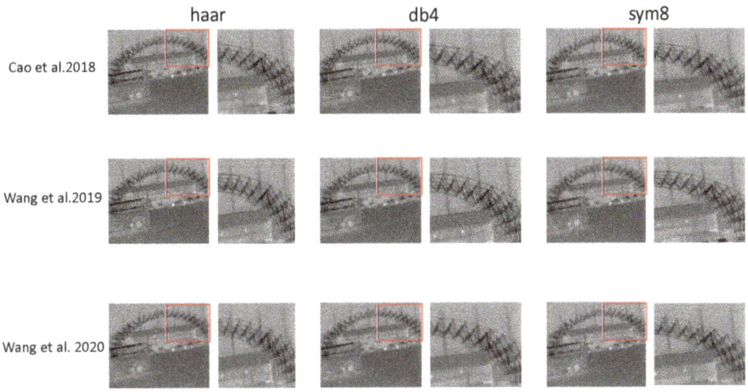

Figure 7. Nonuniformity correction effect of Cao [16], Wang [17], and Wang [18] with respect to distinct wavelet bases. The wavelet bases include "haar", "db4", and "sym8" wavelets. The image "bus" is added with stimulated nonuniformity noise. In order to highlight the correction effect of the algorithm on horizontal stripes and the fuzzy degree of the algorithm on image texture details, the images within the red border are enlarged.

Wang [17] has a good correction impact on the "db4" wavelet, and the texture of the image's details is less blurry. Wang [18] and Cao [16] have severe distortion of image texture detail with the "db4" and "sym8" wavelets, so they selected the "haar" wavelet. The size of the guided filter windows of Cao [16], Wang [17], and Wang [18] are all one-fourth of the width of the processed image sections. Since the simulated image is small, the algorithm in this paper selects the whole frame to calculate the linear correction coefficients without intercepting the image, where the size of the smoothing guide filter window is 8×1, and the size of the horizontal stripe extraction guide filter window is 1×10. The guide filter parameters ε are all 0.16. Figure 8 depicts the effect of image nonuniformity correction using different algorithms.

It can be seen that Cao [16], Wang [17], and Wang [18] have certain effects on horizontal stripe, but they all blur some texture details of the image. Wang [18] has the best nonuniformity correction result on the smooth image, but the blurriest image texture details. Wang [17] and Cao [16] have an average effect on horizontal strip removal, although their harm to picture detail texture is not as severe as Wang's [18]. The damage to image texture details caused by the algorithms of Cao [16], Wang [17], and Wang [18] is mainly due to the fact that they all process the image's horizontal high-frequency information after wavelet transform, and the algorithms eliminate the stripes while also destroying the part of the image with the high-frequency information. In contrast, the algorithm presented in this paper linearly corrects the image in the spatial domain and does not process the image in the frequency domain. Consequently, it destroys fewer image texture details and outperforms the algorithms of Cao [16], Wang [17], and Wang [18] in terms of the nonuniformity correction effect. Finally, in order to better verify the nonuniformity correction effect of this algorithm, real infrared images containing non-uniform strips generated by long-wave infrared focal plane linear array scan detectors are used for experiments to compare the effect of each algorithm. Since the number of columns of a frame of infrared focal plane linear array scan image is about 50,000, in order to show the image's denoising effect, this paper intercepts the scanning image with an image resolution of 3053×4096. The algorithm in this paper uses 1500 columns of scan images to calculate the linear correction coefficients through guided filtering. The size of the smoothing guided filtering window is 12×1. The size of the guided filtering window for horizontal stripe extraction is 1×100, and guided filtering parameters ε are 0.16. The algorithms of Cao [16], Wang [17], and Wang [18] all correct the 3053×4096 images directly.

The real scan image in Figure 9 shows that the scene is the sky, and the horizontal stripes caused by nonuniformity are plainly visible in the clouds. The effects of the Cao [16], Wang [17], and Wang [18] algorithms are not satisfactory in the high-resolution scan image, which contains a large number of horizontal stripes. The correction impact of the proposed algorithm is clearly superior to all others, and the horizontal stripes in the high-resolution cloud image are effectively erased.

3.5. Quantitative Analysis of the Nonuniformity Correction Effect of the Algorithms

In order to better analyze the image correction effect, this paper uses PSNR, roughness, and the energy of vertical gradient for quantitative analysis. In this quantitative analysis, the infrared image in Figures 5 and 6 with added simulated horizontal stripes and 100 randomly selected frames from the FLIRADAS dataset are employed. All images are 8-bit images. The definition of PSNR is described as follows:

$$\mathrm{PSNR} = 10\log_{10}\frac{255^2}{\mathrm{MSE}} \tag{16}$$

where MSE is the mean square error between the original image and the input image. The higher the value of PSNR means the image is closer to the original image, and the better the correction effect. The PSNR results are shown in Table 3.

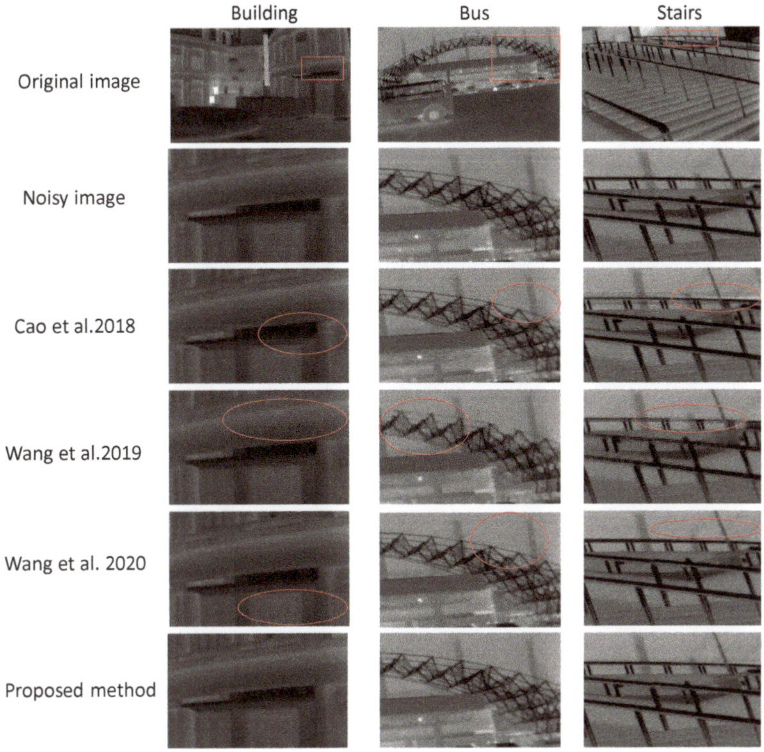

Figure 8. Nonuniformity correction effect of Cao [16], Wang [17], Wang [18], and our proposed algorithm. In order to better show the nonuniformity correction effect of each algorithm, this figure selects local magnified images of building, bus, and stairs to compare the effect of each algorithm. The parts in the red circle in the images are the parts where the non-uniform correction results of the contrast algorithm are not good.

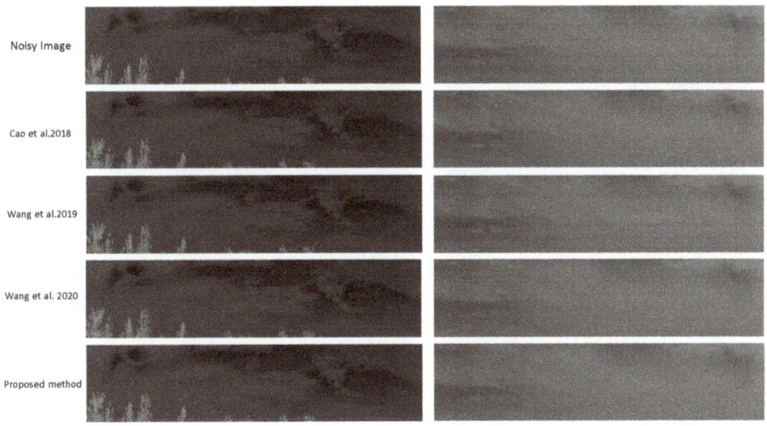

Figure 9. Real infrared focal plane linear array non-uniform correction of Cao [16], Wang [17], Wang [18], and our proposed algorithm. The resolution of the presented image is 1024 × 4096 due to the high resolution of the scanned image, as shown in this figure, in order to more accurately compare the denoising effect.

Table 3. PSNR results of Cao [16], Wang [17], Wang [18], and our method.

Sequence\Method	Cao [16]	Wang [17]	Wang [18]	Proposed Method
bus	40.16	41.18	31.90	**46.24**
people	40.80	42.72	36.19	**45.30**
building	42.35	43.48	39.17	**47.20**
hand	38.49	44.48	43.21	**46.65**
stairs	34.78	38.19	29.77	**38.98**
dataset	39.85	42.84	34.79	**45.74**

It could be seen that our proposed algorithm has the best nonuniformity correction effect with the highest PSNR value. Due to the TV regularization of the wavelet transform DC component, the technique of Wang [18] yields the lowest PSNR of the image. This means that the DC component of the whole frame is changed. The algorithm of Cao [16] has a lower image PSNR in the image "hand", which may come from the fact that this image has more high-frequency details; after the three-stage wavelet transform high-frequency component processing, the image obtains more distortion. The roughness is another indicator to measure the image correction effect, and its definition is as follows:

$$p = \frac{||h * g|| + ||h^T * I||}{||I||} \tag{17}$$

where $||\cdot||$ denotes the L-1 norm, $*$ means the convolution operation, $h = [-1, 1]$, and I denotes the input image. The smoother the image means the lower the roughness of the image. However, this does not imply a superior picture correction effect. The process of smoothing the image may lose the image's high-frequency information. Therefore, this experiment considers that the closer the roughness of the image is to the original image, the better the effect is. The results of roughness are shown in Table 4.

Table 4. Roughness of Cao [16], Wang [17], Wang [18], and our method.

Sequence\Method	Original	Cao [16]	Wang [17]	Wang [18]	Proposed Method
bus	**0.0741**	0.0765	0.0787	0.0652	**0.0740**
people	**0.0942**	0.0962	0.0975	0.0863	**0.0943**
building	**0.0621**	0.0646	0.0662	0.0591	**0.0623**
hand	**0.0877**	0.0939	0.0936	0.0868	**0.0891**
stairs	**0.0977**	0.1009	0.1045	0.0939	**0.0984**
dataset	**0.0834**	0.0851	0.0876	0.0786	**0.0841**

According to the data in Table 4, the roughness of the corrected image generated by our proposed technique is the closest to that of the original image. The image roughness of Wang [18] is the lowest, and the image is the smoothest, which is consistent with the correction effects of different algorithms in Figure 8. Since the nonuniformity noise of the infrared column scan image is in the horizontal direction, the image is not as smooth as the original image in the vertical direction; the energy of vertical gradient is another indicator of the nonuniformity correction effect, and its definition is as follows:

$$E = \sum_{i=1}^{H-1} \sum_{j=1}^{W} \frac{[I(i,j) - I(i+1,j)]^2}{(H-1) \times W} \tag{18}$$

where $I(i,j)$ is the input image, H is the height of the image, and W is the width of the image. The results of the energy of vertical gradient are shown in Table 5.

Table 5. Energy of vertical gradient of Cao [16], Wang [17], Wang [18], and our method.

Sequence\Method	Original	Cao [16]	Wang [17]	Wang [18]	Proposed Method
bus	**90.00**	93.19	92.56	53.86	**87.97**
people	**53.92**	54.66	55.48	37.20	**52.35**
building	**21.62**	23.85	24.50	15.81	**20.95**
hand	**23.50**	32.56	27.12	21.74	**23.45**
stairs	**173.16**	168.90	**176.44**	116.31	162.74
dataset	**75.98**	71.93	78.29	50.43	**72.62**

According to Table 5, the algorithm in this paper has the closest energy of vertical gradient of the corrected image to the original infrared image in most cases. The vertical gradient energy of the corrected image produced by the Wang [18] method is the lowest and the images are smoother. This is consistent with the image in Figure 8.

From Figure 10, we can see that the proposed algorithm has the highest PSNR value of the corrected images, which indicates the best effect. It is worth noting that the PSNR of the corrected image of Wang's [18] algorithm with a variance σ of 0.01 is smaller than that of the noisy image without denoising treatment. The main reason for this is that the algorithm applies TV regularization to the DC component of the image after the wavelet transform, causing the overall pixel value to alter. It can be seen from the effect in Figure 7 that the Wang [18] algorithm has the highest vertical smoothing degree for the image, and a better correction effect for horizontal stripes than the algorithms of Cao [16] and Wang [17]. Therefore, the PSNR of the corrected image varies slowly as the variance increases.

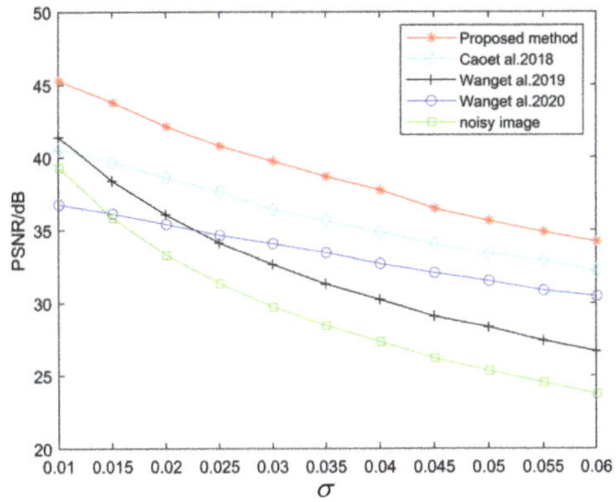

Figure 10. PSNR results of Cao [16], Wang [17], Wang [18], and our method. In order to verify the universality of the nonuniformity correction effect of the algorithm in this paper, 50 images are selected at random from the FLIRADAS dataset, and Gaussian cross-stripe noise with different variance σ is added as shown in Equation (15). The mean PSNR values of the corrected images are calculated using various algorithms to create a relationship curve, as depicted in this figure.

3.6. Computational Time

Due to the fact that infrared focal plane linear array scan detectors image tens of thousands of columns per second, the algorithm must have high performance in real time. Cao [16] and other comparison algorithms perform the wavelet transform on the image. The size of a frame for an infrared focal plane array image is limited, and the image's length and width are around a few hundred pixels; thus, the wavelet transform is not a significant burden for it. For the infrared focal plane linear array scan image, the number of columns in a frame is tens of thousands, so the wavelet transforms applied directly to the whole frame of the line column image are not suitable, while the computational overhead is relatively large. In contrast, the proposed algorithm only requires a portion of the infrared scan image to calculate the linear correction coefficients, which are then used to correct the image nonuniformity of the entire frame. The main cost of the algorithm is the computation of the linear correction coefficients across a portion of the image. Therefore, the proposed algorithm is less time-intensive and operates more in real time. To ensure a fair comparison, all techniques are executed in Matlab R2020b without optimizations or parallel-computing implementation on a PC with a 2.50 GHz Intel Core i7-9750HK processor and 16 GB of RAM. The experimental data in this section are partial images intercepted from real infrared focal plane linear array scan images with an image resolution of 3053 × 8192 and an A/D acquisition accuracy of 14-bit. In order to facilitate subjective evaluation, all 16-bit grayscale images are converted to 8-bit data grayscale images. The proposed algorithm in this paper selects the image with the resolution of 3053 × 1500 for calculating the linear correction model coefficients. The size of the vertical guided filter window for smoothing the image is 12 × 1, and the size of the horizontal guided filter for extracting the horizontal strip is 1 × 100. The artificial guided filter parameters ε are both 0.16. Cao [16] adopts three-level image wavelet decomposition and haar wavelet. Wang [18] chooses the haar wavelet base and Wang's [17] algorithm chooses the "db4" wavelet. The experimental results are shown in Table 6.

Table 6. Computational time comparison of Cao [16], Wang [17], Wang [18], and our method.

Algorithms	Cao [16]	Wang [18]	Wang [17]	Proposed Method
Time\s	1.1014	120.2012	205.6192	1.0143

The algorithm execution time listed in Table 6 is the mean execution time of each algorithm after 10 iterations. An evident conclusion is that, when the image resolution is 3053 × 8192, the algorithm of Cao [16] and the algorithm proposed in this paper have similar minimum times of approximately 1s. In practical application, an infrared scan image frame contains tens of thousands of columns, and the device generates a frame of image in a short time of about 1–2 s. At this point, the processing time for a scanned image utilizing Cao's [16] algorithm will grow dramatically. By intercepting a portion of the image, the algorithm in this paper calculates the linear correction coefficients. The computational time of the proposed algorithm is mainly in the process of solving the linear model coefficients. The time required to generate an image is rather short. Since the infrared scan technology rapidly scans an image frame and the correction coefficients change very little, it can be considered that the correction coefficients do not change in a short time. Consequently, the production of an image frame often requires a simple linear correction. The time required to correct the infrared scan image will not increase significantly. The algorithm proposed by Wang [18] combines image wavelet decomposition with TV regularization. After wavelet decomposition, it takes 114.6317 s to correct the DC component of an image by TV regularization. This is mainly due to the fact that the DC component of the picture wavelet decomposition has a resolution of 1526 × 4096 and the image is huge; thus, the time required to correct it using TV regularization is longer. The algorithm proposed by Wang [17] combines the wavelet transform with the MHE method. It completes eliminating stripes by MHE equalization of the horizontal AC components of the image after wavelet decomposition and guided filtering. In the MHE algorithm, the adaptive selection of parameter s in the optimal Gaussian noise reduction operator $g_s(k) = \frac{1}{s\sqrt{2\pi}}e^{\frac{-k^2}{2s^2}}$ takes 184.4410 s, which is a relatively long time. To conclude, the algorithm of Wang [17] and Wang [18] have a high time complexity and are not suitable for nonuniformity correction of infrared focal plane linear array scan images with high resolution and demand for real-time performance.

4. Conclusions

In this paper, we propose a nonuniformity correction algorithm for an infrared focal plane linear array scan detector based on a single-frame image. Due to the high resolution of the infrared scan image, the algorithm for nonuniformity correction must have a high real-time performance. In order to simultaneously fulfill the denoising effect and high real-time performance of the infrared sequence image, this paper presents a two-step correction approach. In the first stage, a portion of the line-sequence scan image is extracted from the whole frame scan image. On the intercepted picture, nonuniformity correction is accomplished using two one-dimensional guided filters. In the second step, since the corrected image may blur some high-frequency details, the algorithm in this paper calculates the linear correction coefficients using the portion of the image to reduce the damage to the image texture details caused by the guided filtering denoising algorithm. The linear correction coefficients are then used to correct the entire frame image. The proposed algorithm has the following advantages over other nonuniformity correction methods developed in recent years. Based on the real-time analysis of the algorithm, this study only uses a portion of the image to determine the correction coefficients for the nonuniformity correction of the entire frame. It meets the real-time needs of the scan images containing tens of thousands of columns per frame. The designed linear correction model conforms to the structural characteristics corresponding to the imaging of each row of the cycle scan image and the fixed detection unit, as determined by an analysis of the algorithmic principle. The algorithm provided in this paper is processed in the image space domain, does not involve image time-frequency conversion, and does not produce ghosting artifacts. The nonuniformity correction effect is superior to other algorithms. The technique in this paper is based on a single-frame image approach in terms of algorithm robustness. Unlike multi-frame denoising algorithms, the approach does not need to account for the influence of camera shake and scene heat radiation variations on the correction effect, and it is robust. In addition, unlike the conventional correction procedure that uses the black body to correct the detector's nonuniformity, the algorithm proposed in this paper does not require terminating the detector's use, allowing for the detector to be adjusted in real time during use.

Author Contributions: Methodology, B.L.; Formal analysis, B.L.; Resources, Y.Z.; Data curation, W.C.; Writing—original draft, B.L.; Writing—review & editing, B.L., W.C. and Y.Z. All authors have read and agreed to the published version of the manuscript.

Funding: This research received no external funding.

Institutional Review Board Statement: Not applicable.

Informed Consent Statement: Not applicable.

Data Availability Statement: The real infrared image dataset is generated by infrared longwave cooled linear scan detector.It is not a public dataset.The infrared focal array image data set comes from a public data set, https://camel.ece.gatech.edu/, accessed on 11 March 2023.

Conflicts of Interest: The authors declare no conflict of interest.

References

1. Schulz, M.; Caldwell, L. Nonuniformity correction and correctability of infrared focal plane arrays . *Infrared Phys. Technol.* **1995**, *36*, 763–777. [CrossRef]
2. Riou, O.; Berrebi, S.; Bremond, P. Nonuniformity correction and thermal drift compensation of thermal infrared camera. In *Defense and Security, International Society for Optics and Photonics*; SPIE: Bellingham, WA, USA, 2004; pp. 294–302
3. Song, S.; Zhai, X. Research on non-uniformity correction based on blackbody calibration. In Proceedings of the 2020 IEEE 4th Information Technology, Networking, Electronic and Automation Control Conference (ITNEC), Chongqing, China, 12–14 June 2020; Volume 1, pp. 2146–2150.
4. Vollmer, M.; Klaus-Peter, M.A. *Infrared Thermal Imaging: Fundamentals, Research and Applications*; Wiley-VCH: Hoboken, NJ, USA, 2017.
5. Teena, M.; Manickavasagan, A. Thermal infrared imaging. In *Imaging with Electromagnetic Spectrum*; Springer: Berlin/Heidelberg, Germany, 2014; pp. 147–173.
6. Scribner, D.A.; Kruer, M.R.; Gridley, J.C.; Sarkady, K. Physical Limitations to Nonuniformity Correction in Focal Plane Arrays. *SPIE* **1988**, *865*, 185–201.
7. Kim, S. Two-point correction and minimum filter-based nonuniformity correction for scan-based aerial infrared cameras. *Opt. Eng.* **2012**, *51*, 106401. [CrossRef]

8. Hu, X. Study on nonuniformity and calibration method of infrared focal plane detector. *Infrared Laser Eng.* **1999**, *28*, 9–12.
9. Shi, Y.; Zhang, T.; Cao, Z. A New Piecewise Approach for Nonuniformity Correction in IRFPA. *Int. J. Infrared Millim. Waves* **2004**, *25*, 959–972. [CrossRef]
10. Boutemedjet, A.; Deng, C.; Zhao, B. Robust Approach for Nonuniformity Correction in Infrared Focal Plane Array. *Sensors* **2016**, *16*, 1890. [CrossRef] [PubMed]
11. Sheng, M.; Xie, J.; Fu, Z. Calibration-based NUC Method in Real-time Based on IRFPA. *Phys. Procedia* **2011**, *22*, 372–380. [CrossRef]
12. Zhu, R.; Wang, C.; Wei, Q.; Jia, H.; Zhou, W. Development of nonuniformity correction system for infrared detector. *Infrared Laser Eng.* **2013**, *42*, 1669–1673.
13. Tendero, Y.; Landeau, S.; Gilles, J. Non-uniformity Correction of Infrared Images by Midway Equalization. *Image Process. Line* **2012**, *2012*, 134–146. Available online: http://demo.ipol.im/demo/glmtmire/ (accessed on 12 July 2012). [CrossRef]
14. He, K.; Sun, J.; Tang, X. Guided image filtering. *IEEE Trans. Pattern Anal. Mach. Intell.* **2012**, *35*, 1397–1409. [CrossRef]
15. Cao, Y.; Yang, M.Y.; Tisse, C.-L. Effective Strip Noise Removal for Low-Textured Infrared Images Based on 1-D Guided Filtering. *IEEE Trans. Circuits Syst. Video Technol.* **2016**, *26*, 2176–2188. [CrossRef]
16. Cao, Y.; He, Z.; Yang, J.; Ye, X.; Cao, Y. A multi-scale non-uniformity correction method based on wavelet decomposition and guided filtering for uncooled long wave infrared camera. *Signal Process. Image Commun.* **2018**, *60*, 13–21. [CrossRef]
17. Wang, E.; Jiang, P.; Hou, X.; Zhu, Y.; Peng, L. Infrared stripe correction algorithm based on wavelet analysis and gradient equalization. *Appl. Sci.* **2019**, *9*, 1993. [CrossRef]
18. Wang, E.; Jiang, P.; Li, X.; Cao, H. Infrared stripe correction algorithm based on wavelet decomposition and total variation-guided filtering. *J. Eur. Opt.-Soc.-Rapid Publ.* **2020**, *16*, 1–12. [CrossRef]
19. Hardie, R.C.; Hayat, M.M.; Armstrong, E.; Yasuda, B. Scene-based nonuniformity correction with video sequences and registration. *Appl. Opt.* **2000**, *39*, 1241–1250. [CrossRef]
20. Ratliff, B.M.; Hayat, M.M.; Hardie, R.C. An algebraic algorithm for nonuniformity correction in focal-plane arrays. *J. Opt. Soc. Am. Opt. Image Sci. Vis.* **2002**, *19*, 1737–1747. [CrossRef]
21. Zuo, C.; Chen, Q.; Gu, G.; Sui, X.; Ren, J. Improved interframe registration based nonuniformity correction for focal plane arrays. *Infrared Phys. Technol.* **2012**, *55*, 263–269 [CrossRef]
22. Abbass, M.Y.; Sadic, N.; Ashiba, H.I.; Hassan, E.S.; El-Dolil, S.; Soliman, N.F.; Algarni, A.D.; Alabdulkreem, E.A.; Algarni, F.; El-Banby, G.M.; et al. An Efficient Technique for Non-Uniformity Correction of Infrared Video Sequences with Histogram Matching. *J. Electr. Eng. Technol.* **2022**, *17*, 2971–2983. [CrossRef]
23. Ashiba, H.I.; Sadic, N.; Hassan, E.S.; El-Dolil, S.; Abd El-Samie, F.E. New Proposed Algorithms for Infrared Video Sequences Non-uniformity Correction. *Wirel. Pers. Commun.* **2022**, *126*, 1051–1073. [CrossRef]
24. Zhang, X.; Li, H.; Hou, J.; Zhao, D.; Zhou, H.; Zhang, J.; Zhang, Z.; Cheng, K. Non-uniformity correction algorithm based on improved neural network. *Seventh Symp. Nov. Photoelectron. Detect. Technol. Appl. Spie* **2021**, *11763*, 717–726.
25. Li, Y.; Liu, N.; Xu, J. Infrared scene-based non-uniformity correction based on deep learning model. *Optik* **2021**, *227*, 165899. [CrossRef]
26. Guan, J.; Lai, R.; Xiong, A.; Liu, Z.; Gu, L. Fixed pattern noise reduction for infrared images based on cascade residual attention CNN. *Neurocomputing* **2020**, *377*, 301–313. [CrossRef]
27. Luo, Q.; Gao, B.; Woo, W.L.; Yang, Y. Temporal and spatial deep learning network for infrared thermal defect detection. *NDT Int.* **2019**, *108*, 102164. [CrossRef]
28. Tomasi, C.; Manduchi, R. Bilateral filtering for gray and color images. In Proceedings of the Sixth International Conference on Computer Vision (IEEE Cat. No.98CH36271), Bombay, India, 7 January 1998; pp. 839–846.
29. Lu, C.H. Stripe non-uniformity correction of infrared images using parameter estimation. *Infrared Phys. Technol.* **2020**, *107*, 103313. [CrossRef]

applied sciences

MDPI

Article

A Nighttime and Daytime Single-Image Dehazing Method

Yunqing Tang [1,*], Yin Xiang [2] and Guangfeng Chen [1]

1 The College of Mechanical Engineering, Donghua University, Shanghai 201620, China
2 Shanghai Aerospace Control Technology Institute, 1555 Zhongchun Road, Minhang District of Shanghai, Shanghai 201109, China
* Correspondence: 18800235026@163.com

Abstract: In this study, the requirements for image dehazing methods have been put forward, such as a wider range of scenarios in which the methods can be used, faster processing speeds and higher image quality. Recent dehazing methods can only unilaterally process daytime or nighttime hazy images. However, we propose an effective single-image technique, dubbed MF Dehazer, in order to solve the problems associated with nighttime and daytime dehazing. This technique was developed following an in-depth analysis of the properties of nighttime hazy images. We also propose a mixed-filter method in order to estimate ambient illumination. It is possible to obtain the color and light direction when estimating ambient illumination. Usually, after dehazing, nighttime images will cause light source diffusion problems. Thus, we propose a method to compensate for the high-light area transmission in order to improve the transmission of the light source areas. Then, through regularization, the images obtain better contrast. The experimental results show that MF Dehazer outperforms the recent dehazing methods. Additionally, it can obtain images with higher contrast and clarity while retaining the original color of the image.

Keywords: nighttime dehazing; daytime dehazing; mixed-filter; high-light area transmission compensation; regularization

Citation: Tang, Y.; Xiang, Y.; Chen, G. A Nighttime and Daytime Single-Image Dehazing Method. *Appl. Sci.* **2023**, *13*, 255. https://doi.org/10.3390/app13010255

Academic Editors: Przemysław Falkowski-Gilski, Tadeus Uhl and Zbigniew Lubniewski

Received: 19 November 2022
Revised: 4 December 2022
Accepted: 5 December 2022
Published: 25 December 2022

1. Introduction

The suspended particles in haze weaken the spread of light, resulting in the loss of image information and the low contrast and saturation of the collected images, which has a negative effect on visual systems, such as those used for video surveillance and autonomous driving. Therefore, many scholars have researched this aspect, especially when applied to monitoring [1] and image segmentation [2].

Dehazing technology has achieved good results in daytime image processing. There are some methods that are used to directly enhance images, for instance, methods based on histogram equalization [3] or Retinex theory [4–7]. These types of methods enhance the characteristic areas of the image and weaken the areas of no interest according to human perception. However, these methods do not consider the cause of haze in the image, nor do they remove haze in terms of contrast and color. The current methods are basically based on the atmospheric scattering model [8] and use various prior methods that have generated unsatisfactory results for nighttime scenes.

Nighttime scenes are more complicated than daytime scenes. This is mainly because artificial light sources at night make the ambient light spatial. As a result, there are few pieces of research that have investigated dehazing at night, and only a few methods are effective for these images; furthermore, these methods change the color of the original image.

In this paper, we first propose an effective single-image dehazing method called MF Dehazer that can be effectively applied to both nighttime and daytime scenes. The quality of the haze-free images depends on the accuracy of the ambient illumination estimation and the transmission estimation. Secondly, we analyzed the ambient light characteristics of the nighttime images and combined the atmospheric scattering model with the Retinex theory. We also propose a mixed-filter method in order to obtain ambient illumination. Additionally, in order to estimate transmission, the boundary-constrained method was used to estimate the initial transmission and then performed high-light area transmission compensation for the light source area. Finally, we optimized the transmission through regularization. The experimental results demonstrate that regardless of whether it is a nighttime or daytime scene, the image quality obtained using this method is superior when compared to the existing algorithms.

The contribution of our paper can be summarized as follows:

1. Due to the influence of artificial light sources in nighttime haze images, the light is directional and spatial. The mixed-filter method we designed can accurately estimate the direction, shape and color of the image's light.
2. In terms of transmission estimation, we analyzed the reasons for the light source diffusion problem seen in other nighttime dehazing methods. This is because the transmission of the light source area is underestimated. Therefore, we proposed a method to compensate for the high-light area transmission in order to weaken the dehazing effect of the light source areas to solve this problem. Consequently, regularization can improve the contrast of the image and solve the problem of excessive dehazing or poor dehazing effects.
3. Compared to other methods, MF Dehazer can be used effectively for different images, such as images of both distant and close views and nighttime or daytime images. In addition, haze-free images retain their original color and enhance the dark areas, providing a novel approach for removing haze.

2. Background Theory and Related Work

2.1. Atmospheric Scattering Model

Koschmieder's model [8] (the atmospheric scattering model) describes a hazy image as one in which the light intensity $I(x)$ at each image coordinate x is added by direct transmission $D(x)$ and atmospheric light $A(x)$:

The light intensity I received by the imaging system is mainly composed of object-reflected light and atmospheric light. The image can be expressed as follows:

$$I(x) = J(x)t(x) + A_0(1 - t(x)) \tag{1}$$

where x is the pixel, $J(x)$ is the actual reflected light intensity of the object, $I(x)$ is the light intensity received by the imaging system and A_0 is the intensity of the atmospheric light at the time of emission.

In an ideal case, we assume that the atmospheric transmission medium is uniform, and at this time $t(x)$, can be expressed as follows:

$$t(x) = e^{-\beta d(x)} \tag{2}$$

where β is the attenuation coefficient caused by scattering in the medium and $d(x)$ is the transmission distance of the atmospheric light.

Compared to the traditional model, ambient illumination at night is affected by other light sources, such as street lamps and car lights. The estimation of ambient illumination should be a local variable, so the original fixed atmospheric light value A_0 is overwritten as local ambient light A. Hence, the transformation of Equation (1) results in the new defogging model, which is as follows:

$$J(x) = \frac{I(x) - A(x)}{t(x)} + A(x) \tag{3}$$

2.2. Retinex Theory

Retinex theory [4], in short, states that the color of an object is determined by the object's ability to reflect long-wave (red), medium-wave (green) and short-wave (blue) light rather than the absolute intensity of the reflected light. It can be expressed as follows:

$$J(x) = R(x)A(x) \tag{4}$$

where $J(x)$ is the haze-free image, $R(x)$ is the reflectance and $A(x)$ is the ambient illumination.

2.3. Related Works

As previously discussed, most dehazing methods are applied to daytime scenes. The recent methods are basically based on the atmospheric scattering model and remove haze using various prior methods; for example, He et al.'s [9,10] dark channel prior, Fattal et al.'s [11] color-line prior, Zhu et al.'s [12] color attenuation prior and Berman et al.'s [13] haze-line prior. Much of the related work is carried out on the basis of the dark channel a priori. Meng et al. [14] proposed a contextual regularization dehazing method by exploring the inherent boundary constraint on the transmission function. Berman et al. [13] proposed a dehazing method using haze lines. However, the result exceeds the lower bound of the constraint. Subsequently, Berman et al. [15] improved the optimization method of their model. Although in unevenly illuminated areas, such as the sky and lakes, the transmission estimation is inaccurate, resulting in excessive correction or haze residue. In general, the prior methods are liable to failure in different scenes.

In recent years, with the development of Convolutional Neural Networks (CNNs), deep learning has also been applied to dehazing research. Li et al. [16] did not separately estimate the transmission and ambient illumination but directly generated clear images through a lightweight CNN. Chen et al.'s [17] dehazing network focuses on solving the problem of artifacts. Dong et al. [18] embedded the dehazing feature unit into the network and reconstructed the clear image after extracting the features through the encoder module. Shao et al. [19] proposed an end-to-end domain adaptive network to dehaze the real image. Dong et al.'s [20] method is a multi-scale enhanced dehazing network that integrates different levels of features based on back-projection technology. Another type of method is based on Generative Adversarial Networks (GANs). Qu et al. [21] embedded the adversarial network in the enhanced Pix2pix dehazing network and optimized the training model through four loss functions. Deng et al. [22] proposed adding a normalization layer to the dehazing network model to obtain image information.

However, the above-mentioned daytime dehazing methods are not applicable at night. This is because daytime ambient illumination is a fixed value, which is determined by the attenuation and scattering of atmospheric light, while nighttime ambient illumination is a local variable. Daytime dehazing methods usually estimate the ambient illumination with the brightest area of the image. However, in the nighttime scene, the ambient illumination is mainly affected by artificial light sources, such as street lights and car lights. Moreover, the complex environmental conditions make the deep learning dehazing model unable to process the nighttime haze image. Therefore, it is necessary to deal with the nighttime haze image in a targeted manner. Aiming at the characteristics of low contrast, low brightness, and light discoloration of nighttime haze images, Pei et al. [23] proposed using color transfer for preprocessing and removing haze using a dark channel prior and a bilateral filter; however, this method introduces distortion. Zhang et al. [24] performed illumination compensation and color correction for the nighttime haze image and used the dark channel prior to remove the haze; however, this caused artifacts to appear. Then, Zhang et al. [25] proposed a maximum reflectance prior to estimate the ambient illumination, but it changes the original color of the image. Li et al. [26] improved the traditional atmospheric scattering model and introduced a glow layer, and dehazed the new separated haze image and reduced the glow using a dark channel. Although the glow problem can be effectively solved, the color of the output image is unreal. Ancuti et al. [27] optimized their original model and estimated ambient illumination through different window sizes, using a multi-scale fusion method to dehaze. This method can remove nighttime and daytime image haze, but it diffuses the original light source. Zhu et al. [28] also used an image fusion method. They used pixelwise weight maps constructed using global and local exposedness to guide the fusion process of different coefficient gamma operations hazy images.

3. Proposed Dehazing Method

The flow chart of the proposed MF Dehazer is illustrated in Figure 1. The quality of the dehazing effect depends on the accuracy of the two parts of the estimation: the ambient illumination and the transmission. The detailed will follow.

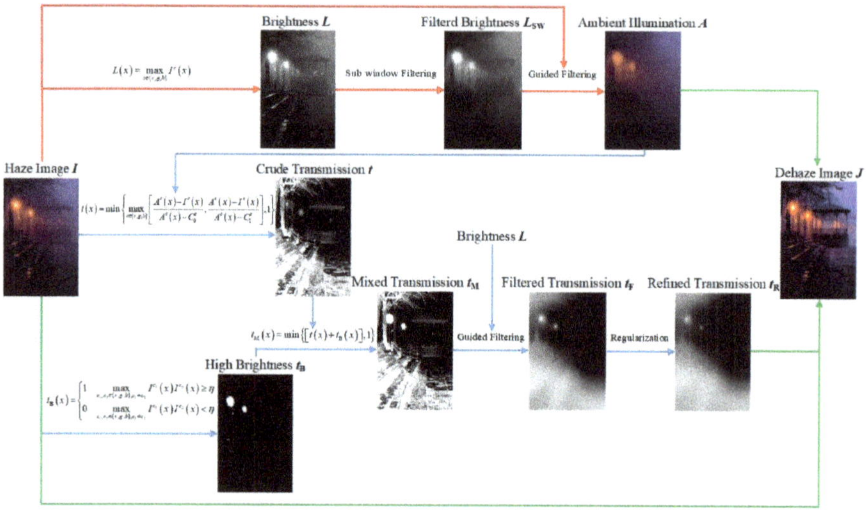

Figure 1. Flow chart of the proposed dehazing method. In this chart, the red line is the ambient illumination estimation method, the blue line is the transmission estimation method, and the green line is the result.

3.1. Mixed-Filter for Ambient Illumination Estimation

According to Retinex theory, we can Substitute (1) into (4):

$$I(x) = A(x)(R(x)t(x) + 1 - t(x)) \tag{5}$$

Ambient illumination, composed of the scattering and reflection of atmospheric light and artificial light sources is a spatially smooth and low-frequency term while $R(x)t(x) + 1 - t(x)$ serves as the high-frequency term. The traditional Retinex algorithm uses Gaussian filtering to estimate the low-frequency components. However, because Gaussian filtering does not retain the edge and it is isotropic, the shape of the light source and the direction of the light cannot be retained when the image is blurred, and the backlight area of the light source is also blurred. Considering this analysis, we propose to initially preserve the edge blur of the brightness of the hazy image. While retaining the shape of the light source and the direction of the light, filter the farther small light source and reflected light, and then pass it as the reference image to guide the filtering as the ambient illumination estimation.

We regard each target pixel as a potential edge and generate multiple local windows (called side windows) in the rectangular area around it. Each window aligns the target pixel with the edge or corner (rather than the center) of the window. In the discrete cases, define eight side windows: upper (U), lower (D), left (L), right ®, upper left (UL), upper right (UR), lower left (DL), and lower right (DR). The eight side windows are mean filtered with the original image, and the side window with the minimum Euclidean distance between the filtering result and the original image is selected as the optimal filtering side window of the target pixel. The process can be expressed as follows:

$$L_n = \frac{1}{N_n} \sum_{j \in \omega_i^n} \omega_{ij} I_j, N_n = \sum_{j \in \omega_i^n} \omega_{ij}, n \in S \tag{6}$$

where L_n is the filter value of the sub-window, ω_{ij} is the weight of the pixel j near the target pixel i based on the kernel function F which is averaging; I_j is the value of pixel j, and $S = U, D, L, R, UL, UR, DL, DR$ is the set of sub-window index.

After obtaining the filter values of the eight sub-windows, in order to preserve the edge information, we use the sub-window with the minimum Euclidean distance to the input intensity as the output of L_{sw}.

$$L_{sw} = argmin_{n \in S} \| I_i - L_n \|_2^2 \tag{7}$$

We assume that the guided image and the output ambient illumination are linear, and then the brightness filtered image L_{sw} is used as the guide image to obtain the linear transform parameter by minimizing the cost function of the target image I :

$$\mathrm{E}(a_k, b_k) = \sum_{x \in \omega_k} \left((a_k L_{sw}(x) + b_k - I(x))^2 + \lambda a_k^2 \right) \tag{8}$$

where ω_k is the filter window, λ is the regularization coefficient with a value of 0.01, which is mainly used to prevent the solution a_k from being too large.

Solve (8) using the least square method as follows:

$$\begin{cases} a_k = \frac{\frac{1}{|\omega|} \sum_{i \in \omega_k} L_{sw} I_i - \mu_k I_k}{\sigma_k^2 + \lambda} \\ b_k = I_k - a_k \mu_k \end{cases} \tag{9}$$

where μ_k and σ_k^2 are the mean and variance of L_{sw} in window ω_k, $|\omega|$ is the number of pixels in the window; I_k is the mean value of all elements in the filter ω_k. After obtaining the coefficients a_k and b_k, the ambient illumination estimation $A^c(x)$ can be obtained by using the mean value of the window containing the pixels as follows:

$$A\ (x) = \frac{1}{|\omega|} \sum_{i \in \omega_k} a_k L_{sw}(x) + b_k \tag{10}$$

Figure 2 shows the comparison of different ambient illumination estimation methods and dehazing effects. He et al.'s [9] method chooses the brightest pixels in the image as the fixed value of the ambient illumination, which is affected by nighttime artificial light sources. If the brightness of the artificial light source is used as the ambient illumination, the light estimation will be higher. As a result, the color of the dehazing image in Figure 2b is too dark, which verifies that the ambient illumination estimation for nighttime dehazing should be a local variable rather than a fixed value. Because of the occlusion, the nighttime ambient illumination is directional and not scattered around. At the same time, the illumination intensity will decrease sharply due to the increase in distance and reflection. Therefore, the reflected light and the distant light source as a low light source should reduce the impact of illuminance estimation. It can be seen from Figure 2a,e that this image light is mainly two red light sources and presents umbrella-shaped illumination on the upper left. The ambient illumination estimation method of Li et al. [26] changes the color of the original light source, and because of its inability to preserve the edge, the light erodes to the backlight side. In comparison, we use the preprocessing image obtained by sub-window filtering as the guide image for ambient illumination estimation. The proposed method retains these two characteristics, and the color after dehazing is closer to the original color.

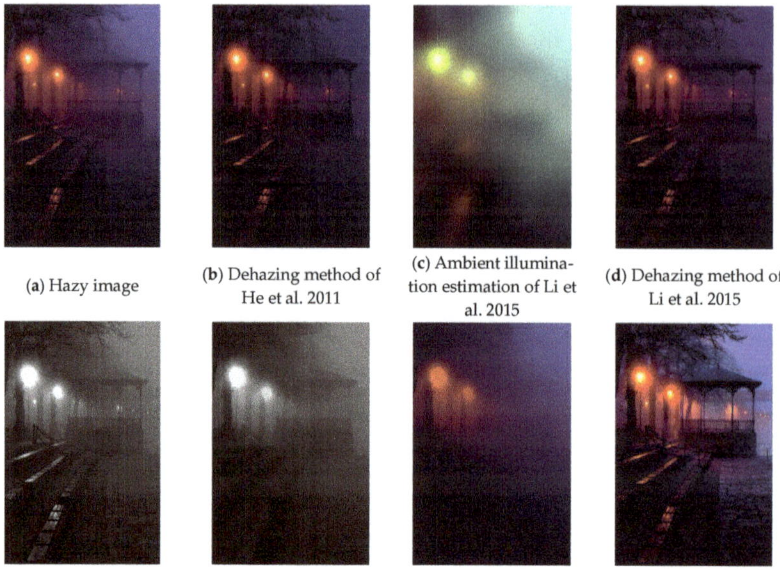

(a) Hazy image (b) Dehazing method of He et al. 2011 (c) Ambient illumination estimation of Li et al. 2015 (d) Dehazing method of Li et al. 2015 (e) Brightness image (f) Our preprocessing image (g) Our ambient illumination estimation (h) MF Dehazer

Figure 2. Comparison of ambient light estimation and dehazing by different methods. In order to ensure a fair comparison, we use the ambient illumination estimation methods of He et al. [9], Li et al. [26] and MF Dehazer and use the proposed transmission estimation method to remove haze to compare the effects of different ambient illumination estimation methods.

3.2. High-Light Area Transmission Compensation

Inspired by Meng et al. [14], we used boundary-constrained initial transmission estimation to replace the dark channel prior. The initial transmission obtained by this method can make the image more structured:

$$t(x) = \min\left\{ \max_{c \in r,g,b} \left\{ \frac{A^c(x) - I^c(x)}{A^c(x) - C_0^c}, \frac{A^c(x) - I^c(x)}{A^c(x) - C_1^c} \right\}, 1 \right\} \tag{11}$$

where C_0^c and C_1^c are the upper and lower boundaries of the three channels.

After the nighttime haze image has been processed for dehazing, the problem of light source diffusion often occurs. It causes the light source to lose its original shape and boundary. This is because the transmission of the light source area is incorrectly estimated. We can draw a conclusion from (3) that with the increased transmission, the pixel value of the image after dehazing will decrease, and when the transmission is closer to 1, the dehazed image will be closer to the hazy image. Thus, the transmission affects the degree of image dehazing. The transmission of the light source area of these methods is estimated to be low, which causes the pixel value to increase after the dehazing of the light source area, and the phenomenon of diffusion occurs. Based on this, we propose high-light area transmission compensation for nighttime scenes to reduce the degree of the dehazing of the light source area to suppress the diffusion. At the same time, considering that the nighttime light sources are mostly colored light sources, at least one of the color channels has a very high value. So the maximum value of the two-channel product is selected as the basis for distinguishing whether it is the light source area as follows:

$$t_B(x) = \begin{cases} 1 & \max\limits_{c_1,c_2 \in r,g,b,c_1 \neq c_2} I^{c_1}(x) I^{c_2}(x) \geq \eta \\ 0 & \max\limits_{c_1,c_2 \in r,g,b,c_1 \neq c_2} I^{c_1}(x) I^{c_2}(x) < \eta \end{cases} \tag{12}$$

where η is the threshold with a value of 0.3, which is used to distinguish whether the light source area is in this image, c_1 and c_2 represents two different color channels in the image, and t_B is the compensation value.

After the transmission is compensated for the high-light area, in order to prevent it from exceeding the boundary, it is restricted using the following:

$$t_M(x) = \min\{(t(x) + t_B(x)), 1\} \tag{13}$$

Finally, guided filtering is used to solve the problem of artifacts in the synthesized transmission after dehazing. Figure 3 shows the effect comparison of whether to use high-light area transmission compensation. This simple method improves the image quality of the light source area so that it retains the shape and boundary of the original light source.

(**a**) Hazy image Road (**b**) Road dehazing with compensation (**c**) Road dehazing without compensation

(**d**) Hazy image Grove (**e**) Grove dehazing with compensation (**f**) Grove dehazing without compensation

Figure 3. Comparison of the high-light compensation effects.

3.3. Regularization and Dehazing

With the purpose of improving the spatial correlation of image pixels to avoid problems such as haze residue or excessive dehazing, we used regularization to find the optimal transmission to improve the image quality. We used the L2-norm as the regularization term and as the data term to prevent over-fitting. We used the L1-norm, which is more robust to outliers, instead of the L2-norm so as to increase the depth of information sensitivity. We find the optimal transmission t_R by minimizing the following objective function:

$$L = \sum_x (t_F(x) - t_R(x))^2 + \lambda \sum_x \sum_{y \in N_x} w(x,y)|t_F(x) - t_F(y)| \tag{14}$$

where the first part is the data term that measures the fidelity by calculating the Euclidean distance between the transmission t_F with the optimized transmission t_R. The second part is the regularization term, where λ is the regularization parameter. N_x is the index set of the pixel x, and $w(x,y)$ is the weighting function. Here, we use the mixed function of color difference and brightness difference as seen in the following:

$$w(x,y) = \alpha \cdot exp\left(\frac{-\|c(x) - c(y)\|^2}{2\sigma^2}\right) + (1+\alpha)\left(|l(x) - l(y)|^\beta + \delta\right) \tag{15}$$

where α is the weight of the color difference and brightness difference, $c(x)$ and $c(y)$ are the color vectors, σ is the prescribed parameter with the value of 0.5, $l(x)$ and $l(y)$ are the brightness vectors, β is the brightness sensitivity parameter (typically 0.3) and δ is a small constant (typically 0.0001) to prevent this term from being 0.

After the optimized transmission is obtained by the least square method, the dehazing image can be obtained using (3). As shown in Figure 4, regularization can effectively improve the structure of image details.

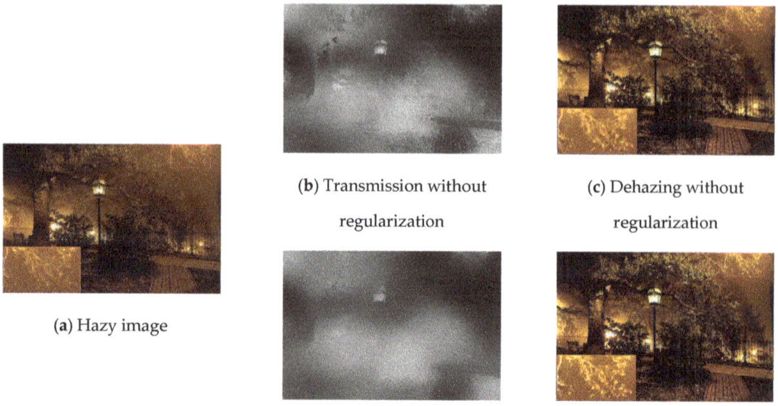

(**a**) Hazy image (**b**) Transmission without regularization (**c**) Dehazing without regularization (**d**) Transmission with regularization (**e**) Dehazing with regularization

Figure 4. Comparison of regularizan.

4. Experimental Results and Discussion

To demonstrate the effectiveness of the proposed MF Dehazer in both day and night conditions, we conducted a series of experiments on the nighttime and daytime hazy images. The images used in the experiment are from the article [9,12,15]. The nighttime experiment and daytime experiment each used ten images of different resolutions. In the nighttime experiment, η is 0.3, and the minimum transmission after guided filtering is 0.2. We compared MF Dehazer with recent nighttime dehazing methods [9,21,24–26]. In the daytime experiment, η is 1 and the minimum transmission after guided filtering is 0.3. We compare MF Dehazer with recent daytime dehazing methods [9,12,14,15,21]. The comparison methods or pictures are provided by the authors.

4.1. Qualitative Comparison

4.1.1. Nighttime Hazy Images

We chose the dark channel prior method of He et al. [9], the dehazing method based on deep learning of Qu et al. [21] and the recent nighttime dehazing methods of Li et al. [26], Zhang et al. [24] and Zhang et al. [25] to carry out the nighttime dehazing experiments. The experimental images are shown in Figure 5. He et al.'s [9] method incorrectly estimates the artificial light as the global ambient illumination, which causes the image to be too dark after dehazing. It proves that the nighttime ambient illumination should be a local variable rather than a simple fixed value. Because of the complex nighttime haze scene, the deep learning model of Qu et al. [21] cannot be effectively applied. It deepens the color of dim nighttime images and introduces unreal colors (Figure 5N3). Li et al.'s [26] method completely changes the color of the image, diffuses the light source, and the color is unstable (Figure 5N6,N9). Zhang et al.'s [24] method diffuses the image light in the light source area (Figure 5N1) and produces noise in the sky and dark areas (Figure 5N2,N3,N8,N10). Although Zhang et al.'s [25] method has better image clarity, it also cannot avoid the problem of noise (Figure 5N2,N3,N9) and light source area diffusion (Figure 5N3,N4). The proposed MF Dehazer effectively solves the defects of the above-mentioned methods. On the basis of retaining the original color, both the dark and bright areas of the image are effectively dehazed, such as the reflection of the lake in Figure 5N1, the clearer and more realistic green space on the left of Figure 5N10, the details and levels of shrubs and leaves in Figure 5N1,N3,N5,N7, and the light source diffusion problem of each image is well solved.

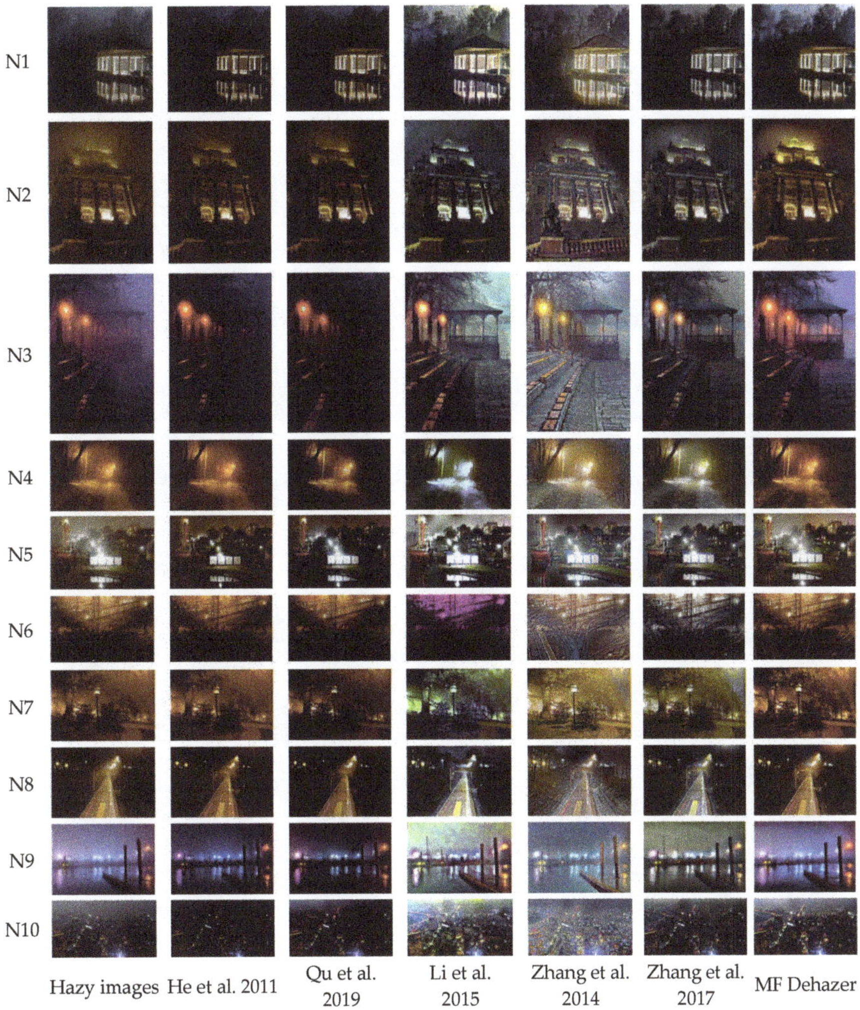

Figure 5. Comparison of nighttime dehazing [9,21,24–26].

4.1.2. Daytime Hazy Images

We chose the different prior methods of He et al. [9], Meng et al. [14], and Zhu et al. [12], the dehazing method based on deep learning of Qu et al. [21] and the recent dehazing methods of Berman et al. [15] to carry out daytime dehazing experiments. When using He et al.'s [9] method, it is easy to overestimate the illumination, causing the image to be too dark (Figure 6D5), and one side is dark when there are far and near scenes (Figure 6D4,D8). Zhu et al.'s [12] method also has the same problem. Meng et al.'s [14] method does not handle the sky area well and produces noise (Figure 6D1,D6,D10) and color artifacts (Figure 6D7). Qu et al.'s [21] method in some areas of complex pictures, such as trees, deepens the color of the image and causes the details to be unrecognizable

(Figure 6D7,D9,D10). Berman et al.'s [15] method produces color artifacts at the junction of the sky area. For example, the sky at the junction with the mountain turns yellow (Figure 6D3) or green (Figure 6D9) as it is affected by the color of the mountain. By contrast, MF Dehazer does not have these defects. The sky, lake and the near and far scenes are handled more naturally; the color is closer to the original image, and it looks visually pleasing.

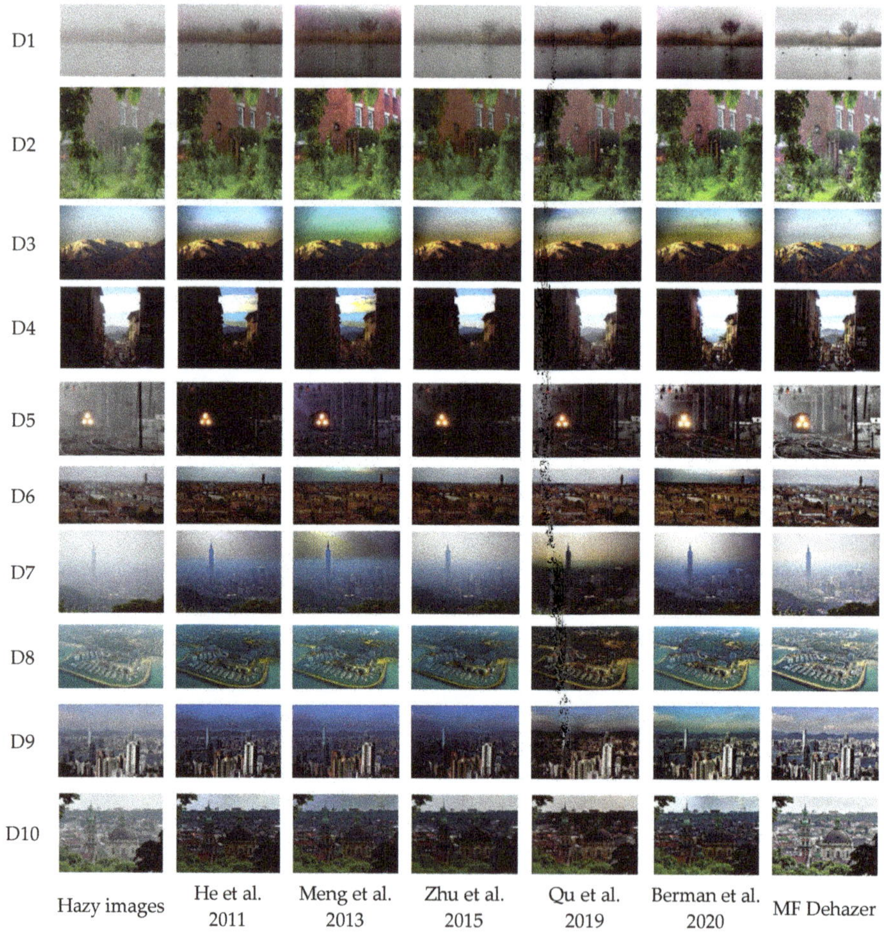

Figure 6. Comparison of daytime dehazing [9,12,14,15,21].

4.2. Quantitative Comparison

Quantitatively, Tables 1 and 2 consider three well-known metrics: PSNR [29], SSIM and CIEDE2000 [30,31]. PSNR is the ratio of the maximum signal amount to the noise intensity, which measures the degree of image distortion. SSIM evaluates image quality from three aspects: brightness, contrast, and structure, which indicates the similarity of image structure information. Higher values indicate better quality for PSNR and SSIM. CIEDE2000 evaluates image quality from five aspects: brightness, chroma, hue, chroma difference, and hue difference, which indicates the chromatic aberration of the image. A lower value indicates better quality for CIEDE2000.

Table 1. Comparosion Index of Nighttime Dehazing Effect.

Fig	Li et al. [26]			Zhang et al. [24]			Zhang et al. [25]			MF Dehazer		
	SSIM	**PSNR**	**CIEDE**	**SSIM**	**PSNR**	**CIEDE**	**SSIM**	**PSNR**	**CIEDE**	**SSIM**	**PSNR**	**CIEDE**
N1	0.648	17.078	19.390	0.662	16.827	24.269	**0.813**	**23.473**	13.374	0.722	21.302	**12.715**
N2	0.785	22.409	20.535	0.604	16.792	24.755	0.853	**24.581**	19.095	**0.875**	23.336	**9.879**
N3	0.617	16.087	18.850	0.390	10.847	38.178	0.773	22.188	16.037	**0.807**	**23.065**	**10.441**
N4	0.835	20.538	24.970	0.509	12.383	31.265	0.802	17.728	19.609	**0.862**	**27.378**	**8.717**
N5	0.721	17.354	19.806	0.791	19.226	19.437	0.886	19.814	20.679	**0.893**	**23.855**	**9.641**
N6	0.505	15.928	38.064	0.464	12.897	34.054	**0.856**	22.522	16.096	0.850	**26.994**	**6.379**
N7	0.754	19.301	24.117	0.578	12.760	29.582	**0.864**	19.481	13.986	0.840	**23.879**	**9.816**
N8	0.789	20.211	15.214	0.557	14.504	29.129	0.919	24.864	13.196	**0.937**	**28.517**	**5.432**
N9	0.544	13.537	24.930	0.516	16.180	21.882	0.785	21.197	20.353	**0.847**	**24.699**	**9.159**
N10	0.624	14.883	21.399	0.488	12.455	32.707	**0.812**	**22.313**	15.336	0.770	20.74	**14.063**
AVG	0.682	17.733	22.727	0.556	14.487	28.526	0.836	21.816	16.776	**0.840**	**24.382**	**9.624**

Table 2. Comparision Index of Daytime Dehazing Effect.

Fig	He et al. [9]			Meng et al. [14]			Qu et al. [21]			Berman et al. [15]			MF Dehazer		
	SSIM	**PSNR**	**CIEDE**	**SSIM**	**PSNR**	**CIEDE**	**SSIM**	**PSNR**	**CIEDE**	**SSIM**	**PSNR**	**CIEDE**	**SSIM**	**PSNR**	**CIEDE**
D1	0.814	13.271	18.816	0.504	8.828	35.924	0.750	9.465	31.349	0.494	8.951	33.463	0.764	**21.993**	**5.157**
D2	**0.908**	17.852	14.675	0.833	18.941	12.504	0.864	17.658	18.575	0.841	19.534	16.859	0.825	**21.877**	**10.616**
D3	0.769	16.842	23.342	**0.869**	16.160	31.645	0.746	15.974	23.298	0.788	15.956	22.663	0.841	**24.477**	**9.325**
D4	0.573	15.308	29.311	**0.915**	18.087	13.385	0.808	20.388	16.390	0.654	20.430	16.078	0.840	**22.710**	**10.406**
D5	0.390	8.001	69.811	0.677	10.590	47.519	0.638	10.650	45.258	0.514	11.511	39.448	**0.680**	**18.441**	**13.277**
D6	**0.831**	16.438	23.480	0.809	14.280	24.589	0.873	18.596	15.247	0.684	13.527	26.763	0.815	**20.339**	**11.820**
D7	0.783	11.012	40.420	0.762	10.053	41.852	0.675	9.198	41.730	0.713	10.267	44.883	0.858	**24.290**	**6.633**
D8	0.767	11.927	53.164	**0.851**	15.569	34.350	0.637	10.239	46.116	0.764	14.753	42.244	0.802	**19.397**	**17.924**
D9	0.746	11.884	50.448	**0.845**	13.223	38.019	0.784	13.746	30.346	0.825	16.084	37.390	0.811	**18.061**	**17.161**
D10	0.676	12.327	39.186	0.728	11.790	36.219	0.746	12.358	31.850	0.632	12.398	40.842	**0.804**	**20.785**	**11.180**
AVG	0.726	13.486	36.265	0.779	13.752	31.601	0.752	13.827	30.016	0.691	14.341	32.063	**0.804**	**21.237**	**11.350**

As shown in Tables 1 and 2, whether it is nighttime or daytime, MF Dehazer is excellent in all metrics, especially in that the PSNR and CIEDE are much higher than in the other methods. It shows that the proposed method has less distortion after dehazing, can effectively reduce noise, and well retains the color of the original image. At the same time, the proposed MF Dehazer has high stability, and the various metrics do not fluctuate. It can adapt to hazy images of different environments and complexity and restore high-definition, high-contrast and high-structure dehazing images.

4.3. Real Haze-Free Image Comparison

Apart from the above experiments, we also tested 80 images from two public image datasets, O-HAZE and I-HAZE [32] for verification. These are dehazing benchmarks that include real hazy and haze-free images. Figure 7 shows a part of the experimental results, and Table 3 shows the average objective indicators of these two datasets. Through the comparison in Figure 7, we can see that our results are close to the real haze-free image. At the same time, it confirms the advantages and disadvantages of the previous methods. All of the indicators in Table 3, except for the PSNR in O-HAZE, are higher than the existing methods. This indicator is lower than other methods because O-HAZE contains a lot of white dense fog scenes. When estimating the ambient illumination through the local filter window, the ambient illumination may be overestimated due to the influence of white dense haze, resulting in haze residue. Therefore, it is a defect of the proposed method to deal with the local white dense haze.

Figure 7. Comparison of O-HAZE and I-HAZE [9,12,14,15,21].

Table 3. Comparision Index Of O-Haze Additionally, I-HazeE.

Method	O-HAZE			I-HAZE		
	SSIM	**PSNR**	**CIEDE**	**SSIM**	**PSNR**	**CIEDE**
He et al.	0.710	16.510	31.420	0.726	14.948	28.058
Meng et al.	0.727	17.379	25.351	0.746	14.345	26.156
Zhu et al.	0.619	17.103	24.220	0.708	15.391	26.316
Qu et al.	0.693	**17.576**	20.813	0.743	15.468	24.855
Berman et al.	0.722	16.544	24.027	0.762	15.713	23.132
MF Dehazer	**0.770**	17.272	**20.231**	**0.792**	**16.135**	**20.958**

5. Conclusions

In this paper, we have proposed a dehazing method for the removal of haze from both nighttime and daytime images. For the ambient illumination estimation, we analyzed the visual properties of hazy images and preserved the illumination direction and shape of the original image using the mixed-filter method. In addition, for the transmission estimation, the proposed method solves the problem of light source diffusion and enhances the image's sense of depth and definition. Both the qualitative and quantitative experimental results demonstrate that MF Dehazer is superior to the recent methods and provides a novel approach for obtaining a better quality dehazing image on the basis of preserving the original image color. However, much more work needs to be undertaken in the future. MF Dehazer shares this limitation because it uses a filter window to estimate the local ambient illumination. If the size of the local dense haze in the image is larger than the window size, it may overestimate ambient illumination and generate haze residue. The method will be improved to overcome these disadvantages.

Author Contributions: Investigation, G.C., Y.X.; methodology, Y.X., Y.T.; validation, Y.X.; data curation, Y.X., G.C.; writing—original draft preparation, Y.T.; writing—review and editing, Y.X., G.C.; project administration, G.C. All authors have read and agreed to the published version of the manuscript.

Funding: This research received no external funding.

Institutional Review Board Statement: Not applicable.

Informed Consent Statement: Not applicable.

Data Availability Statement: Not applicable.

Conflicts of Interest: The authors declare no conflict of interest.

References

1. Duan, J.; Liu, X. Online monitoring of green pellet size distribution in haze-degraded images based on VGG16-LU-net and haze judgement. *IEEE Trans. Instrum. Meas.* **2021**, *70*, 1–16. [CrossRef]
2. Zhang, J.; Lu, Z.; Li, M. Active Contour-Based Method for Finger-Vein Image Segmentation. *IEEE Trans. Instrum. Meas.* **2020**, *69*, 8656–8665. [CrossRef]
3. Xu, H.; Zhai, G.; Wu, X.; Yang, X. Generalized Equalization Model for Image Enhancement. *IEEE Trans. Multimedia* **2013**, *16*, 68–82. [CrossRef]
4. Land, E.H. The Retinex Theory of Color Vision. *Sci. Am.* **1977**, *237*, 108–128. [CrossRef] [PubMed]
5. Galdran, A.; Vazquez-Corral, J.; Pardo, D.; Bertalmío, M. Enhanced Variational Image Dehazing. *SIAM J. Imaging Sci.* **2015**, *8*, 1519–1546. [CrossRef]
6. Wang, Y.; Wang, H.; Yin, C.; Dai, M. Biologically inspired image enhancement based on Retinex. *Neurocomputing* **2016**, *177*, 373–384. [CrossRef]
7. Galdran, A.; Bria, A.; Alvarez-Gila, A.; Vazquez-Corral, J.; Bertalmio, M. On the Duality Between Retinex and Image Dehazing. In Proceedings of the 2018 IEEE/CVF Conference on Computer Vision and Pattern Recognition, Salt Lake City, UT, USA, 18–22 June 2018; pp. 8212–8221. [CrossRef]
8. Koschmieder, H. Theorie der horizontalen sichtweite. *Beitrage zur Physik der Freien Atmosphare* **1924**, *12*, 171–181.
9. He, K.; Sun, J.; Tang, X. Single Image Haze Removal Using Dark Channel Prior. *IEEE Trans. Pattern Anal. Mach. Intell.* **2010**, *33*, 2341–2353. [CrossRef] [PubMed]
10. He, K.; Sun, J.; Tang, X. Guided Image Filtering. *IEEE Trans. Pattern Anal. Mach. Intell.* **2013**, *35*, 1397–1409. [CrossRef] [PubMed]
11. Fattal, R. Single Image Dehazing. *ACM Trans. Graph.* **2008**, *27*, 1–9. [CrossRef]
12. Zhu, Q.; Mai, J.; Shao, L. A Fast Single Image Haze Removal Algorithm Using Color Attenuation Prior. *IEEE Trans. Image Process.* **2015**, *24*, 3522–3533. [CrossRef] [PubMed]
13. Berman, D.; Treibitz, T.; Avidan, S. Non-local Image Dehazing. In Proceedings of the 2016 IEEE Conference on Computer Vision and Pattern Recognition (CVPR), Las Vegas, NV, USA, 27–30 June 2016; pp. 1674–1682. [CrossRef]
14. Meng, G.; Wang, Y.; Duan, J.; Xiang, S.; Pan, C. Efficient Image Dehazing with Boundary Constraint and Contextual Regularization. In Proceedings of the ICCV—IEEE International Conference on Computer Vision, Sydney, Australia, 1–8 December 2013; pp. 617–624. [CrossRef]
15. Berman, D.; Treibitz, T.; Avidan, S. Single Image Dehazing Using Haze-Lines. *IEEE Trans. Pattern Anal. Mach. Intell.* **2018**, *42*, 720–734. [CrossRef] [PubMed]
16. Li, B.; Peng, X.; Wang, Z.; Xu, J.; Feng, D. AOD-Net: All-in-One Dehazing Network. In Proceedings of the 2017 IEEE International Conference on Computer Vision (ICCV), Venice, Italy, 22–29 October 2017; pp. 4780–4788. [CrossRef]
17. Chen, D.; He, M.; Fan, Q.; Liao, J.; Zhang, L.; Hou, D.; Yuan, L.; Hua, G. Gated Context Aggregation Network for Image Dehazing and Deraining. In Proceedings of the 2019 IEEE Winter Conference on Applications of Computer Vision (WACV), Waikoloa, HI, USA, 7–11 January 2019; pp. 1375–1383. [CrossRef]
18. Dong, J.; Pan, J. Physics-Based Feature Dehazing Networks. In Proceedings of the Computer Vision—ECCV 2020, Glasgow, UK, 23–28 August 2020; pp. 188–204.
19. Shao, Y.; Li, L.; Ren, W.; Gao, C.; Sang, N. Domain Adaptation for Image Dehazing. In Proceedings of the 2020 IEEE/CVF Conference on Computer Vision and Pattern Recognition (CVPR), Seattle, WA, USA, 13–19 June 2020; pp. 2805–2814. [CrossRef]

20. Dong, H.; Pan, J.; Xiang, L.; Hu, Z.; Zhang, X.; Wang, F.; Yang, M.-H. Multi-Scale Boosted Dehazing Network With Dense Feature Fusion. In Proceedings of the 2020 IEEE/CVF Conference on Computer Vision and Pattern Recognition (CVPR), Seattle, WA, USA, 13–19 June 2020; pp. 2154–2164. [CrossRef]
21. Qu, Y.; Chen, Y.; Huang, J.; Xie, Y. Enhanced Pix2pix Dehazing Network. In Proceedings of the 2019 IEEE/CVF Conference on Computer Vision and Pattern Recognition (CVPR), Long Beach, CA, USA, 15–20 June 2019; pp. 8152–8160. [CrossRef]
22. Deng, Q.; Huang, Z.; Tsai, C.-C.; Lin, C.-W. HardGAN: A Haze-Aware Representation Distillation GAN for Single Image Dehazing. In Proceedings of the Computer Vision—ECCV 2020, Glasgow, UK, 23–28 August 2020; pp. 722–738.
23. Pei, S.-C.; Lee, T.-Y. Nighttime haze removal using color transfer pre-processing and Dark Channel Prior. In Proceedings of the 2012 19th IEEE International Conference on Image Processing, Orlando, FL, USA, 30 September–3 October 2012; pp. 957–960. [CrossRef]
24. Zhang, J.; Cao, Y.; Fang, S.; Kang, Y.; Chen, C.W. Fast Haze Removal for Nighttime Image Using Maximum Reflectance Prior. In Proceedings of the 2014 IEEE International Conference on Image Processing (ICIP), Paris, France, 27–30 October 2014; pp. 4557–4561. [CrossRef]
25. Zhang, J.; Cao, Y.; Fang, S.; Kang, Y.; Chen, C.W. Fast Haze Removal for Nighttime Image Using Maximum Reflectance Prior. In Proceedings of the 2017 IEEE Conference on Computer Vision and Pattern Recognition (CVPR), Honolulu, HI, USA, 21–26 July 2017; pp. 7016–7024. [CrossRef]
26. Li, Y.; Tan, R.T.; Brown, M.S. Nighttime Haze Removal with Glow and Multiple Light Colors. In Proceedings of the 2015 IEEE International Conference on Computer Vision (ICCV), Santiago, Chile, 7–13 December 2015; pp. 226–234. [CrossRef]
27. CAncuti, C.; Ancuti, C.O.; De Vleeschouwer, C.; Bovik, A.C. Day and Night-Time Dehazing by Local Airlight Estimation. *IEEE Trans. Image Process.* **2020**, *29*, 6264–6275. [CrossRef] [PubMed]
28. Zhu, Z.; Wei, H.; Hu, G.; Li, Y.; Qi, G.; Mazur, N. A Novel Fast Single Image Dehazing Algorithm Based on Artificial Multiexposure Image Fusion. *IEEE Trans. Instrum. Meas.* **2020**, *70*, 1–23. [CrossRef]
29. Wang, Z.; Bovik, A.; Sheikh, H.; Simoncelli, E. Image Quality Assessment: From Error Visibility to Structural Similarity. *IEEE Trans. Image Process.* **2004**, *13*, 600–612. [CrossRef] [PubMed]
30. Sharma, G.; Wu, W.; Dalal, E.N. The CIEDE2000 color-difference formula: Implementation notes, supplementary test data, and mathematical observations. *Color Res. Appl.* **2005**, *30*, 21–30. [CrossRef]
31. Westland, S.; Ripamonti, C.; Cheung, V. *Computational Colour Science Using MATLAB*, 2nd ed.; Wiley: Hoboken, NJ, USA, 2012.
32. Ancuti, C.O.; Ancuti, C.; Timofte, R.; de Vleeschouwer, C. O-HAZE: A Dehazing Benchmark with Real Hazy and Haze-Free Outdoor Images. In Proceedings of the 2018 IEEE/CVF Conference on Computer Vision and Pattern Recognition Workshops (CVPRW), Salt Lake City, UT, USA, 18–22 June 2018; pp. 867–8678. [CrossRef]

applied sciences

MDPI

Article

Reliable Integrity Preservation Analysis of Video Contents with Support of Blockchain Systems

Wan Yeon Lee * and Yun-Seok Choi

Department of Computer Science, Dongduk Women's University, Seoul 02748, Korea
* Correspondence: wanlee@dongduk.ac.kr; Tel.: +82-2-940-4685

Abstract: In this article, we propose an integrity preservation analysis scheme of video contents working on the blockchain systems. The proposed scheme stores the core points of video contents analysis in the blockchain system permanently so that any user can easily verify the results of the proposed analysis procedure and their reliability. The scheme first examines the codec software characteristics of digital camera devices and video editing tools in advance, and stores them in the blockchain systems. Next, the scheme extracts the codec software characteristic from the target video file and compares it with the prepared characteristics in the blockchain system. With a matched characteristic, the scheme finds out the source camera device or the source video editing tool that generates the target video file. We also propose an integrity preservation trace scheme to record the transformation history of video contents into the blockchain systems. This scheme compares the original video and its transformed video frame by frame, and stores the comparison result with a hash value of the transformed video in the blockchain system. Then, the integrity analysis and transformation history of the target file can be easily searched in the blockchain system, where the hash value of the video file is used as the index of searching operation. We implement the proposed scheme into a practical tool upon a commercial blockchain system, Klaytn. Experimental results show that the proposed scheme carries out the integrity analysis of video contents with 100% accuracy, and provides the transformation history of non-original video contents with 100% accuracy when a proper parameter is given. It is also shown that the proposed scheme completes the integrity analysis within at most one second, and the search operation for transformation history within at most four seconds.

Keywords: integrity analysis; integrity preservation trace; video contents; blockchain; digital video forensic

Citation: Lee, W.Y.; Choi, Y.-S. Reliable Integrity Preservation Analysis of Video Contents with Support of Blockchain Systems. *Appl. Sci.* **2022**, *12*, 10280. https://doi.org/10.3390/app122010280

Academic Editors: Przemysław Falkowski-Gilski, Tadeus Uhl and Zbigniew Lubniewski

Received: 29 August 2022
Accepted: 9 October 2022
Published: 12 October 2022

1. Introduction

The video of surveillance digital cameras plays the role of providing crucial evidence for criminal investigation or accident site examinations in many cases. However, many commercial software tools for manipulating video contents are widely distributed, and thus, modifying the video contents becomes very easy. Unless the integrity of video including critical scenes is proved, the video may be useless for evidence or may result in another conflict caused by videos that have been tampered with Moreover, deepfake videos can cause big trouble by spreading false information. Hence, the integrity analysis of video contents has become very important for digital forensics investigations. This article deals with the integrity analysis of video contents and verifies whether the video contents is original or has been tampered with.

A lot of research has progressed in the areas of digital video forensics. The existing techniques of digital video forensics can be classified into three groups: detection of tampered videos, recovery of deleted/hidden videos, and source authentication of videos [1–3]. Studies for the detection of tampered videos are roughly divided into an active intrusive approach and a passive blind approach. In the active intrusive approach, a known identifier trace such as a signature or watermark is embedded imperceptibly into the video contents at the recording time, and it is utilized to discover tampering evidence of the forged video [4]. In the passive blind approach, descriptive features such camera sensor artifacts, coding artifacts and material object features are extracted from the video data without the help of signatures or watermarks, and they are analyzed to detect abnormal patterns of forged multimedia [5,6]. The studies for the recovery of deleted/hidden videos design the mechanism of how to search and extract frame artifacts of deleted video upon unallocated storage space of camera devices, and how to assemble found video fragments [3,7]. The studies for the source authentication of video focus on identifying the specific model of source digital devices or video editing tools by examining unique characteristics of digital video devices, such as parameters of software codecs, lens radial distortion, and sensor imperfections [1,8,9]. The proposed scheme handled in this article belongs to the category of source authentication of videos.

In order to verify whether the target video file is generated by a digital camera device or by a video editing tool, the proposed scheme examines the characteristics of codec software used to compress the image stream into a small-size video file. The scheme prepares the codec characteristics of as many camera devices and editing tools as possible, and stores them in a form of signatures database in the blockchain system. Next, the scheme extracts the codec characteristic from the target video file and compares it with the prepared characteristics in the blockchain system. If the extracted characteristic is matched with one of the digital camera devices, the scheme determines if the target video file is original. On the other hand, if the extracted characteristic is matched with one of the video editing tools, the scheme determines if the target video file is edited, i.e., not original. This approach has a drawback: it cannot discriminate in cases in which video contents are preserved without any tampering during the manipulation of video editing tools. To overcome this drawback, we propose another mechanism that traces the integrity preservation of video contents after the target video file is retouched using a video editing tool. Figure 1 shows the trace procedure of integrity preservation for a video file that has been retouched but not tampered with. In the proposed scheme, the front server extracts the software characteristic from the original video and the multimedia contents of original video and those of retouched video. The backend blockchain system verifies the integrity of the original video and stores the integrity verification result of the original video and the comparison result between the original and retouched videos. Then, a user can easily search for the integrity preservation analysis of the original and the retouched video files.

In the proposed scheme, the back-end blockchain system takes the role of keeping the core parts of the proposed scheme immutable, while the front-end server takes the role of processing heavy computation, such as comparison operation of video contents and hash operation of video files. If the whole source codes are located in the front-end server, the analysis result becomes changeable whenever the source codes are updated. In order to guarantee the reliability of the proposed scheme, the core parts affecting the analysis result are located in the back-end blockchain system so that they are not allowed to change.

To evaluate the feasibility of the proposed scheme, we implement the proposed scheme into a software tool working upon a commercial public blockchain system, *Klaytn* [10]. The software module in the front-end server is implemented with the Javascript language, and the software module in the back-end blockchain system, referred to as smart contract program, is implemented with the Solidity language. In the integrity analysis evaluation, the implemented tool verifies the integrity of two hundred original videos and four hundred retouched videos with 100% accuracy. The implemented tool completes the integrity analysis very quickly, within at most one second, because it examines only the metadata of video files and not their multimedia contents. In the evaluation of integrity preservation trace, we find a noticeable phenomenon: the frames of the original video are not exactly the same as their corresponding frames in the retouched video. Even though there is no manipulation when transforming the original video into the retouched video using a video editing tool, the red, green and blue component values in the original video file are slightly different from those in the retouched video file. So, if any difference between an original video and its retouched video is not allowed, we fail to verify the integrity preservation of the video files which have been retouched but not tampered with in all cases. When a proper difference of red, green and blue components in each pixel is allowed, the implemented tool verifies the integrity preservation of retouched video files with 100% accuracy. When obtaining the comparison result between an original video and its retouched video, the execution time of the implemented tool is affected heavily by the size of multimedia contents, and thus increases proportionally to the file size, When searching the comparison result with a hash value of the input video, the execution time of the implemented tool is affected slightly by the size of multimedia contents, and the search operation is completed within, at most, four seconds in our experiments.

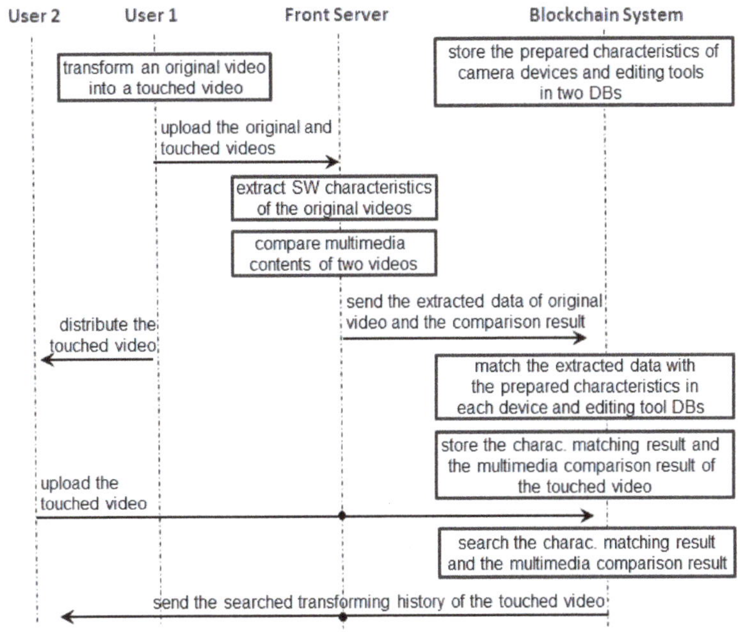

Figure 1. Overall procedure to trace integrity preservation for retouched but non-tampered video file.

The contributions of this article are following;

- This article proposes a trace scheme of multimedia integrity preservation for retouched but not-tampered videos. To the best of our knowledge, this study is the first to record the transforming history from original videos to their retouched videos in order to trace the integrity preservation of the retouched videos.
- To guarantee the reliability, the proposed scheme stores its main code parts and the transforming history of retouched videos into the public blockchain system. When they are stored in a blockchain system, they become immutable, whereas they can be changed at any time when they are stored in a centralized server. The proposed scheme stores its main codes in the form of *smart contract* upon the public blockchain system The smart contract is a digital programming code suitably designed for the public blockchain system.
- To evaluate the practical feasibility of the proposed scheme, we implement the scheme upon a commercial public blockchain system and confirm that the scheme verifies the integrity preservation of retouched videos with 100% accuracy within a reasonable execution time. To avoid the maximum size constraint of a single smart contract, the scheme consists of twenty smarts contracts, and they are linked to their blockchain addresses.

The rest of this article is organized as follows. Section 2 surveys the related previous studies and explains the benefit of the proposed scheme compared to the previous works.

Section 3 describes the proposed integrity analysis and integrity preservation trace scheme in detail. Section 4 explains the implemented software tool of the proposed scheme and deals with the experimental evaluation results of the implemented tool. Section 5 concludes this article with discussions for further study.

2. Related Works

Due to device manufacturing costs, most legacy camera recording systems were not equipped with intentional authentication information such as an invisible watermark. Therefore, the passive blind approach without any authentication information has been more actively studied compared to the active intrusive approach with authentication information. The passive blind approach extracts descriptive features such camera sensor artifacts, coding artifacts and material object features from the video data, and it analyzes them to detect abnormal patterns of forged multimedia [5,6]. Passive blind techniques are classified again into detection of inter-frame forgery and detection of intra-frame forgery [5]. In the inter-frame forgery, some frames are deleted, inserted or duplicated, while some pixel region within a frame is manipulated in the intra-frame forgery. Most of the previous studies [11] for the inter-frame forgery examine the artifacts of optical flow or motion vectors of objects, and those [12–14] for the intra-frame forgery examine the artifacts of noise residue or multiple compressions. All of these passive techniques determine the target video as tampered when the examined artifacts exceed a given threshold. However, it is very difficult to find a proper threshold, and an improper threshold may result in a lot of incorrect detections, such as True Negative detection and False Positive detection.

Recently, deep-learning-based detection techniques [15–17] have been actively investigated to find out the tampered videos. These deep-learning-based methods do not require the threshold decision of abnormal patterns and automatically find out the suspicious static images in the target video with the forgery probability. However, this deep-learning approach requires the prelearning procedure with a well-defined training set. The performance of the deep-learning approach heavily depends on the accuracy of its pre-learning procedure.

The recovery approach of deleted videos focuses on searching artifacts of deleted files upon unallocated storage are of file systems, when the deleted video files remain in the unallocated area until they are overwritten with new video files [18]. The searching technique is called *carving* [19], and it tries to find the unique header or the footer signature of deleted files. Recent carving methods [20–23] considered the case that a deleted video file is split into multiple fragments and the fragments are spread in non-continuous locations. They search for spread fragments and combine them into a single file based on the information of fragment size and fragment location in the video file.

The source authentication approach focuses on identifying the specific model of source digital devices or video editing tools. The traditional source authentication methods [1] examine the unique characteristics of digital video devices such as parameters of lens radial distortion and sensor imperfections in the artifacts of digital multimedia contents. Recent source authentication methods [8,9,24–26] examine software artifacts of video codec in the file metadata, instead of artifacts of digital multimedia contents. These studies consider both software characteristics of video editing tools and digital video cameras. If the extracted metadata are matched with the software characteristics of some camera device, the scheme determines that the given video file has integrity-guaranteed multimedia contents. On the other hand, if the extracted metadata are matched with software characteristics of some video editing tool, the scheme determines that the given video file has integrity-unguaranteed multimedia contents. This approach can quickly verify the name of the video editing tools or camera devices that last touched the target video file. However, this approach is unable to discriminate in cases in which multimedia contents are preserved without any multimedia modification, even though the video is retouched by some video editing tool; for example, file format converting from AVI to MP4. On contrary, the scheme proposed in this article is designed to discriminate in cases in which multimedia contents are preserved without any multimedia modification, even after the video is retouched by a video editing tool.

Due to the immutability of data stored in the blockchain system [27], a few recent studies [28–31] exploit the blockchain system to trace the integrity of video. A blockchain is a digital ledger system that is duplicated and distributed across the network of computer systems, which makes it quite difficult to break. The most expensive resource in the blockchain system is to store a record permanently, because all computers in the blockchain system cooperate to make a consensus and store the record separately in their ledger storage [32]. The method of Yatskiv et al. [28] stores the hash value of each video file in the blockchain system, and protects unauthorized changes by comparing the hash value of the target video with the hash value stored in the blockchain system. The method of Ghimire et al. [29] stores the hash of video segments in chronological order while recording videos in the digital devices. This approach belongs to active intrusive approach and suffers from frequent expensive record operations of the blockchain system, where authentication information is embedded in the blockchain system instead of multimedia contents. The method of Ghimire et al. [30] stores the hash values of captured scenes and suffers from frequent record operations of the blockchain system, similar to the method of Yatskiv et al. [28]. This method segments the continuous stream of captured scenes, calculates their hash value and stores them periodically in the permissioned private blockchain system. The method of Ghimire et al. [31] automatically selects the key frames of target videos with support of the deep learning technique and stores their hash values in the blockchain system in order to guarantee the integrity of the key frames. However, all of these approaches have drawbacks that cannot discriminate in cases where multimedia contents are preserved without any multimedia modification after the video has been retouched by some video editing tool. The scheme proposed in this article is designed to overcome this drawback by storing the transforming history of videos in the blockchain system. Additionally, the proposed scheme requires one-time record operation with short data for a video file to store its integrity analysis result, whereas most of the previous methods [29–31] require frequent record operations with large data on the blockchain system.

3. Proposed Scheme

The proposed scheme consists of two mechanisms. One is to verify the integrity of an original video file by examining the codec software characteristic of the file. The other is to trace the transforming history of a non-original video file by searching the comparison result stored in the blockchain system. The former mechanism is explained in detail in Section 3.1, and the latter mechanism is explained in detail in Section 3.2.

3.1. Analyzing Integrity of Original Video Files

Our previous study [25] deals with the integrity analysis of original video files based on the examination of codec software characteristics. The method given in our previous study prepares two signature databases for video editing tools and digital camera device, respectively. This method extracts the software characteristic from the target video file and compares it with signatures in the two database, as shown in Figure 2a. If any signature in the database of camera devices is matched, the target video file is decided to be original. If any signature in the database of video editing tools is matched, the target video file is decided to not be original. Figure 2b shows a signature example in the database of video editing tools for the editing tool named Movavi Video Editor.

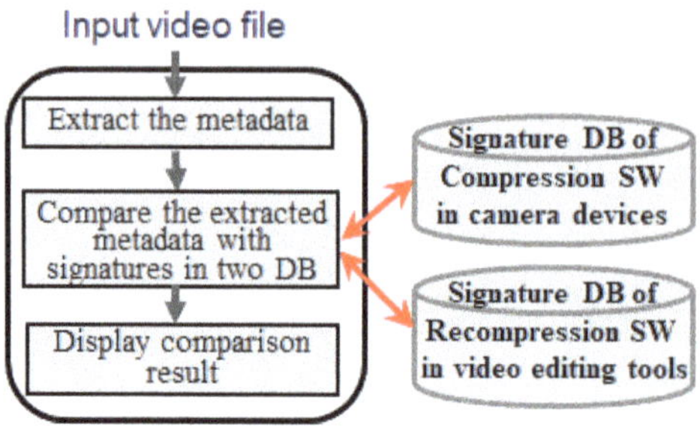

(a)


```
{ "item_name": "Movavi Video Editor",
  " id_seq " :
    [ ["ftyp", "free", "mdat", "moov", "mvhd",
      "trak", "tkhd", "edts", "elst", "mdia", "mdhd", "hdlr", "minf", vmhd",
      "dinf", "dref", "stbl", "stsd", "stts", "stss", "stsc", "stsz", "stco",
      "trak", "tkhd", "edts", "elst", "mdia", "mdhd", "hdlr", "minf", "smhd",
      "dinf", "dref", "stbl", "stsd", "stts", "stsc", "stsz", "stco",
      "sgpd", "sbgp", "udta", "meta" ]
    ]
  "check_value " :
    [ { "id": "meta", "value": "mdtaencoder" } ]
}
```


(b)

Figure 2. Outline of Integrity Analysis: (**a**) Modules of integrity analysis function and (**b**) signature example of video editing tool.

However, the method given in our previous study has a drawback in that the integrity analysis result becomes changeable whenever its source codes are modified. The integrity analysis result of the method is not permanent, and thus, this method does not provide sufficient reliability. To enhance the reliability of integrity analysis result, the mechanism proposed in this article upgrades the method of our previous study [25]. In order to fix the software source codes permanently, the proposed mechanism stores the source codes in the public blockchain system. Considering the expensive record costs of a blockchain system, the proposed mechanism minimizes the data amount stored in the blockchain system. The core parts of the source code are stored in their original forms and the rest non-essential source parts are stored as a shortened form of their entire hash value. That is, the two signature databases shown in Figure 2a and the signature comparison procedure are stored in their original forms in the blockchain system, while the rest of the source codes are stored as their entire hash value. Separation of core source parts prevents the reliability corruption caused by modifications of the rest of the source parts. Even when the rest of the source codes are updated, the key operation of integrity analysis is still immutable, and its analysis result preserves its reliability. Additionally, it is easy to confirm the comparison procedure between the extracted signature and the signatures in the databases, where the signature extracted from the target file is always displayed as a form of log and the signature databases are open to any user through the public blockchain system. If the whole source code, including the databases, is stored as a form of its hash value, the record costs of the blockchain system are reduced, but it becomes difficult to check the signature comparison procedure between the extracted signature and the signatures in the databases.

Figure 3 shows the enhanced procedure (This procedure is not depicted in Figure 1 for the sake of simplicity) to check the source code reliability of the proposed scheme compared to our previous study. The core parts of source codes are constructed as digital programming codes, called smart contracts, in the public blockchain system. The remaining parts of the source codes are constructed in the front server, and their hash value is stored in the blockchain system. Then, the core parts of source codes are fixed permanently due to the systemic nature of blockchain systems, and the remaining parts of the source codes can be checked to see whether they have been updated or not.

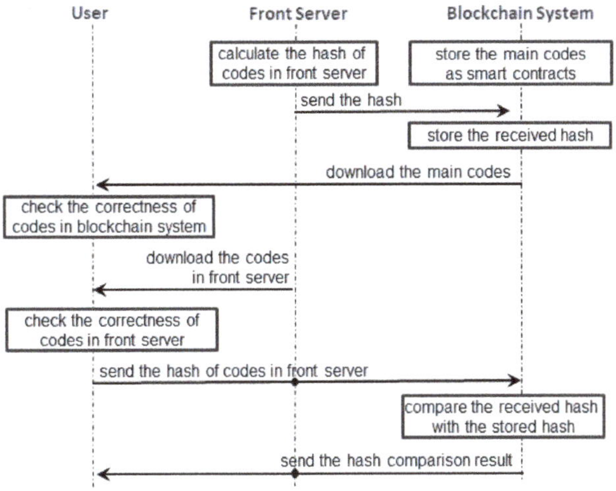

Figure 3. Procedure to check the source code reliability of the proposed scheme.

Figure 4 shows the detailed structure of the proposed mechanism. The signature databases are stored in their original form in the public blockchain system, and the comparison operation with the extracted signature is constructed as a smart contract program on the blockchain system. The rest of the software modules, such as those extracting the signature from the input file and displaying the ongoing analysis results, are constructed as a stand-alone tool working on a local computer, called the front-end server. The proposed mechanism calculates the hash value of the whole source code in the front-end server and compares it with the hash value stored in the blockchain system. In this structure, whenever the source code in the front-end server is updated, its hash value is stored with its time stamp separately in the blockchain system. Then, it can be checked to see which update version the source codes in the front-end server belong to.

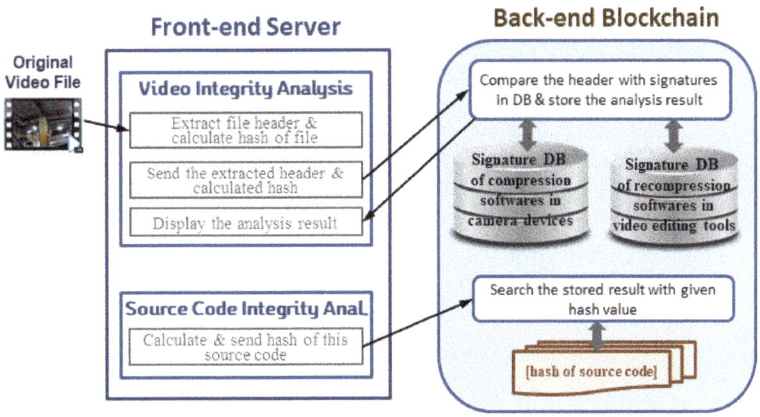

Figure 4. Detailed structure of integrity analysis machanism.

3.2. Integrity Preservation Trace of Reretouched Video Files

Although the mechanism explained in Section 3.1 can quickly verify the name of the video editing tools or camera devices that last touched the target video file, it is unable to discriminate in cases in which video contents are preserved without any modification during retouching of the original video file by some video editing tool. It is common to convert the file format of an original video file into another format suitable for running on portable multimedia devices; for example, from AVI file format to MP4 file format. In this case, the mechanism explained in Section 3.1 determines the converted video file as non-original, but the integrity of the multimedia contents in the converted video file is preserved in its original form.

To overcome this drawback, the proposed scheme provides another mechanism, which is not considered in our previous study [25]. The mechanism compares multimedia contents of the original video and those of the retouched video frame by frame. When all corresponding frames of the original video and the retouched video are the same, this mechanism decides that contents of the two videos are equal. If some corresponding frames of the original video and the retouched video are different, this mechanism decides that contents of the two videos are not equal. The comparison result and the integrity analysis result of the original file are stored in the public blockchain system with a hash value of the retouched video file. Then, the integrity of the retouched video file can be traced with its hash value on the blockchain system. The retouched video file is determined as original only when the comparison result is equal and its corresponding source video file has integrity-guaranteed video contents.

Figure 5 shows the flow chart of the procedure to compare the multimedia contents of two videos. The scheme opens the original video and the retouched video and extracts their Group of Picture (GoP) sequentially. When the frame rates per second (FPS) of the two videos are different, the comparison of each corresponding frame pair in two videos may result in an incorrect comparison result. So, the scheme compares only the first frame of extracted GoPs, called the I-frame, which is not affected by the different FPSs. The scheme calculates the maximum or average difference of the first frames in two extracted GoPs. If the calculated maximum or average difference is smaller than the given threshold values, the scheme extracts the next GoPs of the two videos and continues to compare the first frames of the extracted GoPs until there are no more GoPs in the two videos. If the numbers of GoPs in the two videos are different, and thus the two videos do not reach their ends simultaneously, the scheme returns the result "Not Same". Only when the two videos reach their ends simultaneously with passes of comparing all GoPs, will scheme return the result "Same".

Figure 6 shows the detailed structure of the integrity preservation trace mechanism. The function modules for "Video Integrity Analysis" and "Source Code Integrity Analysis" are equal to those in Figure 4. Added function modules are "Integrity Trace Support" and "Stored Integrity Analysis Search". The proposed mechanism first checks the integrity of the input video file in the module "Video Integrity Analysis". If the integrity of the file is not proved, the following comparison procedure is skipped. If the integrity is proved, the proposed mechanism compares the frames of an original video and the frames of its retouched video one by one sequentially in the module "Integrity Trace Support". As well as the comparison result, the integrity analysis result of the original file and the hash value of the retouched video file are stored as a form of the single group variable into the blockchain system. In the module "Stored Integrity Analysis Search", the mechanism searches for the comparison result of the retouched video and the integrity analysis result of its corresponding original video. This module calculates the hash value of the retouched video file and utilizes it as the index of the searching operation.

The core parts of source code are stored in their original forms, and the rest of the non-essential source parts are stored as a shortened form of their entire hash value, similar to the mechanism explained in Figure 3. The core parts are the signature comparison procedure and the search procedure of stored results, and they are constructed with a program language designed for the blockchain system, called smart contract.

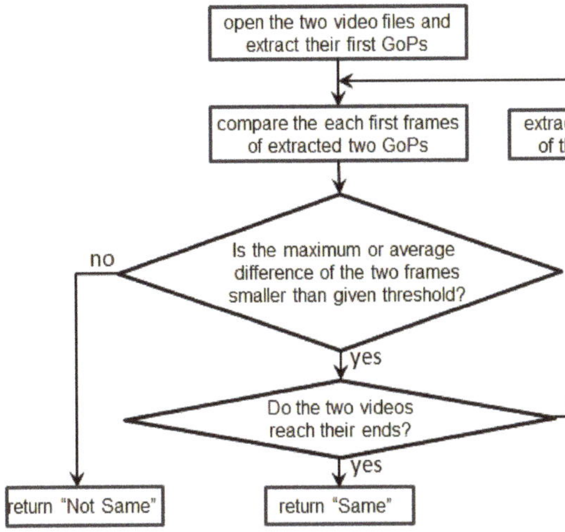

Figure 5. Flow chart of comparison procedure.

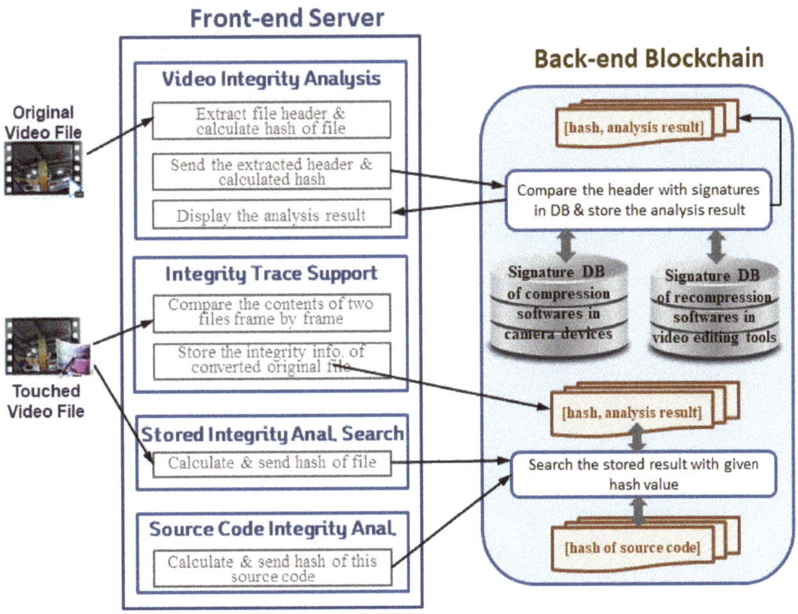

Figure 6. Detailed structure of integrity preservation trace.

4. Evaluation

For evaluation, we implement the proposed scheme upon a commercial public blockchain system. We choose the *Klaytn* platform [10], although the Ethereum platform [27] is the most popular commercial blockchain system, because the Klaytn platform is much faster and cheaper than the Ethereum platform. Instead of the main network of the Klaytn platform, we use its non-commercial test network, called *Baobab*, where its operation costs are free.

4.1. Implemented Software Tool

The proposed scheme is implemented into a software tool (The source codes of our implemented tool can be found at the github site, https://github.com/wanlee02/Klaytn_WhereCamFrom, accessed on 15 August 2022.). The software modules in the front-end server are implemented with the Javascript and HTML languages and managed by the Nodejs package. The software modules in the back-end blockchain system are implemented with the Solidity language, constructed in the form of smart contracts, and managed by the Truffle package. As mentioned in Section 3, the software codes in the blockchain systems are constructed as digital programming codes called smart contracts. Each smart contract has the maximum size constraint: 24 Kbytes on bytecode. So the software codes are divided into 20 modules and each module is implemented into a single smart contract. The 20 smart contract modules are connected with links to their addresses in the blockchain system. Five smart contracts store signature databases of camera devices and five smart contracts store signature databases of video editing tools. The remaining 10 smart contracts compare the extracted signature with two databases and searching the stored result with the transferred hash value.

Figure 7 shows the graphic user interface for the main function of the integrity analysis mechanism, explained in Section 3.1. It is the output display of Chrome browser that is operated by the software modules in the front-end server with an IP address of 201.121.139.90 and a port number of 8080. The button "File Open" starts the uploading process of an input video file. The button "Analysis" starts the integrity analysis process of the input video file. The software modules in the front-end server just extracts the meta data of the input video file, excluding multimedia contents, and transmits it to the software modules in the back-end blockchain system. Then the software modules in the back-end blockchain system extracts the code characteristic from the transmitted metadata, compares it with the prepared signatures, and determines whether the integrity of the input video file is original, non-original or unknown. The integrity analysis result is displayed as *Original*, *Edited* or *Unknown* as shown in the bottom tree boxes of Figure 7, where *Edited* represents non-original video contents. The result *Unknown* occurs when the extracted codec characteristic is not found in the two signature databases, i.e., the signature databases are insufficiently prepared in advance. The decision of integrity analysis is transmitted to the software modules in the front-end server. The matched name of digital camera device or video editing tool is also transmitted. The module in the front-end server displays the transmitted decision as shown in Figure 7.

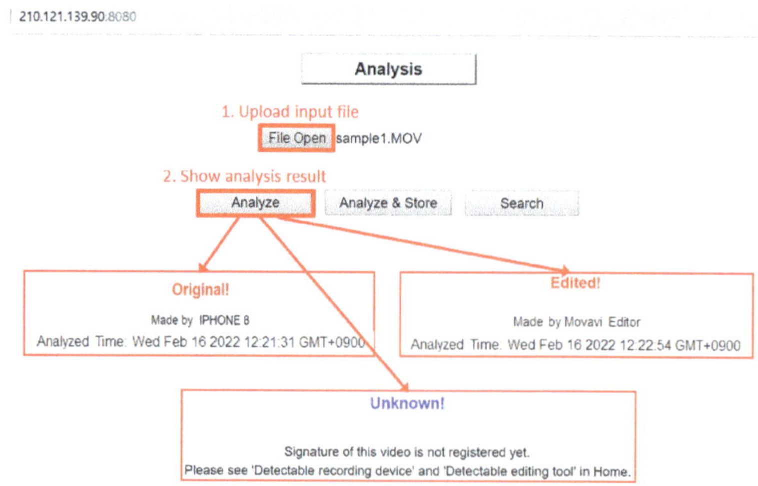

Figure 7. Graphic User Interface for the integrity analysis mechanism.

Figure 8 shows the graphic user interface for supplementary functions of the integrity analysis mechanism. The button "Analysis & Store" stores the integrity analysis result in the blockchain systems with a hash value of the input file. When storing the integrity analysis result is completed, the integrity analysis requests of the same input file can be replaced with the fast search operation, instead of extracting long metadata from the input file and comparing the metadata with prepared signatures. The button "Search" starts the search operation for the stored integrity analysis result with a 32-bytes hash value of the input file.

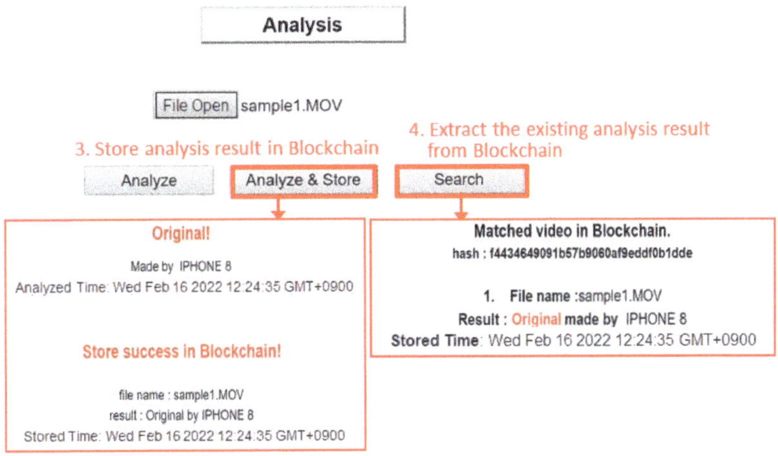

Figure 8. Storing and searching the integrity analysis result on the blockchain system.

Figure 9 shows the graphic user interface for the function of integrity preservation trace between an original file and its retouched file, explained in Section 3.2. The button "File Open" provides the uploading process of the original video file and its retouched non-original video file. The button "Compare" starts the process of comparing each pair of corresponding frames between the original video and the non-original video one by one and sequentially. This function operates in the front-end server and takes a long time proportional to the running time of the two video files. To avoid the confusion caused by different frame rates per second between the original video and its retouched video, our software module compares only the first frame in each group of pictures (GoP) instead of comparing all frames in each group of pictures. The comparison result is displayed either *Same* or *Not Same*, as shown in the bottom of Figure 9. If the comparison result is *Same*, the software modules in the front-end server extract the meta data of the two files and transmit them to the software modules in the blockchain system. As explained in Figure 7, the software modules in the blockchain system verify the integrity of the two files with the transmitted meta data. The integrity analysis results of the two files are sent back to the front-end server and displayed on the Chrome browser, as shown in the left bottom box of Figure 9. If the comparison result is *Not Same*, the software modules in the front-end server provide the specific time when the two videos run differently for the first time, and depict their two corresponding frames, as shown in the bottom right box of Figure 9. Note, in the bottom right box of Figure 9, the right frame contains a hat image marked as a circle, whereas the left frame does not.

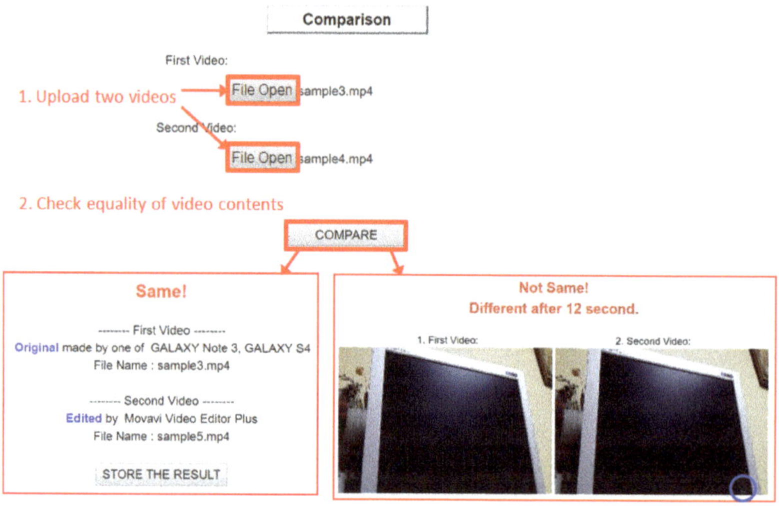

Figure 9. Graphic User Interface for the integrity preservation trace.

Figure 10 shows the procedure used to store the integrity preservation trace in the blockchain system. The button "Analysis" in the left box of Figure 10 starts the storing procedure, where the comparison result of the two files, the comparison time, their integrity analysis results and their hash values are transmitted from the front-end server to the back-end blockchain system. These values are stored as a group variable in the blockchain system. The right box in Figure 10 shows the result of blockchain operation transmitted from the back-end blockchain system to the front-end server.

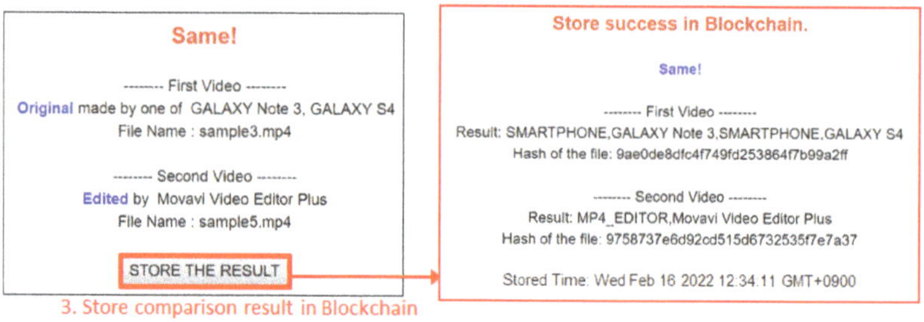

Figure 10. Procedure to store the integrity preservation trace into the blockchain system.

Figure 11 shows the procedure to search the integrity preservation trace of retouched video files. The button "File Open" starts the uploading process of an input video file. The button "OK" starts the operation to search for the transforming history of the input files stored in the blockchain system. The software module in the front-end server calculates the hash value of the input file and transmits it to the software modules in the blockchain system. The software module in the blockchain system searches for the stored integrity preservation trace with the transmitted hash value as the index key. The search result is transmitted to the software module in the front-end server, and displayed as shown in the bottom box of Figure 11.

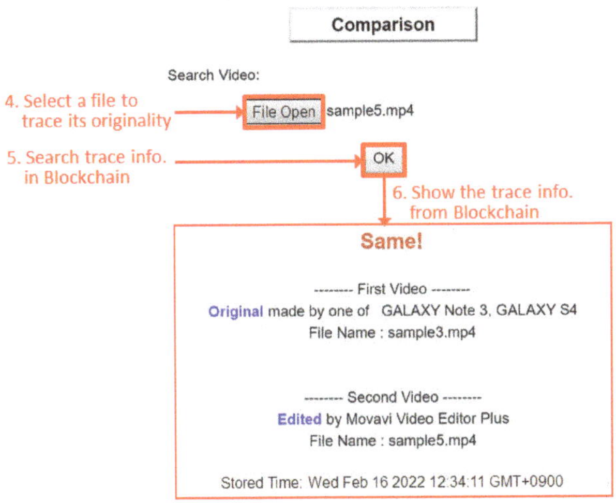

Figure 11. Procedure to search the integrity preservation trace with a hash of input file.

4.2. Experiment Results

To examine the performance of the implemented tool, we prepare two hundred original video files and their retouched video files. The two hundred original videos are generated using USB web cameras with Windows 10, Galaxy smartphones with Android and Apple smartphones with iOS. These two hundred video files are retouched with five video editing tools: Movavi Video Editor, VapMix, GomMix, FFmpeg and Pinnacle Studio. We first transform these two hundred original videos without any modification into their corresponding retouched videos, referred to as non-modified videos. Next, we transform these two hundred original videos with some modification into their corresponding retouched videos, referred to as non-modified videos. One hundred files among the two hundred original files are transformed with frame deletion, referred to as inter-frame forgery videos, and the other one hundred files are transformed with image modification within frames, referred to as intra-frame forgery videos. The position of the deleted frame and the location of modified images are selected randomly.

Table 1 shows the integrity analysis accuracy and the execution time of the proposed scheme. The proposed scheme detects the exact name of camera devices or the name of video editing tools in all cases. Based on the detected name, the scheme classifies the input videos into original videos and retouched videos. However, the scheme does not classify the group of retouched videos: non-manipulated, inter-frame forgery and intra-frame forgery. The execution time of the proposed scheme is less than one second in all cases, whereas the maximum execution time of the scheme is 443 ms. The average execution time for integrity analysis of original videos is 221 ms; for non-modified videos, it is 227; for inter-frame forgery videos, it is 228 ms; for intra-frame forgery videos, it is 232 ms.

Table 1. Accuracy and execution time of integrity analysis.

Measure	Original Video	Non-Modified	Inter-Frame Forgery	Intra-Frame Forgery
Accuracy	100.0%	100.0%	100.0%	100.0%
Avg exec. time	221 ms	227 ms	228 ms	232 ms

Table 2 shows the comparison accuracy between the original video and the retouched video in the proposed scheme. In Table 2, "Allowed Difference Ratio" means the maximally allowed difference ratio of red, green and blue components in each pixel of the retouched video against those in each pixel of the original video. Even though there is no modification when transforming the original video into the retouched video with video editing tools, the red, green and blue component values in the original video file are slightly different from those in the retouched video file. This variance comes from the re-compression procedure of codecs in the video editing tool, which is different the compression procedure of original codecs in the video recording device. Therefore, if any difference between original video and retouched video is not allowed, the first frame of the original video is determined to not be the same as the first frame of its corresponding retouched video in all cases. As result, when "Allowed Difference Ratio" is 0.0%, the accuracy of comparing original video and retouched videos is 0.0% for all transformations of non-modified, inter-frame forgery and intra-frame forgery videos, as shown in Table 2.

Table 2. Comparison accuracy between original videos and retouched videos.

Allowed Difference Ratio	Non-Modified	Inter-Frame Forgery	Intra-Frame Forgery
0.0%	0.0%	0.0%	0.0%
0.025%	99.0%	99.0%	96.0%
0.05%	100.0%	100.0%	100.0%
0.10%	100.0%	100.0%	100.0%
0.20%	100.0%	100.0%	100.0%
0.30%	100.0%	100.0%	100.0 %
0.40%	100.0%	100.0%	100.0%
0.50%	100.0%	100.0%	99.0%
0.60%	100.0%	100.0%	98.0%

When the "Allowed Difference Ratio" is 0.025%, the accuracy becomes close to 100.0% but not equal to 100.0%. Comparison between original video and retouched video is correct for all static scenes, where objects rarely move in subsequent frames. However, comparison becomes incorrect for some dynamic scenes, where objects moves very quickly in subsequent frames. When the "Allowed Difference Ratio" is set to 0.05%, the accuracy measures of all transforming operations are 100.0%. Until the "Allowed Difference Ratio" is set to 0.40%, the accuracy measures are 100.0%. It is worth mentioning that if the "Allowed Difference Ratio" is set too high, some intra-frame forgery may not be detected during the comparison. For example, when "Allowed Difference Ratio" is set to 0.50%, the proposed scheme determines one intra-frame forgery video as a non-manipulated retouched file among the 100 intra-frame forgery videos. Similarly, when the "Allowed Difference Ratio" is set to 0.60%, the scheme misses two intra-frame forgery videos.

Finally, we examine the processing speed of the proposed scheme against the video file size. Table 3 shows the consumed execution time on the Samsung notebook with a i7-core processor with 16 Gbytes memory. The average execution time of videos against the range of file size is depicted. The execution time of analyzing the integrity of input videos is slightly proportional to the file size, but almost close to about 220 ms. The processing time of analyzing the integrity is rarely affected by the file size, because this operation examines only the file header fields, but not the multimedia contents of the video. This processing time is mostly dominated by the speed of the smart contract program in the blockchain system. The execution time of comparing video frames between original videos and retouched videos heavily depends on the file size, as shown in the column "Comparing" of Table 3. This execution time is determined by the hardware performance of the front-end server, because it is carried out on the front-end server. The execution time for storing the comparison result is manifestly proportional to the file size, where the storing operation requires the hash function to compress the input video file into a 32-bytes hash value. The processing speed of storing the comparison result per mega bytes is about 212 (ms)/(Mbytes). This execution time is also affected by the hardware performance of the front-end server. The execution time taken when the comparison result is also manifestly proportional to the file size, because the search operation also calculates the hash value of an input file. The processing speed of searching the comparison result per mega bytes is about 74 (ms)/(Mbytes). The maximum execution time of the search operation is about 3971 ms against the input file with 1.98 Gbytes. The search operation is completed within, at most, four seconds in this experiment. The procedure to obtain the comparison result and store it into the blockchain system is called once for each video, whereas the search operation may be called repeatedly for each video.

The operations of comparing videos and storing result are set for recording the transforming history of a retouched video, while the operation of analyzing integrity is separate and the operation of searching result is separate. The operation of analyzing integrity is completed within, at most, half of a second. The operation of the searching result is completed within, at most, five seconds. The operations of comparing videos and storing the result are completed within, at most, fifteen minutes.

Table 3. Execution time against video size on the blockchain system.

Range of Video File Size (Mbytes)	Analyzing Integrity	Comparing Videos	Storing Result	Searching Result
10∼50	212 ms	19,233 ms	1744 ms	637 ms
50∼100	218 ms	38,568 ms	2824 ms	938 ms
100∼200	221 ms	71,201 ms	3424 ms	1180 ms
200∼500	223 ms	136,119 ms	4026 ms	1315 ms
500∼1024	226 ms	293,744 ms	5725 ms	1974 ms
1024∼1536	233 ms	586,983 ms	7332 ms	2876 ms
1536∼2048	234 ms	722,726 ms	9154 ms	3527 ms

The coin fees required for analyzing integrity and searching result are zero, because these operations do not require data to be permanently stored in the blockchain. On the contrary, the coin fee required for storing the result is about 0.018756 Klay, where the coin fee is the amount of consumed gas $\times$ fee per gas (250×10^{-9} Klay). This coin fee is required whenever storing the transforming history of a retouched video. The total coin fee required for distributing and storing our 20 smart contracts in the blockchain system is about 2.365831 Klay, This operation is only executed once.

4.3. Hardware and Blockchain Platform Dependency

The execution time of the proposed scheme depends on both the hardware specification of the front-end server and the platform of the back-end blockchain system. To check the effect of hardware specification of the front-end server, we replace the notebook with a low-performance IoT device, Raspberry Pi 3 with 1.2 GHz Quad Core and 1 Gbytes memory [33,34]. In the Raspberry Pi 3, the execution time of analyzing integrity is similar to that in Table 3. The operation required for analyzing integrity in the front-end servers is to just extract the file header fields, and thus it is rarely affected by the hardware performance of the front-end server. The execution time of comparing video frames in the Raspberry Pi 3 is increased by about 48%, compared to that in Table 3. This operation is heavily affected by the hardware performance of the front-end server. The execution times of comparing video frames and searching for the result in the Raspberry Pi 3 are similar to those in Table 3. These operations are rarely affected by the hardware performance of the front-end server, but are heavily affected by the platform performance of the blockchain system.

Next, we replace the blockchain platform with other public blockchain platforms such as Ethereum [27] and Tron [35] in order to check the effect of the blockchain platform. We use their non-commercial test networks instead of their commercial main networks: Ropsten network for Ethereum platform and Shasta network for Tron platform. We do not consider the private blockchain platform such as Hyperledger Fabric [36] and Quorum [37], because the hardware and network specifications of private blockchain platforms are not fixed or determined by the private platform operator. The private blockchain platform may be not suitable for the proposed scheme because the disclosure of smart contracts and stored data in the blockchain system requires the permission of the network operator in the private platform. In the Ethereum platform, the execution times of analyzing integrity and comparing videos are similar to those in Table 3. The execution time of the searching result is increased by about 740% compared to that in Table 3, and the operation to store results takes about 650 s. In the Tron platform, the execution times of analyzing integrity and comparing videos and searching result are also similar to those in Table 3. The execution times for storing the result and searching for the result are decreased by about 32% and 16%, respectively, compared to that in Table 3. Through these experiments, it is confirmed that the execution speed of comparing video frames depends on the hardware performance of the front server, and the execution speeds of the storing result and the searching result depend on the platform performance of the public blockchain system.

The proposed scheme is applicable to any public blockchain platform supporting two basic functions: storing data permanently and searching for the stored data. When the proposed scheme employs other public blockchain platforms instead of Klaytn, Ethereum and Tron, the metrics of storing time, searching time and the required blockchain operation fee are changed according to the employed blockchain platform. For example, when the proposed scheme employs IoT-oriented blockchain platforms implemented with Raspberry Pi devices [38–40], the storing time and searching time of the scheme may be increased due to their relatively low hardware and network performance, while the required operation fee depends on the operation fee policy of platform designer or operator. These IoT-oriented blockchain platforms provide many benefits, such as smaller size of generated blocks, lower energy consumption and lower implementation costs. However these benefits do not affect the performance of the proposed scheme. So comparative evaluation about other benefits and drawbacks of the employed blockchain platform, except storing time, searching time, and the required operation fee, is beyond the scope of this study.

5. Conclusions and Discussions

In this article, we propose an integrity preservation analysis scheme of video contents with the support of blockchain systems. To guarantee the reliability of the proposed scheme, core source codes of the scheme are stored in their original form in the public blockchain system immutably, so that any user can easily verify the core parts of the scheme. The rest of the codes are stored as a short form of their hash value in the blockchain system so that any user can check their modification time.

The scheme first examines the software characteristics of digital camera devices and video editing tools in advance, and stores them in the blockchain systems. Next, the scheme extracts the software characteristic from the target video file and compares it with the prepared characteristics in the blockchain system. With a matched characteristic in the blockchain system, the scheme determines the source camera device or the source video editing tool that generates the target video file. In addition, the proposed scheme stores the transformation history of retouched video contents into the blockchain system. The scheme compares the original video and its retouched video frame by frame, and stores the comparison result with a hash value of the retouched video in the blockchain system. Then, the transformation history of the non-original video file can be easily searched in the blockchain system, where the hash value of the video file is used as the index of the searching operation. We implement the proposed scheme into a practical tool on a commercial blockchain system. Experimental results show that the proposed scheme performs the integrity analysis of video files with 100% accuracy and provides the transformation history of retouched video files with 100% accuracy when a proper parameter is given. It is also shown that the proposed scheme completes the integrity analysis within at most one second, and the search operation for transformation history of retouched video files within at most four seconds.

Although this article deals with only the integrity preservation trace for non-modified transforming from original videos to retouched video, the proposed scheme can be extended to the integrity preservation for modified transforming. Examples of modified transforming with integrity preservation are adding the text subtitle to video frames for language translation and generating a short highlight video from a long video. Additionally, this article addresses only the comparison of image contents, but the proposed scheme is applicable to the comparison of audio contents. This study empirically determines the allowed difference ratio used for comparison between original and retouched videos. We leave the problem of systematically determining the allowed difference ratio for further study. We will also investigate the cost minimization of a blockchain system required to store the core parts in further study.

Author Contributions: Conceptualization, W.Y.L.; Implementation, W.Y.L. and Y.-S.C.; Testing and Validation, W.Y.L. and Y.-S.C.; Writing—Original Draft Preparation, W.Y.L.; Writing—Review and Editing, W.Y.L.; Supervision, W.Y.L. All authors have read and agreed to the published version of the manuscript.

Funding: This research was supported by Basic Science Research Program through the National Research Foundation of Korea (NRF) funded by the Ministry of Education (NRF-2021R1F1A1045676).

Institutional Review Board Statement: Not applicable.

Informed Consent Statement: Not applicable.

Data Availability Statement: Not applicable.

Conflicts of Interest: The authors declare no conflict of interest.

References

1. Lanh, T.V.; Chong, K.S.; Emmanuel, S.; Kankanhalli, M.S. A Survey on Digital Camera Image Forensic Methods. In Proceedings of the IEEE International Conference on Multimedia and Expo (ICME), Beijing, China, 2–5 July 2007; pp. 16–19.
2. Singh, R.D.; Aggarwal, N. Video Content Authentication Techniques: A Comprehensive Survey. *Multimed. Syst.* **2018**, *24*, 211–240. [CrossRef]
3. Pahade, R.; Singh, B.; Singh, U. A Survey on Multimedia File Carving. *Int. J. Comput. Sci. Eng. Surv.* **2015**, *6*, 27–46. [CrossRef]
4. Husain, F. A Survey of Digital Watermarking Techniques for Multimedia Data. *Int. J. Electron. Commun. Eng.* **2011**, *2*, 37–43.
5. Kk, S.; Mehtre, B. Digital video tampering detection: An overview of passive techniques. *Digit. Investig.* **2016**, *18*, 8–22.
6. Lee, S.; Song, J.E.; Lee, W.Y.; Ko, Y.W.; Lee, H. Integrity Verification Scheme of Video Contents in Surveillance Cameras for Digital Forensic Investigations. *IEICE Trans. Inf. Syst.* **2015**, *E98-D*, 95–97. [CrossRef]
7. Poisel, R.; Tjoa, S. Forensic Investigations of Multimedia Data: A Review of the State-of-the-Art. In Proceedings of the International Conference on IT Security Incident Management and IT Forensics (IMF), Stuttgart, Germany, 10–12 May 2011; Volume 6, pp. 48–61.
8. Song, J.; Lee, K.; Lee, W.Y.; Lee, H. Integrity Verification of the Ordered Data Structures in Manipulated Video Content. *Digit. Investig.* **2016**, *18*, 1–7. [CrossRef]
9. Sim, S.G.; Kim, E.S.; Kim, D.S.; Lee, S.W.; Lee, W.Y. Apparatus and Method for Verifying the Integrity of Video File. U.S. Patent US15/976,754, 10 May 2018.
10. Klaytn Foundation. Available online: https://docs.klaytn.foundation/ (accessed on 15 August 2022).
11. Kingra, S.; Aggarwal, N.; Singh, R.D. Inter-frame forgery detection in H.264 videos using motion and brightness gradients. *Multimed. Tools Appl.* **2017**, *76*, 25767–25786. [CrossRef]
12. Singh, R.D.; Aggarwal, N. Optical Flow and Pattern Noise-based Copy-Paste Detection in Digital Videos. *Multimed. Syst.* **2021**, *27*, 449–469. [CrossRef]
13. Kobayashi, M.; Okabe, T.; Sato, Y. Detecting Forgery From Static-Scene Video Based on Inconsistency in Noise Level Functions. *IEEE Trans. Inf. Forensics Secur.* **2010**, *5*, 883–892. [CrossRef]
14. He, P.; Jiang, X.; Sun, T.; Wang, S. Detecting of double compression in MPEG-4 videos based on block artifact measurement. *Neurocomputing* **2017**, *228*, 84–96. [CrossRef]
15. Wang, Q.; Zhang, R. Double JPEG compression forensics based on a convolutional neural network. *EURASIP J. Inf. Secur.* **2016**, *23*, 30879–30906. [CrossRef]
16. Chuah, J.H.; Khaw, H.Y.; Soon, F.C.; Chow, C.O. Detection of Gaussian Noise and Its Level using Deep Convolutional Neural Network. In Proceedings of the IEEE Region 10 Conference (TENCON), Penang, Malaysia, 5–8 November 2017; pp. 2447–2450. [CrossRef]
17. Bae, W.; Nam, S.H.; Yu, I.J.; Kwon, M.J.; Yoon, M.; Lee, H.K. Dual-path Convolutional Neural Network for Classifying Fine-Grained Manipulations in H.264 Videos. *Multimed. Tools Appl.* **2021**, *80*, 30879–30906. [CrossRef]
18. Carrier, B. *File System Forensic Analysis*; Addison-Wesley: Boston, MA, USA, 2005.
19. Yoo, B.; Park, J.; Lim, S.; Bang, J.; Lee, S. A study on multimedia file carving method. *Multimed. Tools Appl.* **2012**, *61*, 243–261. [CrossRef]
20. Na, G.H.; Shin, K.S.; Moon, K.W.; Kong, S.G.; Kim, E.S.; Lee, J. Frame-based recovery of corrupted video files using video codec specifications. *IEEE Trans. Image Process.* **2013**, *23*, 317–326.

21. Alghafli, K.; Martin, T. Identification and recovery of video fragments for forensics file carving. In Proceedings of the IEEE International Conference for Internet Technology and Secured Transactions (ICITST), Barcelona, Spain, 5–7 December 2016; Volume 11, pp. 267–272.
22. Yang, Y.; Xu, Z.; Liu, L.; Sun, G. A security carving approach for AVI video based on frame size and index. *Multimed. Tools Appl.* **2017**, *76*, 3293–3312. [CrossRef]
23. Lee, W.Y.; Kim, K.H.; Yang, H.; Ko, Y.W. Automatic reconstruction of deleted AVI video files composed of scattered and corrupted fragments. *Multimed. Tools Appl.* **2020**, *79*, 28355–28367. [CrossRef]
24. Gloe, T.; Fisher, A.; Kirchner, M. Forensic Analysis of Video File Formats. *Digit. Investig.* **2014**, *11*, S68–S76. [CrossRef]
25. Lee, W.Y. Practical Video Authentication Scheme to Analyze Software Characteristics. *IEICE Trans. Inf. Syst.* **2021**, *E104-D*, 212–215. [CrossRef]
26. Huanman, C.Q.; Orozco, A.L.S.; Villalba, L.J.G. Authentication and Integrity of Smartphone Videos through Multimedia Container Structure Analysis. *Future Gener. Comput. Syst.* **2020**, *108*, 15–33. [CrossRef]
27. Wood, G. Ethereum: A Secure Decentralized Generalized Transaction Ledger *Ethereum Proj. Yellow Pap.* **2014**, *151*, 1–32.
28. Yatskiv, V.; Yatskiv, N.; Bandrivskyi, O. Proof of Video Integrity Based on Blockchain. In Proceedings of the International Conference on Advanced Computer Information Technologies (ACIT), Ceske Budejovice, Czech Republic, 5–7 June 2019; Volume 9, pp. 431–434. [CrossRef]
29. Ghimire, S.; Choi, J.Y.; Lee, B. Using Blockchain for Improved Video Integrity Verification. *IEEE Trans. Multimed.* **2020**, *22*, 108–121. [CrossRef]
30. Mercan, S.; Cebe, M.; Aygun, R.S.; Akkaya, K.; Toussaint, E.; Danko, D. Blockchain-based Video Forensics and Integrity Verification Framework for Wireless Internet-of-Things Devices. *Secur. Priv.* **2021**, *4*, 1–17. [CrossRef]
31. Khan, G.; Jabeen, S.; Khan, M.Z.; Khan, M.U.G.; Iqbal, R. Blockchain-enabled Deep Semantic Video-to-video Summarization for IoT Devices. *Comput. Electr. Eng.* **2020**, *81*, 106524. [CrossRef]
32. Lee, W.Y. Cost Mimization of Solidity Smart Contract on Blockchain Systems. *Int. J. Adv. Smart Converg.* **2020**, *9*, 157–163.
33. Raspberry Pi Foundation. Available online: https://www.raspberrypi.org/ (accessed on 1 October 2022).
34. Tirumala, S.S.; Nepal, N.; Ray, S.K. Raspberry Pi-based Intelligent Cyber Defense Systems for SMEs and Smart-homes: An Exploratory Study. *EAI Endorsed Trans. Smart Cities* **2022**, *6*, e4. [CrossRef]
35. Tron Developer Hub. Available online: https://developers.tron.network/ (accessed on 1 October 2022).
36. Hyperledger Fabric. Available online: https://www.hyperledger.org/projects/fabric (accessed on 1 October 2022).
37. Build on Quorum. Available online: https://consensys.net/goquorum/ (accessed on 1 October 2022).
38. Chaabane, F.; Ktari, J.; Frikha, T.; Hamam, H. Low Power Blockchained E-Vote Platform for University Environment. *Future Internet* **2022**, *14*, 269. [CrossRef]
39. Ktari, J.; Frikha, T.; Ben Amor, N.; Louraidh, L.; Elmannai, H.; Hamdi, M. IoMT-Based Platform for E-Health Monitoring Based on the Blockchain. *Electronics* **2022**, *11*, 2314. [CrossRef]
40. Gururaj, H.L.; Kumar, V.R.; Goundar, S.; Elngar, A.A.; Swathi, B.H. The Convergence of Internet of Things and Blockchain Technologies. In *AI/Springer Innovations in Communication and Computing*; Springer: Berlin/Heidelberg, Germany, 2022; pp. 57–75. [CrossRef]

applied sciences

MDPI

Article

Performance Evaluation of a Multidomain IMS/NGN Network Including Service and Transport Stratum

Sylwester Kaczmarek * and Maciej Sac *

Faculty of Electronics, Telecommunications and Informatics, Gdansk University of Technology, Narutowicza 11/12, 80-233 Gdansk, Poland
* Correspondence: kasyl@eti.pg.edu.pl (S.K.); maciej.sac@eti.pg.edu.pl (M.S.)

Abstract: The Next Generation Network (NGN) architecture was proposed for delivering various multimedia services with guaranteed quality. For this reason, the elements of the IP Multimedia Subsystem (IMS) concept (an important part of 4G/5G/6G mobile networks) are used in its service stratum. This paper presents comprehensive research on how the parameters of an IMS/NGN network and traffic sources influence mean Call Set-up Delay (*E*(*CSD*)) and mean Call Disengagement Delay (*E*(*CDD*)), a subset of standardized call processing performance (CPP) parameters, which are significant for both network users and operators. The investigations were performed using our analytical traffic model of a multidomain IMS/NGN network with Multiprotocol Label Switching (MPLS) technology applied in its transport stratum, which provides transport resources for the services requested by users. The performed experiments allow grouping network and traffic source parameters into three categories based on the strength of their effect on *E*(*CSD*) and *E*(*CDD*). These categories reflect the significance of particular parameters for the network operator and designer (most important, less important and insignificant).

Keywords: IMS; NGN; MPLS; call processing performance; quality of service; performance evaluation; traffic model

Citation: Kaczmarek, S.; Sac, M. Performance Evaluation of a Multidomain IMS/NGN Network Including Service and Transport Stratum. *Appl. Sci.* **2022**, *12*, 11643. https://doi.org/10.3390/app122211643

Academic Editor: Juan-Carlos Cano

Received: 10 September 2022
Accepted: 12 November 2022
Published: 16 November 2022

1. Introduction

The Next Generation Network (NGN) [1,2] is a packet-based network dedicated to providing various broadband multimedia services (e.g., VoIP telephony, IPTV or VoD services) with Quality of Service (QoS) and support for generalized mobility. The basic principle of the NGN is the independence of its service-related functions from underlying transport-related technologies. This principle is reflected in the NGN architecture, which includes a service stratum (based on elements of the IP Multimedia Subsystem (IMS) [3–6]; thus, the term "IMS/NGN network" is widely used) and a transport stratum (where any transport technology supporting transfer of IP packets can be applied). There are various technologies considered for the IMS/NGN transport stratum. Access networks can be based on—among others—Ethernet, xDSL, fiber, Wi-Fi or 4G/5G/6G mobile networks. Core transport networks may be based on Ethernet, Flow-State-Aware (FSA), Multiprotocol Label Switching (MPLS) [7,8] and other technologies.

Ensuring proper values of standardized QoS parameters is crucial for the operation and commercial success of the IMS/NGN architecture. From the point of view of the service stratum and services provided to users, which are of our interest, call processing performance (CPP) parameters defined by ITU-T [9,10] are very important. They include, among others, Call Set-up Delay (*CSD*) and Call Disengagement Delay (*CDD*). These parameters result from the operation of the service stratum and the response times of the transport stratum regarding the allocation and release of transport resources for user services (calls). For MPLS technology, which is considered in this paper, transport stratum response times contain response times of MPLS domains resulting from their structure and operation. Consequently, both the service stratum and the transport stratum of an

IMS/NGN network have to be properly designed so that strictly determined values of CPP parameters are not exceeded, which would result in dissatisfaction of users with the provided services.

To achieve this aim, it is necessary to determine how network and traffic source parameters affect guaranteed CPP parameters, which is the goal of this paper. We use our analytical traffic model of a multidomain IMS/NGN network with an MPLS-based transport stratum, proposed in [11], to thoroughly examine the impact of all input parameters on mean Call Set-up Delay ($E(CSD)$) and mean Call Disengagement Delay ($E(CDD)$) calculated for various types of successful call scenarios. The performed research and discussion enable the indication of parameters that are very important, less important and irrelevant from the point of view of the network operator and designer.

The rest of the paper is organized as follows. Section 2 contains a review of the related work. Section 3 describes the structure and operation of our analytical traffic model of a multidomain IMS/NGN network with an MPLS-based transport stratum in terms of the performed investigations (presented in Section 4). Section 5 summarizes the paper.

2. Related Work

Our review of the work related to the analysis and design of IMS/NGN networks indicates that standards organizations (e.g., ITU-T, ETSI, 3GPP, etc.) do not provide any guidelines for traffic modeling and traffic engineering which can be used to achieve appropriate values of CPP parameters. Their scope focuses on functionality [1,2], architecture [2,3] and communication protocols [3]. Although the keywords "IMS" and "NGN" lead to many scientific papers, only a small number of them concern the subject of traffic modeling and engineering in an IMS/NGN network. Examples of such papers in which an analytical approach to network analysis and design is used are described in the next part of this section.

In Ref. [12], a trade-off between delay and throughput in IMS session setup is discussed. A scheme for decreasing the excessive signaling in IMS session setup is proposed. To assess this proposition, an analytical traffic model is used. The model focuses only on IMS elements and does not consider reservation of the transport resources necessary for the services. In particular, there is no Resource and Admission Control Function (RACF) unit proposed by the ITU-T for controlling transport resources.

An analytical traffic model allowing estimation of packet delay variation is proposed in Ref. [13]. Despite the authors' declarations, the proposed model is general and not closely related to the IMS and NGN architecture.

The paper [14] proposes an analytical model for IMS/NGN with G/G/1/N queuing models based on diffusion approximation. The traffic model allows evaluation of average time for establishing multimedia sessions. The authors do not, however, specify the detailed network architecture, number of domains and considered service scenarios. Moreover, no information about verification of the analytical model is provided.

A similar study with the same types of queuing models was performed in Ref. [15]. The described analytical traffic model considers only one IMS domain and does not include resource control mechanisms defined by the ITU-T (RACF unit).

In Ref. [16], an analytical traffic model is presented that allows assessment of bandwidth requirements for IMS architecture. The model contains only the basic elements and a single domain of IMS.

The application of MPLS technology in an NGN is discussed in Ref. [17]. However, this paper is an overview and does not contain any propositions for IMS/NGN performance analysis and resource design.

Papers [18,19] concern the subject of performability of IMS in virtualized containerized environments. They assess availability of a single IMS domain with respect to an extensive set of input parameters. Moreover, an automated procedure aimed at supporting the performability management of IMS deployments that must satisfy the optimal trade-off among high availability requirements, capacity load and deployment costs is proposed in Ref. [19].

Paper [20] proposes an analytical model based on an open queuing network of G/G/m queues to estimate mean response time of a chain of Virtualized Network Functions (VNFs). The presented analytical model is verified using simulations. The proposition is very general and can be applied to model various systems, including IMS/NGN networks.

The described papers demonstrate that the available analytical approaches to IMS/NGN network design and analysis do not cover these aspects completely. They are either very general (not related closely to the IMS/NGN architecture) or consider only the elements of the service stratum (in most cases, it is a single domain of pure IMS without taking into account elements of the transport stratum, including RACF) and only selected network functionalities (services and elements). Moreover, in the majority of cases, there is no information about verification of the abovementioned analytical models using simulations or real networks.

Therefore, we decided to create our own analytical traffic model of an IMS/NGN network which comprehensively models the operation of this network and enables assessment of $E(CSD)$ and $E(CDD)$. The model takes into account the elements of both the IMS-based service stratum and the MPLS-based transport stratum with an extensive set of input variables and service scenarios. Our model is proposed and verified in Ref. [11]. The description of this model is provided in the next section in terms of the performed investigations, which are presented in Section 4.

3. Traffic Model

Figure 1 presents the structure of the considered multidomain IMS/NGN network, which is divided into two domains belonging to two operators (the terms "operator" and "domain" will be used interchangeably with similar meaning). To distinguish between elements of IMS/NGN service and the transport stratum administered by different operators, we designate them with the numbers 1 and 2, respectively. The service stratum of each domain controls service requests sent by user terminals (User Equipment, UE) and includes elements derived from the IMS concept [3–5]:

- Call Session Control Function (CSCF) servers:
    - Proxy-CSCF, P-CSCF—the first contact point for UE;
    - Serving-CSCF, S-CSCF—the main server handling all calls in its domain;
    - Interrogating-CSCF, I-CSCF—the server handling messages coming from other domains.
- Service User Profile Functional Entity/Service Authentication and Authorization Functional Entity, SUP-FE/SAA-FE—the element storing user profiles and performing AAA functions.

In each domain, the transport stratum includes one access network with several access areas and one MPLS core network. The technology of access networks in domains 1 and 2 is not determined—it may be, for example, xDSL, fiber, Wi-Fi or 4G/5G/6G mobile networks. The resources of all access and core networks are controlled by dedicated Resource and Admission Control Function (RACF) units (RACF A1, RACF C1, RACF A2 and RACF C2; Figure 1).

There are two main assumptions for the operation of MPLS core networks: a static bandwidth reservation mode [8] and a quota-based approach [21] are used for Label Switched Paths (LSPs, logical channels transporting aggregated data). LSPs are created based on routing tables, which are determined by a routing algorithm working in the background. Consequently, routing has a negligible impact on the performance of the IMS/NGN service stratum, which is described by CPP parameters. The assumption of a static bandwidth reservation mode means that LSPs are initially allocated with some bandwidth, which can be modified by RACF through the Label Edge Router (LER) beginning the LSP. The quota-based approach indicates that during network operation, LSP bandwidth is allocated with some reserves so that a part of resource allocation requests generated from the service stratum results only in updating the resource state database of RACF without changing the physical LSP bandwidth. A similar situation takes place for resource release requests generated from the service stratum—they are partially handled by updating the RACF database, and the LSP bandwidth is decreased only when its utilization is low.

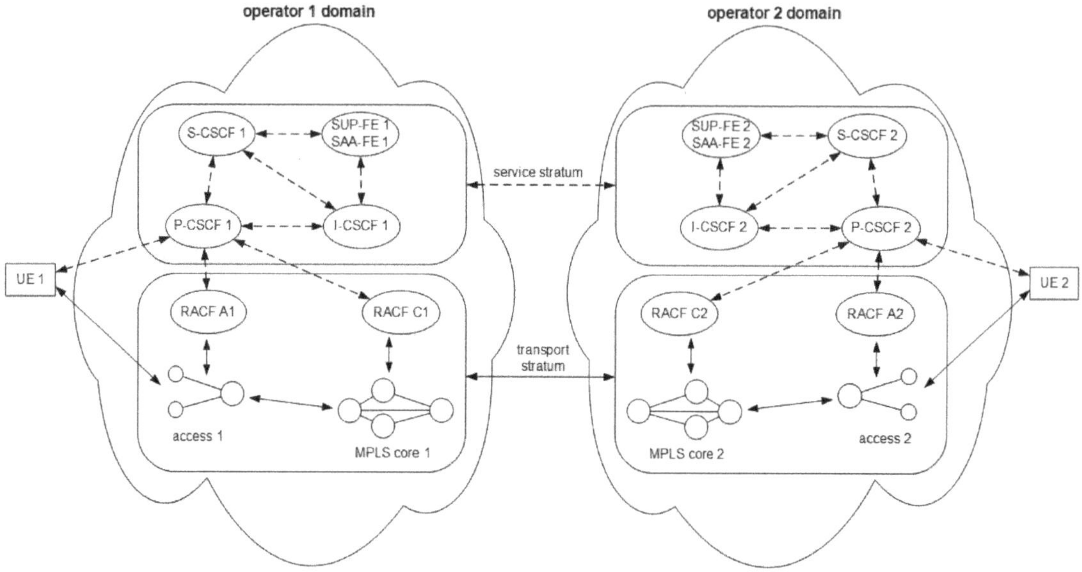

Figure 1. Structure of the multidomain IMS/NGN network with an MPLS-based transport stratum.

The set of service scenarios considered in the modeled network includes UE registration in domains 1 and 2 (scenarios a1 and a2) as well as various type of calls, whose parameters are presented in Table 1. Figure 2 depicts an example of a successful call scenario (inter-operator call originating in domain 2—scenario f2). The most important events for this scenario are as follows:

- Originating UE (UE 2) sends call set-up request (SIP INVITE) to terminating domain (domain 1) (messages 1–11).
- P-CSCF 1 starts transport resources reservation in access and core networks on behalf of UE 1 (messages 14–15) and informs P-CSCF 2 about initiating this process (messages 13 and 16–18).
- P-CSCF 2 starts transport resources reservation in access and core networks on behalf of UE 2 (messages 19–20) and confirms receiving message 18 (messages 21–28).
- P-CSCF 2 informs P-CSCF 1 about successful resource reservation (messages 29–32), and this information is confirmed (messages 33, 35–37).
- After resources are reserved in both domains, SIP INVITE message (34) is sent to UE 1.
- UE 1 starts ringing (messages 39–44), it is confirmed by P-CSCF 2 (messages 45–52).
- UE 1 answers (53–62), which involves updates in packet filtration in access networks (messages 54–55 and 60–61) and ACK confirmation (messages 63–68).
- UE 2 sends a call disengagement request (SIP BYE), which is forwarded to UE 1 and causes resource release in access and core of domains 2 and 1 (messages 69–77).
- Resource release is confirmed (messages 78–82).

Table 1. Call scenarios in the modeled IMS/NGN network.

Name	Type of Call	Originating Domain	Same Access Areas	Necessary Resources	Successful	Remarks
b1	Intra-operator	1	Yes	Access 1	Yes	
b2	Intra-operator	2	Yes	Access 2	Yes	
c1	Intra-operator	1	Yes	Access 1	No	
c2	Intra-operator	2	Yes	Access 2	No	
d1	Intra-operator	1	No	Access 1, core 1	Yes	
d2	Intra-operator	2	No	Access 2, core 2	Yes	
e1	Intra-operator	1	No	Access 1, core 1	No	
e2	Intra-operator	2	No	Access 2, core 2	No	
f1	Inter-operator	1	No	All networks	Yes	
f2	Inter-operator	2	No	All networks	Yes	
g1	Inter-operator	1	No	All networks	No	Variant 1
g2	Inter-operator	2	No	All networks	No	Variant 1
h1	Inter-operator	1	No	All networks	No	Variant 2
h2	Inter-operator	2	No	All networks	No	Variant 2

It is very important that, according to the abovementioned quota-based approach used in MPLS core networks, not all resource control operations performed by RACF C1 and RACF C2 involve communication with MPLS LERs ("MPLS core 1" and "MPLS core 2" blocks; Figure 2) for changes of LSP bandwidth. Blue arrows with dashed lines are used in Figure 2 to illustrate this optional communication between RACF units and controlled MPLS networks.

The aim of our research is to evaluate selected CPP parameters in a multidomain IMS/NGN network with an MPLS-based transport stratum, which is depicted in Figure 1. For this reason, an analytical traffic model was developed [11] which uses definable queuing models (M/M/1 or M/G/1; Figure 3) to reflect the operation of CSCF servers and RACF C1 and RACF C2 units as well as all links between network elements. The choice of queuing models will be commented on later in this section. Modeling of the structure in Figure 1 comes down to the network of queuing systems and selected elements represented by random variables (RACF A1, RACF A2, SUP-FE 1/SAA-FE 1, SUP-FE 2/SAA-FE 2 and MPLS domains; details are provided later). Unfortunately, this network does not meet the assumptions of Jackson's theorem, so it cannot be used to determine the probabilities of the state distributions of this network. Since we are interested in the mean values of delays ($E(CSD)$ and $E(CDD)$), these distributions are not necessary.

It is assumed in our model that message processing times for CSCF servers, RACF C1 and RACF C2 depend on the message type and are different for each network element. For links, message processing times (message transmission times) depend on the message lengths and link bandwidths. Random variables are used to model the response times for the remaining network elements presented in Figure 1: RACF A1, RACF A2, SUP-FE 1/SAA-FE 1 and SUP-FE 2/SAA-FE 2. For the RACF A1 and RACF A2 units, the response times include the time necessary to configure the controlled access networks.

The response times of particular MPLS domains ("MPLS core 1" and "MPLS core 2"; Figure 1) are also taken into account as random variables in the described analytical model. They include the time of processing the request by the LER beginning the LSP and changing the LSP bandwidth, which requires interaction with all MPLS routers on the path. Thus, the influence of the structure and operation of MPLS domains on the analyzed output parameters is reflected by these random variables.

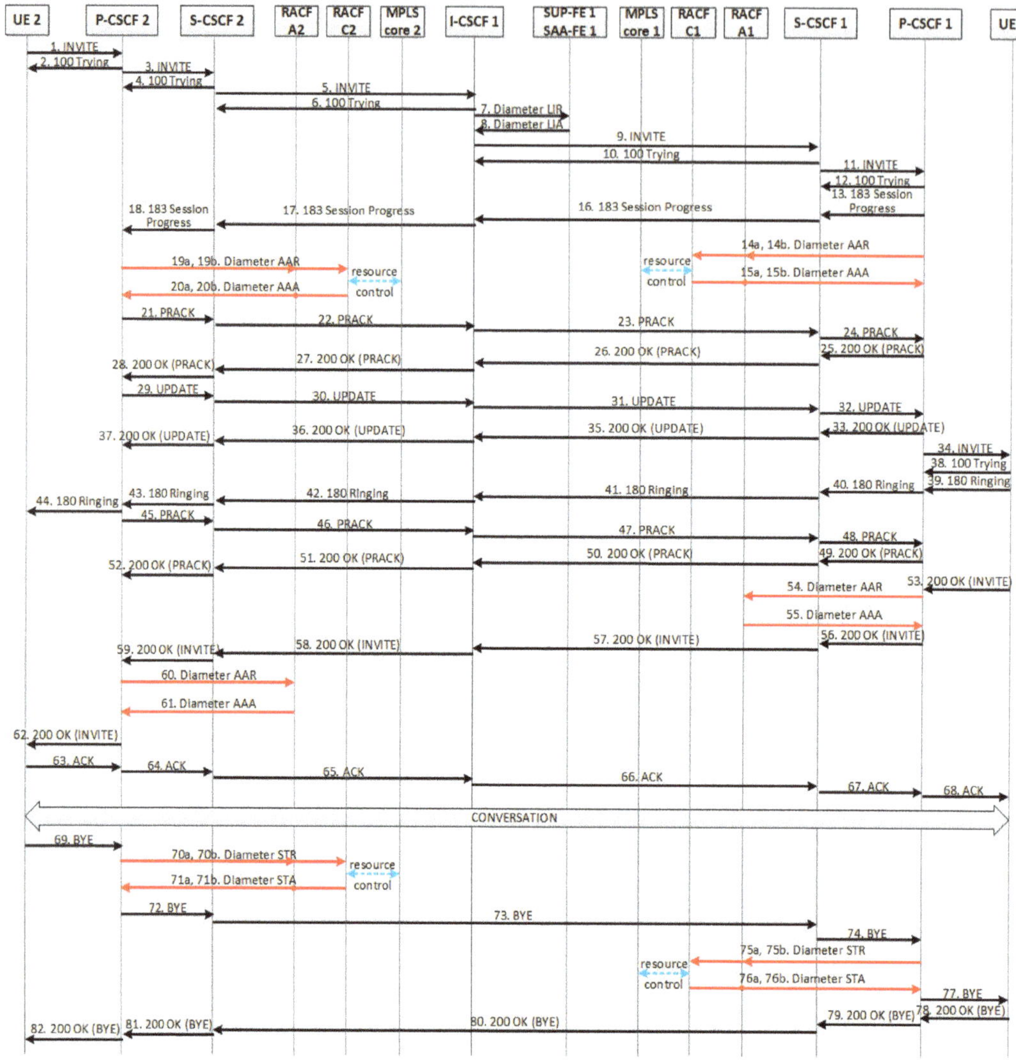

Figure 2. Message flow for the f2 call scenario (black arrows—communications using SIP protocol; red arrows—Diameter protocol; blue arrows—dedicated MPLS resource control protocol).

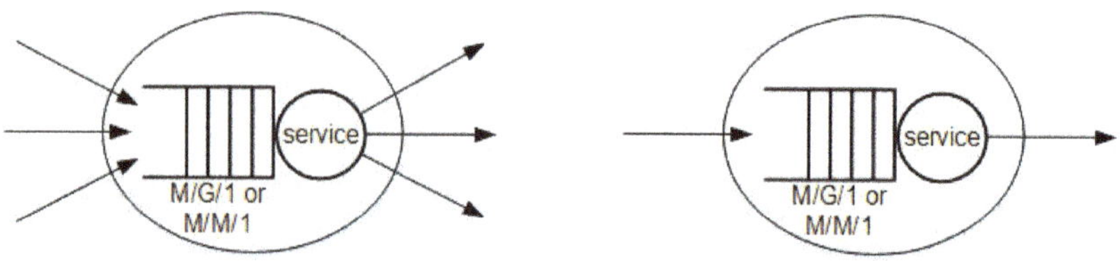

Figure 3. Model of CSCF servers, RACF C1, RACF C2 (**left**) and links (**right**).

To conform to ITU-T standards [9,10] defining CPP parameters, the delays introduced by UE 1 and UE 2 (Figure 1) are not considered in calculations of these parameters.

The output variables of our traffic model are mean values of Call Set-up Delay (*E*(*CSD*)) and Call Disengagement Delay (*E*(*CDD*)) evaluated individually for all successful call scenarios presented in Table 1 (scenarios b1, b2, d1, d2, f1 and f2). To distinguish between the evaluated CPP parameters, the names of scenarios are used in their indexes—e.g., $E(CSD)_{\text{b1}}$ represents mean *CSD* for scenario b1. The calculations of the output parameters are based on the input variables of the traffic model, which can be divided into the following groups:

- Intensities of requests generated in both domains:
  - UE registration request intensity (**lambdaR** vector);
  - Intra-operator call set-up request intensity (**lambda1d** vector);
  - Inter-operator call set-up request intensity (**lambda2d** vector).
- Processing times for network elements containing message queues (CSCF servers, RACF C1 and RACF C2):
  - Times of processing SIP INVITE message by CSCF servers in both domains (**TINV** vector);
  - Time of message authorization and request type determination by RACF C1 and RACF C2 (**TA** vector);
  - Time of performing elementary database operations by RACF C1 and RACF C2 (**Tproc** vector);
  - Time of processing a response from LER by RACF C1 and RACF C2 (**Tresp** vector).
- Response times of network elements modeled as random variables (MPLS domains, RACF A1, RACF A2, SUP-FE 1/SAA-FE 1 and SUP-FE 2/SAA-FE 2):
  - Mean response time of "MPLS core 1" and "MPLS core 2 (**ETR** vector);
  - Mean time of processing requests by RACF A1 and RACF A2 (**EXA** vector);
  - Mean time of processing requests by SUP-FE 1/SAA-FE 1 and SUP-FE 2/SAA-FE 2 (**EY** vector).
- Parameters of the transport stratum:
  - Ratio of calls involving multiple access areas to all intra-operator calls generated in domains 1 and 2 (**rC** vector);
  - Probability of transport resource unavailability in all access and core networks (**pb** vector);
  - Probability of a successful bandwidth reservation or increase without the necessity of increasing LSP bandwidth in MPLS core networks of operators 1 and 2 (**p11** vector);
  - Probability of a bandwidth release or decrease without the necessity of decreasing LSP bandwidth in MPLS core networks of operators 1 and 2 (**p21** vector).
- Link parameters:
  - Lengths (**d** vector);
  - Bandwidths (**b** vector).
- Types of queuing models for CSCF servers, RACF C1, RACF C2 and links (M/M/1 or M/G/1).

In our analytical model of a multidomain IMS/NGN network with an MPLS-based transport stratum, the *E*(*CSD*) and *E*(*CDD*) times for successful call scenarios are calculated based on the definitions given by ITU-T [9,10]. The calculations are performed by summation of mean values of component delays, which correspond to sending messages through links and message sojourn in different network elements. These delays result from the set of messages exchanged in a particular call scenario as well as network elements "visited" by these messages. For *m* "visited" links and *n* "visited" network elements:

$$CPP_x = \sum_{i=1}^{m} T_{link-i} + \sum_{j=1}^{n} T_{NE-j}, \quad (1)$$

where:

- CPP_x is the $E(CSD)$ or $E(CDD)$ for the scenario x = b1, b2, d1, d2, f1 or f2;
- T_{link-i} is a delay corresponding to sending a message through the i-th link;
- T_{NE-j} is a delay corresponding to message sojourn in the j-th network element.

The delays related to sending messages through links (T_{link-i}) can be further decomposed:

$$T_{link-i} = T_{link-w-i} + T_{link-t-i} + T_{link-p-i}, \tag{2}$$

where:

- $T_{link-w-i}$ represents mean message waiting time in a communication queue storing messages when links are busy (to obtain the value of this delay, the M/M/1 or M/G/1 queuing model is applied);
- $T_{link-t-i}$ is a message transmission time calculated by dividing the message length by the link bandwidth (message lengths are set according to the values measured experimentally in Ref. [22]; link bandwidths are taken from the **b** vector);
- $T_{link-p-i}$ is a propagation time proportional to link length (taken from the **d** vector) and equal to 5µs/km for optical links.

For network elements whose response times are modeled as random variables, message sojourn times (T_{NE-j}) are simply included in calculations of $E(CSD)$ and $E(CDD)$ times by taking mean values of these random variables (**ETR, EXA** and **EY** vectors). Message sojourn times in the network elements containing queues can be calculated as follows:

$$T_{NE-j} = T_{NE-w-j} + T_{NE-p-j}, \tag{3}$$

where:

- T_{NE-w-j} represents mean message waiting time in the CPU queue buffering messages when the CPU is busy (to obtain the value of this delay, the M/M/1 or M/G/1 queuing model is applied);
- T_{NE-p-j} is a message processing time.

For each CSCF server, which is the j-th network element in (1), message processing times are calculated in the following way:

$$T_{NE-p-j} = a_{msg} \cdot T_{INV}, \tag{4}$$

where:

- a_{msg} is a factor determining the time of processing a message *msg* in relation to the time of processing the SIP INVITE message (this parameter depends only on the message type and does not depend on the CSCF server type);
- T_{INV} is the SIP INVITE message processing time by a particular CSCF server, and values of this time for all CSCF servers are stored in the **TINV** vector.

The message processing times (T_{NE-p-j}) of RACF C1 and RACF C2 are obtained by appropriately summing elements of the **TA**, **Tproc** and **Tresp** vectors. Components of these sums are dependent on the type of resource operation and the necessity of adjusting LSP bandwidth. Detailed calculations are described in Ref. [23].

It is worth noting that the method of constructing our analytical model is universal. Therefore, it can be used to develop new analytical models, which would allow analysis of other aspects of the IMS/NGN network operation or even analysis of other network types. For example, IMS/NGN resource virtualization can be investigated in which new instances of individual IMS servers are launched in order to guarantee quality (defined by values of $E(CSD)$ and $E(CDD)$) as the load increases. Then, new virtual machines for a specific type of server can be added, and when the load decreases, they can be removed, as suggested in Ref. [24]. In this way, the use of resources allocated for service control in the service stratum can be optimized.

Implementation of the analytical traffic model for the IMS/NGN network with service stratum resource virtualization requires modification of the previously presented network model (Figure 1) and the set of service scenarios (Table 1). They must take into account the fact that there are many instances of individual IMS servers in the network working in parallel, to which messages generated in particular service scenarios are distributed. Then, according to the service scenarios, messages handled by these servers are sent to subsequent network elements that may be implemented in one or more instances. Thus, in addition to changing the number of network elements, the number of links and the way of handling messages in the network also change.

The general calculation methodology presented in this section (Formulas (1)–(4)) is suitable for computing mean values of *CSD* and *CDD* times in the modified IMS/NGN network structure involving virtualization of service stratum resources. Its advantage is the uncomplicated method of obtaining final results by summation of the mean values of message sojourn times in links and network elements, which is dedicated to engineering applications. Of course, the detailed formulas for *E*(*CSD*) and *E*(*CDD*) times in the service stratum resource virtualization case need to be modified to take into account the fact that message streams are split into multiple substreams at certain points in the network, served by multiple instances of individual IMS servers, and the resulting messages are then recombined into a single stream. The average time of all the abovementioned operations, taken into account during the calculation of the *E*(*CSD*) and *E*(*CDD*) times, can be computed as the weighted average of the sojourn time of individual message substreams in the appropriate links and instances of IMS servers. The weights of particular message substreams correspond to their share in the total stream of exchanged messages.

To analyze various configurations of IMS/NGN service stratum virtualization, the described steps should be repeated to create a set of analytical traffic models, each of which includes different considered numbers of IMS server instances.

It is very important that the analytical traffic model of the multidomain IMS/NGN described in this paper was successfully verified by simulations (service stratum and MPLS-based transport stratum separately; [23,25,26]). For the IMS/NGN service stratum and the MPLS-based transport stratum, simulation models were developed [23,25], in which standardized communication scenarios were implemented. The simulators take into account the most important elements of the IMS/NGN network and communication links for exchanging messages between these elements. Network elements handle messages according to definable processing times depending on the element and message type. Messages waiting to be handled are stored in FIFO queues. Similarly, in the case of communication links, FIFO queues are used, and the time of message handling depends on its length and definable link bandwidth. The exchange of messages between the modeled network elements complies with the standards, and the introduced delays result from the operation of elements and links. In this way, the implemented simulation models reflect the phenomena taking place in the real IMS/NGN network and can be a reference for examining the quality of the analytical results.

Our simulation studies described in Ref. [27] demonstrated that for exponential intervals between generated requests, message inter-arrival time distributions at the inputs of IMS/NGN network elements and links are generally not exponential but close to multimodal. This results from the overlapping of many correlated messages sent in particular service scenarios. The nature of message service time distributions is similar, and it results from the assumed method of modeling the operation of network elements and links. Each network element and link processes its own set of messages (resulting from the implemented service scenarios), where the processing times of particular messages depend on their type and the type of element/link.

The simulation model of the service stratum was used to evaluate the quality of the analytical results for overall IMS/NGN network responses (*E*(*CSD*) and *E*(*CDD*) times for all types of successful call scenarios) [26]. In the analytical research, various types of queuing models were used to calculate mean message waiting times for service in the IMS/NGN network elements and links. A wide range of queuing models were examined, regardless of how well they fit to message inter-arrival and service time distributions previously measured using simulations in Ref. [27]. The set of tested queuing models included M/M/1, M/G/1, G/G/1 approximations based on moments, PH/PH/1 and their special cases (PH/M/1 and M/PH/1) with algorithms for fitting phase-type distributions to arrival and service distributions based on moments or whole histograms. The conducted research demonstrated that M/M/1 and M/G/1 models are sufficient for analysis of *E*(*CSD*) and *E*(*CDD*) times. They offer good conformity with reference simulation results and lead to fast calculations without the need for additional experimental data. It is worth noting that the compliance of the analytical and simulation results for individual service systems (mean message waiting times) was not analyzed. The aim of the performed research was to assess overall IMS/NGN network responses, which are important for users and operators.

The experiments presented in Ref. [11] indicate that the overall analytical traffic model containing the IMS/NGN service stratum and the MPLS-based transport stratum works correctly. Consequently, it can be used in thorough investigations on how parameters of traffic sources and an IMS/NGN network with an MPLS-based transport stratum influence CPP parameters. The results of these investigations are presented in the next section.

4. Results and Discussion

4.1. Research Assumptions

The performed experiments are described by the input data sets depicted in Table 2. For CSCF servers, RACF C1, RACF C2 and links, M/M/1 queuing models were applied. For each data set, elements of two selected input variables (vectors) corresponding to domain 1 were widely modified according to Table 2, while the remaining elements of the selected vectors (corresponding to domain 2) as well as the elements of other vectors were left unchanged. Such an approach allowed investigating how modifications of particular input variables in domain 1 would affect CPP parameters in both domains. The obtained results can be used to determine changes in all output parameters when more than one input variable in domain 1 or input variables in both domains are manipulated.

4.2. Results

The results of the conducted experiments are summarized in Table 3, where the following terminology is used:

- + indicates that a particular output variable increases when values of the selected parameter of domain 1 are higher (l represents linear increase; n—nonlinear increase);
- − indicates that a particular output variable decreases when values of the selected parameter of domain 1 are higher (l represents linear decrease; n—nonlinear decrease);
- 0 represents no impact of the selected parameter of domain 1 on a particular output variable;
- ≈0 represents a negligible impact of the selected parameter of domain 1 on a particular output variable.

Analysis of Table 3 can lead to the conclusion that there are groups of output variables which react identically to modifications of all investigated parameters of domain 1. For these output variables (e.g., $E(CSD)_{\mathrm{d1}}$, $E(CSD)_{\mathrm{f1}}$, $E(CDD)_{\mathrm{d1}}$ and $E(CDD)_{\mathrm{f1}}$), the columns of Table 3 are the same. From the twelve columns of Table 3 corresponding to different types of CPP parameters, only four are unique. Consequently, there are four groups of CPP parameters which are influenced by changes in the parameters of domain 1 in the same way. Of course, within these four groups, there are differences in the values of particular CPP parameters. For example, *E*(*CDD*) times are always lower than *E*(*CSD*) times for the same types of call scenarios (call disengagement process is less complicated than call set-up process), and mean *CSD* and *CDD* times for inter-operator calls are higher than those for intra-operator calls (the former involve much more communication between elements).

Table 2. Input data sets (D1 represents values of input variables for domain 1; D2—domain 2; when not specified, the same values of input variables were applied for both domains).

Input Variable	Data Set 1	Data Set 2	Data Set 3	Data Set 4	Data Set 5	Data Set 6	Data Set 7	Data Set 8
lambdaR [1/s]	D1: 0–350 D2: 50	50	50	50	50	50	50	50
lambda1d [1/s]	D1: 0–100 D2: 50	50	50	50	50	50	50	50
lambda2d [1/s]	50	D1: 0–100 D2: 50	50	50	50	50	50	50
TINV [ms]	0.5	D1: 0–0.8 D2: 0.5	0.5	0.5	0.5	0.5	0.5	0.5
TA [ms]	0.5	0.5	D1: 0.05–2.5 D2: 0.5	0.5	0.5	0.5	0.5	0.5
Tproc [μs]	50	50	D1: 5–600 D2: 50	50	50	50	50	50
Tresp [ms]	0.5	0.5	0.5	D1: 0.05–2.9 D2: 0.5	0.5	0.5	0.5	0.5
ETR [ms]	5	5	5	D1: 0–1000 D2: 5	5	5	5	5
EXA [ms]	10	10	10	10	D1: 0–400 D2: 10	10	10	10
EY [ms]	10	10	10	10	D1: 0–400 D2: 10	10	10	10
rC	0.5	0.5	0.5	0.5	0.5	D1: 0–1 D2: 0.5	0.5	0.5
pb	0	0	0	0	0	D1: 0–0.1 D2: 0	0	0
p11	0.4	0.4	0.4	0.4	0.4	0.4	D1: 0–1 D2: 0.4	0.4
p21	0.4	0.4	0.4	0.4	0.4	0.4	D1: 0–1 D2: 0.4	0.4
d [km]	200	200	200	200	200	200	200	D1: 1–1000 D2: 200
b [Mbit/s]	50	50	50	50	50	50	50	D1: 1–200 D2: 50

In the next part of this section, detailed CPP results (Figures 4–11) are presented for the abovementioned four groups of CPP parameters—as their representatives, $E(CSD)_{b1}$, $E(CSD)_{f1}$, $E(CSD)_{b2}$ and $E(CSD)_{f2}$ are chosen. Moreover, changes in all output variables according to the groups of the input variables described in Section 3 are discussed.

The registration and call set-up request intensities (**lambdaR**, **lambda1d** and **lambda2d**; Figures 4 and 5) increase the load of network elements and links, resulting in nonlinear growth of the analyzed CPP parameters. As registration and intra-operator calls do not involve elements of multiple domains, modifications of **lambdaR** and **lambda1d** in domain 1 affect only mean *CSD* and mean *CDD* times for scenarios in which communication is performed in domain 1 (scenarios b1, d1, f1 and f2). When **lambda2d** (call set-up request intensity for inter-operator calls which use both domains 1 and 2) is increased, the CPP parameters for scenarios b2 and d2 are also affected, but this effect is much weaker than for other types of call scenarios (Figure 5).

Table 3. Impact of changing input variables in domain 1 on CPP parameters in both domains.

Input Variable (domain 1)	$E(CSD)_{b1}$	$E(CSD)_{d1}$	$E(CSD)_{f1}$	$E(CDD)_{b1}$	$E(CDD)_{d1}$	$E(CDD)_{f1}$	$E(CSD)_{b2}$	$E(CSD)_{d2}$	$E(CSD)_{f2}$	$E(CDD)_{b2}$	$E(CDD)_{d2}$	$E(CDD)_{f2}$
lambdaR	+n	+n	+n	+n	+n	+n	0	0	+n	0	0	+n
lambda1d	+n	+n	+n	+n	+n	+n	0	0	+n	0	0	+n
lambda2d	+n	+n	+n	+n	+n	+n	+n	+n	+n	+n	+n	+n
TINV	+n	+n	+n	+n	+n	+n	0	0	+n	0	0	+n
TA	0	+n	+n	0	+n	+n	0	0	+n	0	0	+n
Tproc	0	+n	+n	0	+n	+n	0	0	+n	0	0	+n
Tresp	0	+n	+n	0	+n	+n	0	0	+n	0	0	+n
ETR	0	+l	+l	0	+l	+l	0	0	+l	0	0	+l
EXA	+l	+l	+l	+l	+l	+l	0	0	+l	0	0	+l
EY	0	0	0	0	0	0	0	0	+l	0	0	0
rC	≈0	≈0	≈0	≈0	≈0	≈0	0	0	≈0	0	0	≈0
pb	≈0	≈0	≈0	≈0	≈0	≈0	≈0	≈0	≈0	≈0	≈0	≈0
p11	0	−n	−n	0	−n	−n	0	0	0	0	0	−n
p21	0	−n	−n	0	−n	−n	0	0	0	0	0	−n
d	+l	+l	+l	+l	+l	+l	0	0	+l	0	0	+l
b	−n	−n	−n	−n	−n	−n	0	0	−n	0	0	−n

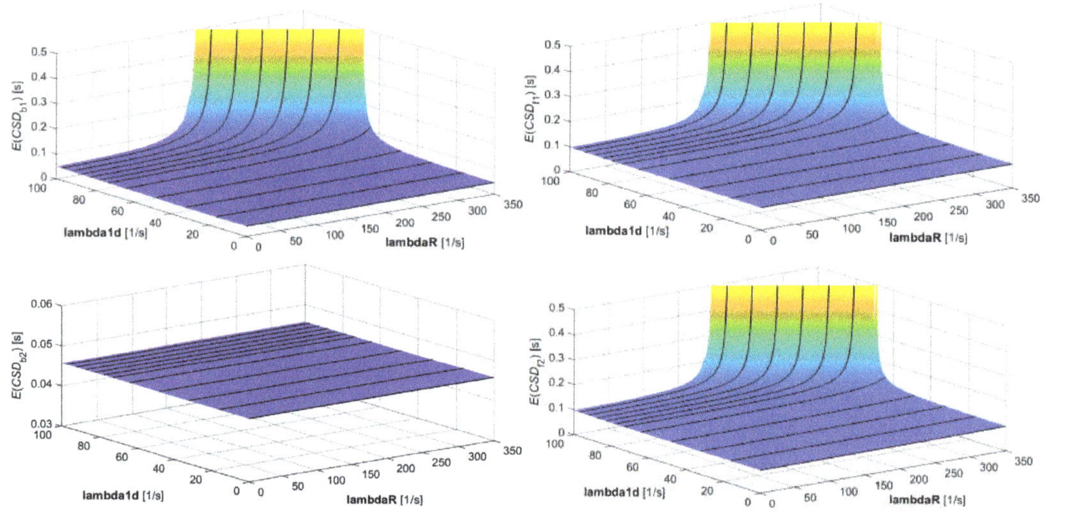

Figure 4. Selected results for data set 1 from Table 1.

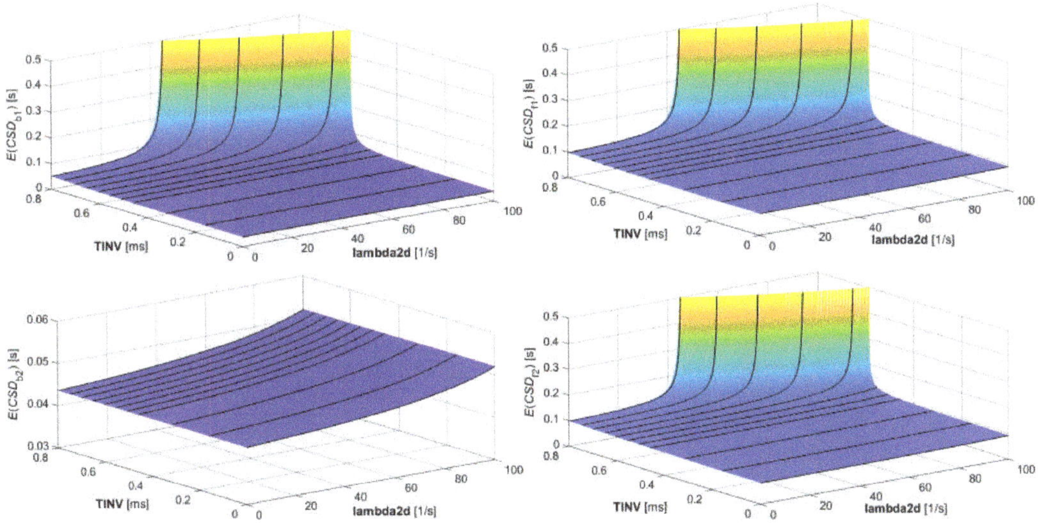

Figure 5. Selected results for data set 2 from Table 1.

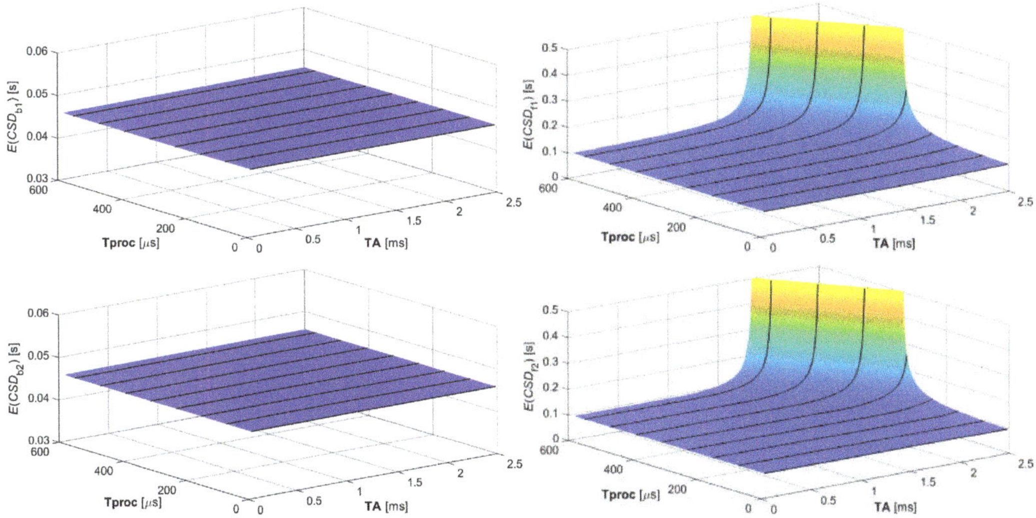

Figure 6. Selected results for data set 3 from Table 1.

The message processing times by CSCF servers and RACF C1 (**TINV**, **TA**, **Tproc** and **Tresp**; Figures 5–7) affect loads of these network elements. Higher values of these parameters nonlinearly increase mean *CSD* and mean *CDD* times (similarly to request intensities). When elements of the **TINV** vector are modified in domain 1, the CPP parameters for all scenarios in which messages are exchanged in this domain are affected (scenarios b1, d1, f1 and f2). Changes in the message processing times (**TA**, **Tproc** and **Tresp**) of RACF C1 influence only mean *CSD* and mean *CDD* times for call scenarios with resource reservation in the MPLS core network of operator 1 (scenarios d1, f1 and f2).

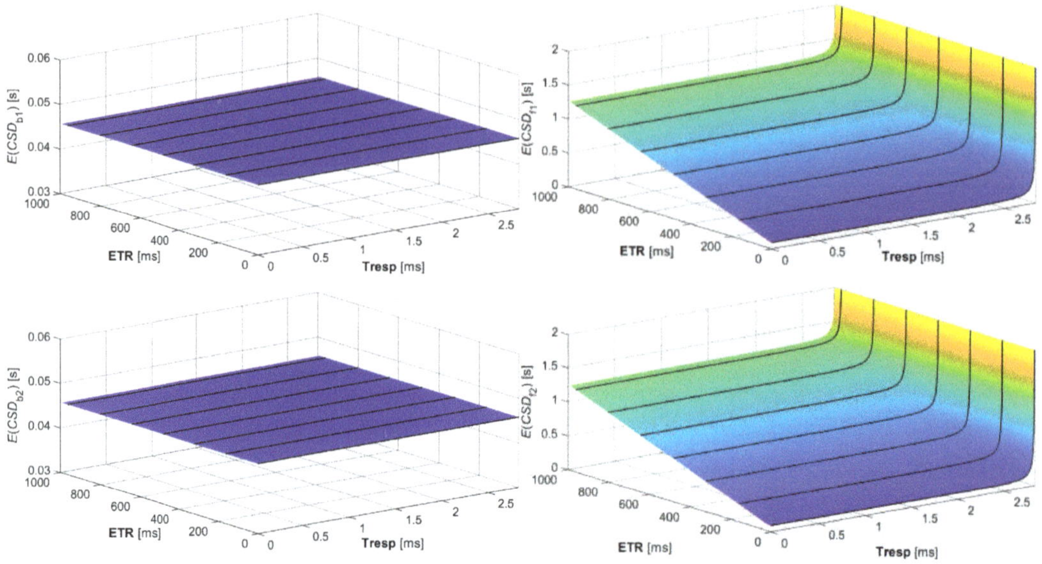

Figure 7. Selected results for data set 4 from Table 1.

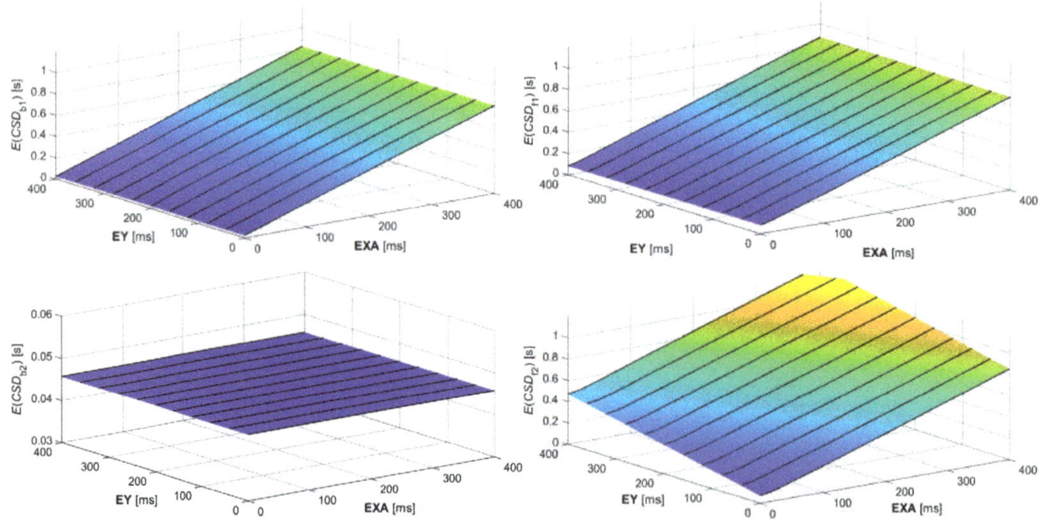

Figure 8. Selected results for data set 5 from Table 1.

For network elements whose response times are specified by random variables, larger mean values of these random variables (**ETR**, **EXA** and **EY**) result in proportionally higher *E*(*CSD*) and *E*(*CDD*) times (Figures 7 and 8). The set of affected CPP parameters is, however, dependent on the parameter modified in domain 1. Changes in the mean time of processing requests by RACF A1 (**EXA**) have an impact on the *E*(*CSD*) and *E*(*CDD*) times for all scenarios with resource reservation in the access network of operator 1 (scenarios b1, d1, f1 and f2). The mean response time of the operator 1 MPLS network (**ETR**) influences only CPP parameters for call scenarios with resource allocation in this network (scenarios d1, f1 and f2). A very interesting case is that of modifying the mean time of processing requests by SUP-FE 1/SAA-FE 1 (**EY**). This network element is only used during the set-up of inter-operator calls originated in domain 2, so increasing its processing time increases only $E(CSD)_{f2}$ and has no impact on other CPP parameters.

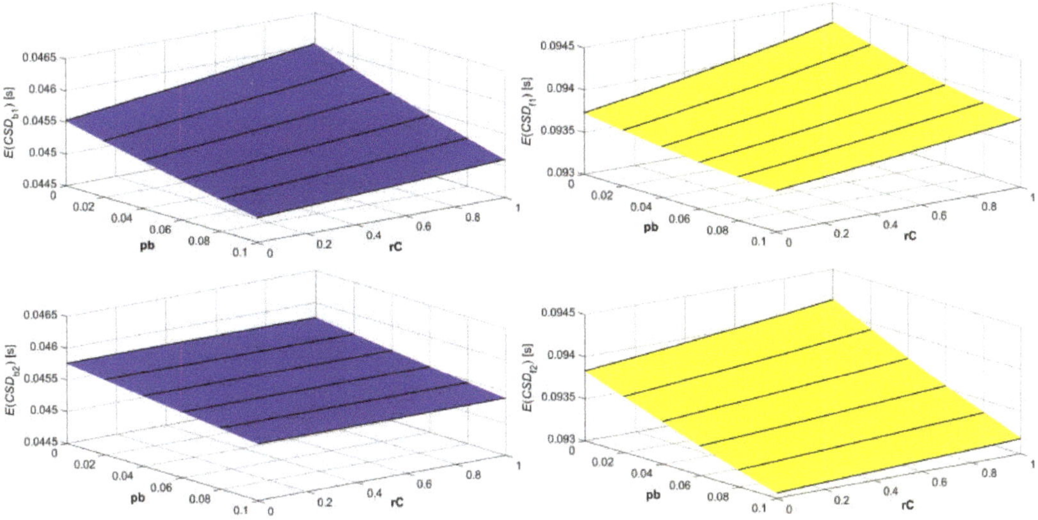

Figure 9. Selected results for data set 6 from Table 1.

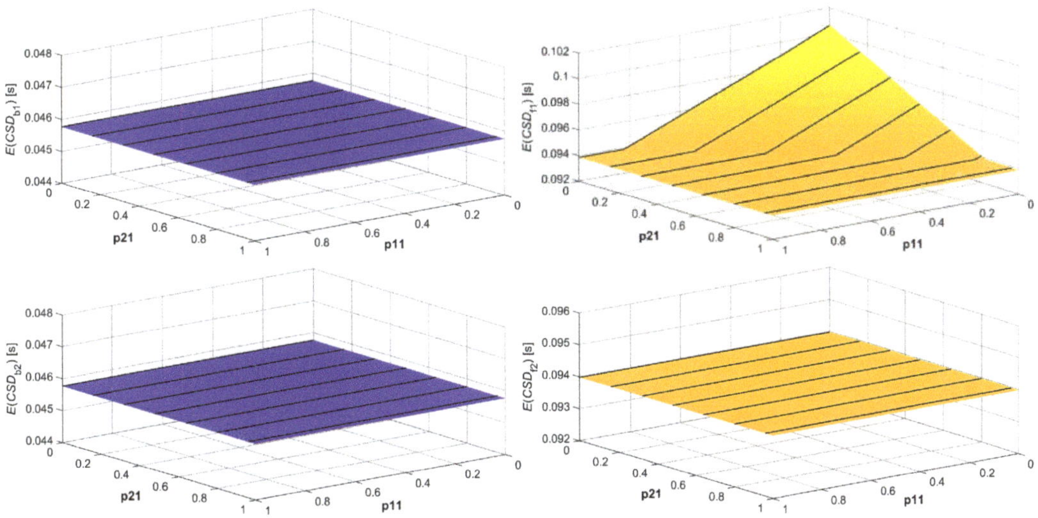

Figure 10. Selected results for data set 7 from Table 1.

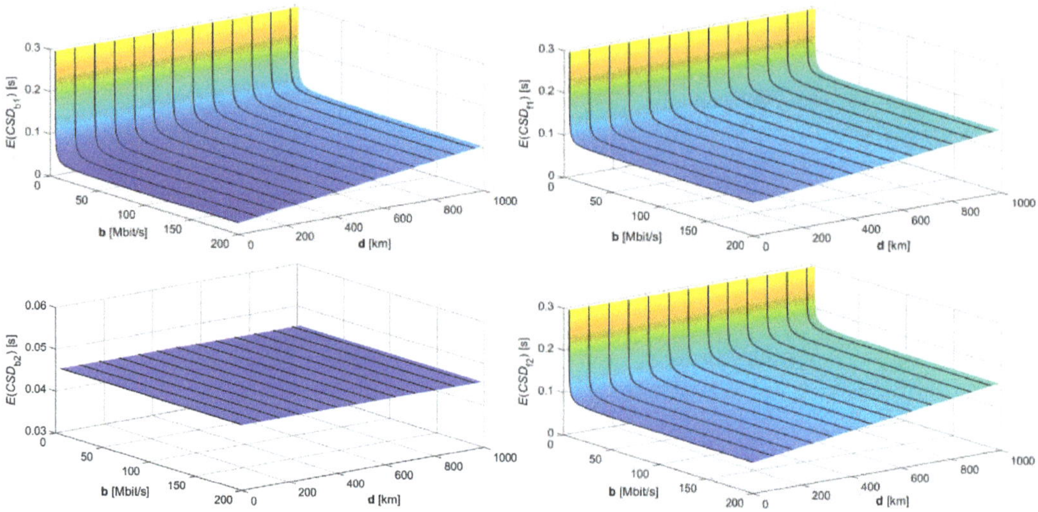

Figure 11. Selected results for data set 8 from Table 1.

Modification of the **rC** and **pb** parameters in domain 1 (Figure 9) has a very small influence on CPP parameters. Increasing the ratio of calls involving multiple access areas to all intra-operator calls generated in domain 1 (**rC**) makes resource reservations in the MPLS core network of operator 1 more common, which results in more messages exchanged in domain 1 and, consequently, a higher load of network elements. As a result, the *E*(*CSD*) and *E*(*CDD*) times for scenarios b1, d1, f1 and f2 are slightly increased. A higher probability of transport resource unavailability in the access and core networks of operator 1 (**pb**) leads to more unsuccessful call scenarios, which involves sending and processing less messages compared to successful ones. This results in a slightly lower load of all network elements and links in both domains and lower values of all CPP parameters.

Higher values of the **p11** and **p21** probabilities describing the operation of the MPLS core network of operator 1 (Figure 10) result in more resource operations (bandwidth reservation or increase—**p11**; bandwidth release or decrease—**p21**) performed only by updating the resource state in the RACF C1 database (without communication with LERs). This decreases mean *CSD* and *CDD* times for scenarios d1 and f1, in which resource operations in the core network of operator 1 are necessary. A similar decrease can be observed in the values of $E(CDD)_{f2}$. Interestingly, the values of $E(CSD)_{f2}$ are not affected by modifications of the **p11** and **p21** probabilities. It results from the fact that during the set-up of inter-operator calls generated in domain 2 (scenario f2), there are two concurrent processes (Figure 2):

1. Resource reservation in the destination domain 1 (messages 14–15);
2. Communication from domain 1 to domain 2 (messages 13 and 16–18), resource reservation in the originating domain 2 (messages 19–20) and communication from domain 2 to domain 1 (confirmation of resource reservation; messages 29–32).

Typically, the second process is much more time-consuming and has a decisive influence on $E(CSD)_{f2}$. Such a situation took place for the examined sets of input variables. The time necessary for resource reservation in the core network of operator 1 does not have an impact on $E(CSD)_{f2}$, independently of the applied values of the **p11** and **p21** parameters.

The last data set (Figure 11) examines the influence of link parameters in domain 1 on the analyzed CPP parameters. Higher link lengths (**d**) increase mean *CSD* and mean *CDD* times linearly due to distance-dependent propagation times. Such a situation applies for all call scenarios with message exchange in domain 1 (scenarios b1, d1, f1 and f2). The same CPP parameters are affected by the values of the link bandwidth (**b**) in domain 1.

Increasing **b** in domain 1 decreases mean *CSD* and mean *CDD* times nonlinearly. Overly low values of bandwidth can lead to link overload and cause *E*(*CSD*) and *E*(*CDD*) times to rise to infinity.

4.3. Discussion

The presented results allow drawing synthetic conclusions on how modifying the IMS/NGN network and traffic source parameters in domain 1 affects all analyzed CPP parameters:

- The intensities of requests significantly affect *E*(*CSD*) and *E*(*CSD*) times:
    - Registration and intra-operator call set-up requests generated in domain 1 (**lambda** and **lambda1d**) increase only CPP parameters for scenarios performed in domain 1 (b1, d1, f1 and f2);
    - Inter-operator call set-up requests generated in domain 1 (**lambda2d**) increase all investigated CPP parameters.
- The processing/response times of network elements increase *E*(*CSD*) and *E*(*CSD*) times dependent on the type of network element and its usage in particular call scenarios:
    - Changing the times of processing SIP INVITE messages by CSCF servers in domain 1 (**TINV**) and the mean time of processing requests by RACF A1 (**EXA**) affects CPP parameters for all call scenarios performed in domain 1 (b1, d1, f1 and f2);
    - Modification of the message processing times of RACF C1 (**TA**, **Tproc** and **Tresp**) and the mean response time of "MPLS core 1" (**ETR**) affects only CPP parameters for call scenarios with resource reservation in the core network of domain 1 (d1, f1 and f2);
    - Changing the mean time of processing requests by SUP-FE 1/SAA-FE 1 (**EY**) influences only $E(CSD)_{f2}$, as this network element is used solely in establishing inter-operator calls originated in domain 2.
- Link parameters in domain 1 affect *E*(*CSD*) and *E*(*CSD*) times for all call scenarios performed in domain 1 (b1, d1, f1 and f2):
    - Link lengths (**d**) have an influence on propagation delays and increase the analyzed CPP parameters linearly;
    - Sufficient link bandwidths (**b**) must be assured, as below a certain bandwidth level, *E*(*CSD*) and *E*(*CSD*) times rise rapidly.
- Other parameters of our traffic model (**rC**, **pb**, **p11** and **p21**) do not have a significant impact on the analyzed CPP parameters.

Based on the abovementioned remarks, it is possible to indicate the influence of the IMS/NGN network and traffic source parameters in domain 2 on the analyzed *E*(*CSD*) and *E*(*CDD*) times using the rule of symmetry. For example, larger values of **lambdaR** and **lambda1d** in domain 1 increase the analyzed CPP parameters for scenarios b1, d1, f1 and f2. Similar changes in **lambdaR** and **lambda1d** values in domain 2 will affect the *E*(*CSD*) and *E*(*CDD*) times for scenarios b2, d2, f2 and f1.

As a result of identifying the relations between the input and output parameters of our traffic model, the performed research (Table 3, Figures 4–11) allows grouping the parameters of the IMS/NGN network with an MPLS-based transport stratum into three categories, which correspond to their importance for the network operator and designer. The first group contains input variables with a significant impact on CPP parameters, which must definitely be taken into account during network analysis and design. This group includes most of the input variables of our model: **lambdaR**, **lambda1d**, **lambda2d**, **TINV**, **TA**, **Tproc**, **Tresp**, **EXA**, **ETR**, **d** and **b**. The second group consists of network parameters that generally have little effect on *E*(*CSD*) and *E*(*CDD*) times: **EY**, **p11** and **p21**. The **EY** input variable affects only one CPP parameter of the twelve analyzed, while the **p11** and **p21** probabilities have a very limited impact on several output parameters. The third group includes input parameters with a negligible influence on mean *CSD* and *CDD* times: **rC** and **pb**.

It is worth noting that from the group of input variables that most affect *E*(*CSD*) and *E*(*CDD*) times, mainly the request intensities (**lambdaR**, **lambda1d** and **lambda2d**), which have a direct impact on the offered traffic load, change during normal operation of the IMS/NGN network. The results of our research (Figures 4 and 5) can be used to indicate at which threshold values of these intensities *E*(*CSD*) and *E*(*CDD*) times become unacceptable. As the intensities approach these thresholds, additional instances of IMS servers can be launched, as described in Section 3, which will result in decreasing *E*(*CSD*) and *E*(*CDD*). For a later decrease in request intensities, these additional instances may be turned off, thus optimizing the resources of the IMS/NGN network service stratum.

5. Conclusions

This paper presents investigations on the impact of IMS/NGN network and traffic source parameters on *E*(*CSD*) and *E*(*CDD*) times. These times belong to the set of call processing performance parameters, which proper values must be ensured in the network to achieve user satisfaction with the provided services. In the performed research, our analytical traffic model of a multidomain IMS/NGN network based on the MPLS technology in the transport stratum was used. The model allows the calculation of *E*(*CSD*) and *E*(*CDD*) times for all types of successful call scenarios with respect to an extensive set of input variables regarding the service and transport stratum.

During the investigations, particular input variables in domain 1 were modified, and the effect of these modifications on all *E*(*CSD*) and *E*(*CDD*) times was examined. The obtained results can be used to determine changes in all output parameters when any combinations of input variables in domains 1 and/or 2 are modified. The performed experiments also allowed indicating the network and traffic source parameters with a significant influence on *E*(*CSD*) and *E*(*CSD*), which should be definitively taken into account during IMS/NGN network analysis and design. They include the intensities of registration requests as well as intra- and inter-operator call set-up requests; the processing times of CSCF servers, RACF C1 and RACF C2; the mean response time of RACF A1, RACF A2 and MPLS domains as well as link lengths and bandwidths.

Our future work will include extensions of the analytical traffic model. Our aim is to take into account more details regarding the structure and operation of MPLS core networks, which affect, for example, the probability of transport resource unavailability in these networks. We are also planning to include modeling of user (voice) traffic in MPLS transport networks and apply segment routing [28,29] for transport resource optimization. Another planned research pursuit is to use the experience gained from the experiments described in this paper to create a set of analytical traffic models enabling the analysis of IMS/NGN service stratum resource virtualization, which is mentioned in Section 3. In such a solution, there exist several instances of the same types of IMS servers working in parallel, where the number of active instances depends on the network load, which results from, among others, the values of request intensities. Increasing the number of active IMS server instances will reduce the *E*(*CSD*) and *E*(*CDD*) times for the same values of request intensities.

Author Contributions: Conceptualization, S.K. and M.S.; methodology, S.K. and M.S.; software, M.S.; validation, S.K. and M.S.; formal analysis, S.K.; investigation, S.K. and M.S.; data curation, S.K. and M.S.; writing—original draft preparation, M.S.; writing—review and editing, S.K.; visualization, M.S.; supervision, S.K. All authors have read and agreed to the published version of the manuscript.

Funding: This research received no external funding.

Institutional Review Board Statement: Not applicable.

Informed Consent Statement: Not applicable.

Data Availability Statement: Not applicable.

Conflicts of Interest: The authors declare no conflict of interest.

References

1. *ITU-T Recommendation Y.2001*; General Overview of NGN. ITU: Geneva, Switzerland, 2004.
2. *ITU-T Recommendation Y.2012*; Functional Requirements and Architecture of Next Generation Networks. ITU: Geneva, Switzerland, 2010.
3. *3GPP TS 23.228: IP Multimedia Subsystem (IMS)*, version 17.3.0.; Stage 2 (Release 17); 3GPP: Valbonne, France, 2021.
4. Chang, Y.-C.; Li, J.-W.; Lv, J.-H. An IP Multimedia Subsystem Services Proxy Gateway Based on a JAVA Dynamic Module System. *Appl. Sci.* **2018**, *8*, 2060. [CrossRef]
5. Bahaa, A.; Shehata, M.; Gasser, S.M.; El-Mahallawy, M.S. Call Failure Prediction in IP Multimedia Subsystem (IMS) Networks. *Appl. Sci.* **2022**, *12*, 8378. [CrossRef]
6. Liao, L.; Leung, V.C.M.; Chen, M. Virtualizing IMS Core and Its Performance Analysis. In *Cloud Computing. CloudComp 2014. Lecture Notes of the Institute for Computer Sciences, Social Informatics and Telecommunications Engineering*; Leung, V., Lai, R., Chen, M., Wan, J., Eds.; Springer: Cham, Switzerland, 2015; Volume 142.
7. Rosen, E.; Viswanathan, A.; Callon, R. *IETF RFC 3031: Multiprotocol Label Switching Architecture*; IETF: Wilmington, NC, USA, 2001.
8. *ITU-T Recommendation Y.2175*; Centralized RACF Architecture for MPLS Core Networks. ITU: Geneva, Switzerland, 2008.
9. *ITU-T Recommendation Y.1530*; Call Processing Performance for Voice Service in Hybrid IP Networks. ITU: Geneva, Switzerland, 2007.
10. *ITU-T Recommendation Y.1531*; SIP-Based Call Processing Performance. ITU: Geneva, Switzerland, 2007.
11. Kaczmarek, S.; Sac, M. Analytical Traffic Model for a Multidomain IMS/NGN Network Including Service and Transport Stratum. In Proceedings of the 30th International Conference on Software, Telecommunications and Computer Networks, SoftCOM 2022, Split, Croatia, 22–24 September 2022.
12. Oveis, A.H.; Asadi, M.A.S.; Homami, S.M.S. A trade-off between delay and throughput in IMS network session setup. In Proceedings of the 22nd Iranian Conference on Electrical Engineering, ICEE 2014, Tehran, Iran, 20–22 May 2014; pp. 1614–1618.
13. Alassane, B.A.; Karim, K. An analytical jitter model in IMS network. In Proceedings of the 2016 IEEE 7th Annual Ubiquitous Computing, Electronics & Mobile Communication Conference, UEMCON 2016, New York, NY, USA, 20–22 October 2016.
14. Hasanov, M.H.; Ibrahimov, B.G.; Mardanov, N.T. Research and Analysis Performance Indicators NGN/IMS Networks in the Transmission Multimedia Traffic. In Proceedings of the 2019 Wave Electronics and its Application in Information and Telecommunication Systems, WECONF 2019, St. Petersburg, Russia, 3–7 June 2019.
15. Ibrahimov, B.G.; Ismaylova, S.R. The effectiveness NGN/IMS networks in the establishment of a multimedia session. *Am. J. Netw. Commun.* **2018**, *7*, 1–5.
16. Romanov, O.I.; Nesterenko, M.M.; Veres, L.A.; Hordashnyk, Y.S. IMS: Model and calculation method of telecommunication network's capacity. In Proceedings of the 2017 International Conference on Information and Telecommunication Technologies and Radio Electronics, UkrMiCo 2017, Odessa, Ukraine, 11–15 September 2017.
17. Rodríguez, M.A.V.; Muñoz, E.C. Review of Quality of Service (QoS) mechanisms over IP Multimedia Subsystem (IMS). *Ing. Y Desarro.* **2017**, *35*, 262–281. [CrossRef]
18. Di Mauro, M.; Galatro, G.; Longo, M.; Postiglione, F.; Tambasco, M. Performability Management of Softwarized IP Multimedia Subsystem. In Proceedings of the NOMS 2020—2020 IEEE/IFIP Network Operations and Management Symposium, Budapest, Hungary, 20–24 April 2020; pp. 1–6.
19. Di Mauro, M.; Galatro, G.; Longo, M.; Postiglione, F.; Tambasco, M. IP Multimedia Subsystem in a containerized environment: Availability and sensitivity evaluation. In Proceedings of the 2019 IEEE Conference on Network Softwarization (NetSoft), Paris, France, 24–28 June 2019; pp. 42–47.
20. Prados-Garzon, J.; Ameigeiras, P.; Ramos-Munoz, J.J.; Andres-Maldonado, P.; Lopez-Soler, J.M. Analytical modeling for Virtualized Network Functions. In Proceedings of the 2017 IEEE International Conference on Communications Workshops (ICC Workshops), Paris, France, 21–25 May 2017; pp. 979–985.
21. Zhang, Z.; Duan, Z.; Hou, Y. On Scalable Design of Bandwidth Brokers. *IEICE Trans. Commun.* **2001**, *E84-B*, 2011–2025.
22. Abhayawardhana, V.S.; Babbage, R. A traffic model for the IP Multimedia Subsystem (IMS). In Proceedings of the IEEE 65th Vehicular Technology Conference, VTC 2007, Dublin, Ireland, 22–25 April 2007.
23. Kaczmarek, S.; Łuczaj, Ł.; Sac, M. Simulation Model of IMS/NGN with Transport Stratum Based on MPLS Technology. *Telecommun. Rev. Telecommun. News* **2016**, 777–784. [CrossRef]
24. Femminella, M.; Giacinti, F.; Reali, G. Optimal deployment of open source application servers providing multimedia services. *IEEE Netw.* **2014**, *28*, 54–63. [CrossRef]
25. Kaczmarek, S.; Sac, M. Verification of the Analytical Traffic Model of a Multidomain IMS/NGN Using the Simulation Model. In *Information Systems Architecture and Technology: Proceedings of the 36th International Conference on Information Systems Architecture and Technology—ISAT 2015—Part II. Advances in Intelligent Systems and Computing, Karpacz, Poland, 20-22 September 2015*; Grzech, A., Borzemski, L., Świątek, J., Wilimowska, Z., Eds.; Springer: Cham, Switzerland, 2016; Volume 430, pp. 109–130.
26. Kaczmarek, S.; Sac, M. Performance Models of a Multidomain IMS/NGN Service Stratum. In Proceedings of the 29th International Conference on Software, Telecommunications and Computer Networks, SoftCOM 2021, Hvar, Croatia, 23–25 September 2021.

27. Kaczmarek, S.; Sac, M. Approximation of message inter-arrival and inter-departure time distributions in IMS/NGN architecture using phase-type distributions. *J. Telecommun. Inf. Technol.* **2013**, 9–18. Available online: https://www.itl.waw.pl/czasopisma/JTIT/2013/3/9.pdf (accessed on 10 September 2022).
28. Ventre, P.L.; Salsano, S.; Polverini, M.; Cianfrani, A.; Abdelsalam, A.; Filsfils, C.; Camarillo, P.; Clad, F. Segment Routing: A Comprehensive Survey of Research Activities, Standardization Efforts, and Implementation Results. *IEEE Commun. Surv. Tutor.* **2021**, *23*, 182–221. [CrossRef]
29. Desmouceaux, Y.; Pfister, P.; Tollet, J.; Townsley, M.; Clausen, T. SRLB: The Power of Choices in Load Balancing with Segment Routing. In Proceedings of the 2017 IEEE 37th International Conference on Distributed Computing Systems, ICDCS 2017, Atlanta, GA, USA, 5–8 June 2017; pp. 2011–2016.

applied sciences

MDPI

Article

Improving Streaming Video with Deep Learning-Based Network Throughput Prediction

Arkadiusz Biernacki

Department of Computer Networks and Systems, Silesian University of Technology, Akademicka 16, 44-100 Gliwice, Poland; arkadiusz.biernacki@polsl.pl

Abstract: Video streaming represents a significant part of Internet traffic. During the playback, a video player monitors network throughput and dynamically selects the best video quality in given network conditions. Therefore, the video quality depends heavily on the player's estimation of network throughput, which is challenging in the volatile environment of mobile networks. In this work, we improved the throughput estimation using prediction produced by LSTM artificial neural networks (ANNs). Hence, we acquired data traces from 4G and 5G mobile networks and supplied them to two deep LSTM ANNs, obtaining a throughput prediction for the next four seconds. Our analysis showed that the ANNs achieved better prediction accuracy compared to a naive predictor based on a moving average. Next, we replaced the video player's default throughput estimation based on the naive predictor with the LSTM output. The experiment revealed that the traffic prediction improved video quality between 5% and 25% compared to the default estimation.

Keywords: traffic prediction; artificial neural networks; adaptive video

Citation: Biernacki, A. Improving Streaming Video with Deep Learning-Based Network Throughput Prediction. *Appl. Sci.* **2022**, *12*, 10274. https://doi.org/10.3390/app122010274

Academic Editors: Zbigniew Lubniewski, Tadeus Uhl and Przemysław Falkowski-Gilski

Received: 15 September 2022
Accepted: 9 October 2022
Published: 12 October 2022

1. Introduction

Highly variable packet delays, packet losses and fluctuating mobile network bandwidth make streaming a video challenging. Although modern mobile networks offer high bandwidth, the nature of wireless transmission—inherent variability in signal strength, interference, noise, and user mobility—results in high throughput variability. To overcome this unstable network environment, content providers encode video into multiple versions at different bit rates and split them into segments. A video player periodically downloads the segments, concatenates them and delivers a continuous video stream to a user. During the playback, the player can request a different version of the segments to adapt the video bit rate to the current network conditions. This video delivery concept is called dynamic adaptive streaming over HTTP (DASH).

The rate-adaptation algorithm is a significant element of a DASH system because the player has to estimate network conditions, including available throughput to ensure proper quality of experience (QoE). The measurement of network throughput, usually based on the moving average, often differs from real throughput. Because of this difference, streaming algorithms can be too conservative and download a video of much lower quality than the throughput allows, or too aggressive and download content of too high quality in given network conditions. A solution to this problem is to predict the future network throughput. As the video segment is usually several seconds long, the prediction is short-term and does not rely on seasonality. Prediction can improve the efficiency of bit rate selection, which translates to lower startup delay and offers better mid-stream bit rate adaptation. Researchers estimate that the applied prediction achieves between 68% and 89% of the optimal quality. The exact numbers are dependent on the playback algorithms employed in a DASH player [1].

In our work, we predict throughput using traces from 4G and 5G networks from static and mobile usage scenarios. In the next step, we replace the default methods of throughput estimation in three DASH algorithms with the obtained predicted results. Finally, we show that the DASH algorithms achieve better performance when equipped with prediction compared with reliance only on an average of network measurements. Our approach does not require any modification to network protocols or the DASH standard and its components. The solution only affects users' players where a more elaborate algorithm based on LSTM artificial neural networks (ANNs) replaces the default throughput estimation. With the information provided by the player's measurement module, the algorithm can estimate the expected throughput for the next few seconds. Because an LSTM ANN requires some computation needed to learn and continually update its knowledge, this process increases the usage of memory and CPU of a video player.

Taking the above into account, the contributions of the paper are: the prediction of 5G network throughput at the application level in static and mobile scenarios; the comparison of the prediction quality with the 4G trace; the usage of stacked uni- and bidirectional LSTM ANNs versions for the prediction, and the application of the prediction made by an LSTM ANN to improve quality of video for three DASH players.

In the next section, we describe the related works. In Section 3, we present the video streaming model, adaptation strategies of video playback and video quality measurement. Section 4 shows the prediction model, LSTM ANNs, used for the prediction, data characteristics, prediction results and the prediction influence on the video quality. The last section concludes the paper.

2. Related Works

The literature on network throughput prediction is extensive and usually focuses on the Internet traffic, which aggregates flows generated by a whole spectrum of applications. The authors of [2] show a comprehensive review of prediction methods and compare some of them. A number of research studies used linear regression, autoregressive integrated moving average (ARIMA), ARIMA with an explanatory variable (ARIMAX), multiple linear regression, support vector regression, GARCH processes and many other techniques derived from statistics. These methods are conventionally called traditional and are considered descriptive in nature, aimed at analyzing datasets with finite, countable, and explainable predictors. The authors of [3] made a comprehensive comparison of the traditional methods using a 3G/HSPA data set.

The newer prediction schemes based on machine learning are sometimes used in accordance or opposition to the traditional methods, and there is an ongoing debate about their accuracy and computational requirements [4]. ANNs, considered a non-traditional approach, play a role in network traffic prediction mainly due to their universality and accuracy. Historically, the popularity gained simple ANNs with a single hidden layer because versions with multiple layers, called deep ANNs, were harder to train. After researchers overcame the training issues, deep ANNs became popular and gradually replaced single-layered ANNs as a tool for the solution of more complex problems. Because deep ANNs added more complexity and nonliterary to the network traffic model, their prediction accuracy increased. Oliveira's comparative study [5] discusses both approaches in the context of network traffic prediction.

From the variety of deep ANNs, LSTM ANNs are one of the possible choices and are applied readily due to their capability to capture long-range dependencies characteristic of aggregated network traffic, i.e., traffic composed of multiple flows. For example, an LSTM ANN was used to identify recurrent patterns of various metrics to predict traffic in cellular networks in [6]. The work focuses on long-horizon prediction counted in days. In [7], the authors enhanced their LSTM-based prediction with an autocorrelation function, which captured the traffic cycles on a daily scale with a prediction span between 5 and 60 min. In [8], the authors proposed to remove extreme values by scaling a traffic trace. Their LSTM ANN had four hidden layers with a time window set to 32 units or more. The authors did not provide a prediction horizon; however, the data granularity was more than one minute.

In the case of multimedia applications, the prediction of a single traffic flow is more important and challenging than aggregated traffic [9]. A single flow, especially if transmitted through an unstable mobile environment, can have much variability and sudden changes compared with an aggregated trace smoothed by statistical multiplexing. Hence, the models of the aggregated traffic are usually not suitable for describing a single flow generated by a DASH system streaming video segments in an ON–OFF fashion, which contributes to significant bursts in the transmission. As single-flow prediction is a more challenging task, there are fewer works dedicated to this topic, and their authors focus mainly on a short-time prediction span. In [10], the authors applied LSTM and gated recurrent units (GRU) ANNs to predict traffic flow obtained from a sensor network for a five-minute time horizon. They achieved better results compared with the ARIMA processes. A comparative study [11] showed that for cellular network traffic, LSTM provides better prediction than a stacked autoencoder. Unfortunately, the authors did not provide information about the traffic traces used in the experiment.

Except for the historical throughput, an LSTM ANN sometimes operates on additional data to improve the results; for example, the authors of [12] used a complex LSTM ANN with five hidden layers and a multivariate dataset containing information about all downlink and uplink traffic for a given base station. Other works use a meta-learning scheme, including a set of predictors, each optimized to predict a particular kind of traffic. The study by He et al. [9] applied this technique to improve traffic scheduling in base stations with a prediction horizon set to 100 ms. The authors of [13] stated that, for the prediction of flow throughput, the best result was achieved by a random forest technique, which outperformed a deep ANN. They tested different time windows spanning from 0.03s to 4s on nearly 1 million flows containing traces from miscellaneous networks and applications. In addition to flow history data, the study used other attributes related to TCP: window size, window scale or segments retransmissions. Azari et al. in [14] analyzed the effect of different parameters of an LSTM ANN and ARIMA on prediction accuracy. The simulation demonstrated the superiority of LSTM over ARIMA, especially for longer time series. The paper [15] employs the same methods and concludes that historical traffic is the most important feature when predicting future network throughput. However, the authors also state that relying only on the throughput alone leads to worse performance compared to the methods using more inputs. Another short-term prediction is proposed in [16], where He et al. evaluated the ARIMA and LSTM for the prediction of both single and multiple flows taking into account video and non-video traffic. Using a prediction span between 10 and 100 ms, they found that LSTM ANNs achieved good results at the cost of more training computations.

DASH algorithms usually implement a simple throughput prediction based on an average of past measurements. Advanced solutions try to optimize video playback by employing more elaborate prediction techniques. Some researchers state that the playback algorithms fed with an accurate estimation of network throughput outperform those that do not apply predictive methods, while others observe that prediction alone is not sufficient and should be combined with other techniques, e.g., buffer control [1]. In [17], the authors focus on inaccuracies in throughput measurements occurring due to idle periods between the chunks causing sub-optimal adaptation decisions. The authors show that a simple prediction scheme based on adaptive filtering can improve the quality of the streaming video. Raca et al. in [18] compared several prediction schemes, among them support vector machines, random forest and LSTM ANN, and applied them to a single DASH algorithm. Interestingly, the authors reported that a DASH player supported by the best predictor achieved nearly the same performance as in the case of the ideal prediction. In addition to application data, the model relied on network data, reaching as deep as the physical layer, which may be a reason behind the extraordinary high accuracy of the prediction. To improve the video quality, in [19], the authors propose active queue management (AQM). The AQM uses LSTM to predict the router's queue delay, which temporarily stores DASH packets. Increasing queue delay triggers removal of the packets in the queue. This, in turn, forces traffic sources to reduce their transmission rate and induces a video player to decrease the video-bit rate before the network stalls, which allows for smoother video playback. In [20], we proposed employing (F)ARIMA processes, shallow ANNs and an recursive auto-encoder to enhance the performance of a DASH system.

In this work, we operate on traces gathered from 4G and 5G networks. To date, the analysis of the traffic generated by the latter network has limited coverage in the literature. We rely only on application-level throughput measurements; therefore, the proposed solution does not require modifications to network infrastructure or the TCP stack. Consequently, our work shows the impact of the prediction on video QoE without the support of any other information, which is hard to gather and requires profound changes in a DASH system. Contrary to the majority of the works, which aim at either short or long time horizons, our prediction span is 4 s, which is suitable for a DASH system.

3. Theoretical Background

3.1. Video Streaming Model

We model the video stream as a set of consecutive data segments $\{c_i\}$, where c_i is the amount of data measured in bytes and $i \in \{1, 2, \dots N\}$. Every segment contains L seconds of video encoded with a given bit rate; hence, the total length of the video is $N \times L$ seconds. A video player obtains from a server a video segment c_i encoded with a bit rate $q_i \in \{Q_{\min}, \dots, Q_{\max}\}$ (bit/s), where $Q_{\min}$ and $Q_{\max}$ are the lower and upper bound, respectively, of the bit rate levels. A user considers a video with a higher bit rate as having better quality compared to its lower bit rate version.

A video player downloads video segments and places them in its buffer, where they wait for a playback. At the time t_i, the player starts to download the video segment c_i. Its downloading time is a function of the bit rate q_i and averaged network throughput r_i. At the time t_{i+d}, the video player completes the download of the segment c_i. At time t_{i+1}, where $t_{i+d} \leq t_{i+1}$, the video player starts to download the next segment c_{i+1}. At the time t_i, the player's buffer stores $b_i \in [0, B_{\text{full}}]$ time units of video. The maximum buffer capacity B_{full} depends on the video player implementation. If r_i (bit/s) is the average network throughput at time $\Delta t_i = t_{i+d} - t_i$, then:

$$t_{i+d} = t_i + \frac{Lq_i}{r_i},$$

where

$$r_i = \frac{1}{\Delta t_i} \int_{t_i}^{t_{i+d}} r \, dt. \tag{1}$$

During the video playback, the amount of data in the player's buffer fluctuates. The buffer increases by L seconds after a player downloads the segment c_i, while it decreases when a user watches the video. Hence, the evolution of the buffer is described as:

$$b_{i+1} = \max[(b_i - t_{i+1} + t_i), 0] + L.$$

If $b_i < t_{i+1} - t_i$, the buffer becomes empty, which freezes the video.

3.2. Adaptation Strategies

At the time t_{i+1}, a video player, which has only information of the past network throughput $\{r_t, t \leq t_i\}$, selects the bit rate q_{i+1}. To estimate the future throughput $\{r_t, t > t_i\}$, the player may use a predictor defined as $\{\hat{r}_t, t > t_i\}$. Except the throughput, a player may also take into account its buffer level. Having the information about the network and a buffer state, the player selects the next segment c_{i+1} quality, see Figure 1, as follows:

$$q_{i+1} = f(b_i, \{\hat{r}_t, t > t_i\}).$$

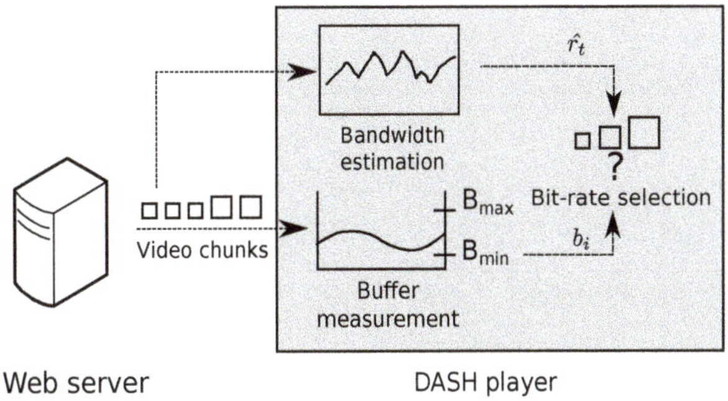

Figure 1. Adaptation strategy for video. A video player chooses the quality of video, taking into account the prediction of network throughput and its buffer state.

In our work, we presented three exemplary methods of the video bit rate selection. The first method is based on the simplified version of the algorithm implemented in the MSS video player, which is described in [21]. The second approach employs the implementation of ExoPlayer—a video player for Android developed by Google [22]. The third solution is described in [23], and we will further refer to it as *Tian2016*.

The first algorithm selects a video bit rate as follows

$$q_{i+1} = \begin{cases} q_i^+ & b_i > B_{\max} & \text{(2a)} \\ q_i^- & b_i < B_{\min} & \text{(2b)} \\ q_i^- & b_i \in [B_{\min}, B_{\max}] \land (q_i > \hat{r}_i) & \text{(2c)} \\ q_i & b_i \in [B_{\min}, B_{\max}] \land \hat{r}_i - d < q_i < \hat{r}_i & \text{(2d)} \\ q_i^+ & b_i \in [B_{\min}, B_{\max}] \land (q_i < \hat{r}_i - d) & \text{(2e)} \end{cases}$$

The buffer-based part of the algorithm selects the video bit rate considering the player's buffer state. When the buffer level reaches $B_{\max}$, the system increases the bit rate (2a) by one level, denoted as q_i^+. Similarly, when the playback drains the buffer below the threshold $B_{\min}$ (2b), the player reduces the video quality by one level, denoted as q_i^-. When the buffer level is between the thresholds $B_{\min}$ and $B_{\max}$, the player estimates throughput $\hat{r}_i$ and selects a corresponding quality level. If the estimated throughput $\hat{r}_i$ is lower than the video bit rate q_i, the player decreases video quality (2c). The bit rate remains unchanged (2d), when the estimated throughput is not sufficiently high. In the case when the bit rate surpasses the throughput by at least d bits/s, the player increases the video quality (2e). The parameter d marks a stripe of the throughput for which the player continues to download videos of the same quality. Hence, the parameter d stabilizes video quality and prevents frequent switches, which degrades content reception. The constants $Q_{\min}$ and $Q_{\max}$ define the lowest and the highest bit rates of video segments. We implemented this idea in the open-source software *dash.js* [24] and presented it in Figure 2. Taking the experiment's parameters from [25], we set the player's buffer size B_{full} to 35 s and the thresholds $B_{\min}$ and $B_{\max}$ to 40% and 80% of the buffer size.

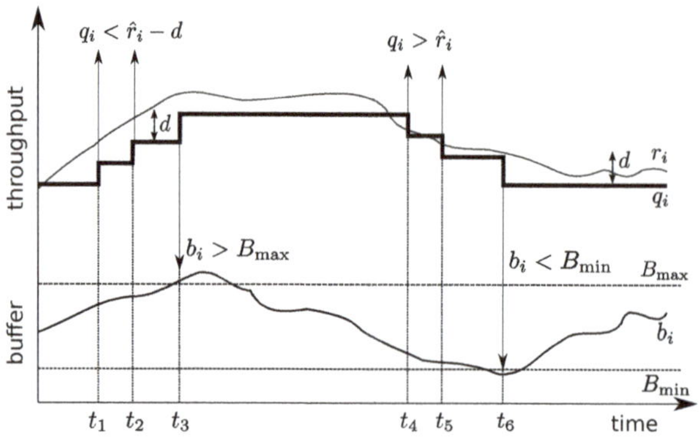

Figure 2. An example operation of a simplified version of the MSS algorithm. At times t_1 and t_2, network throughput is high enough to increase the video bit rate (2e). At t_3, the buffer level, which is above the threshold, allows increasing the video bit rate again (2a). At times t_4 and t_5, network throughput drops, which causes the video bit rate to decrease (2c). At time t_6, the buffer level drops below the threshold, and the video bit rate is reduced again (2b). After t_6, network throughput is not high enough to increase the video-bit rate (2d).

The second algorithm is a part of the ExoPlayer library. Its input parameters are a buffer size and throughput estimation. The algorithm calculates the video quality that fits in the currently available bandwidth according to the formula:

$$q_{i+1} = \begin{cases} \max(Q) & q_{i+1} \le \lambda \hat{r}_i & \text{(3a)} \\ q_i & \hat{r}_i > r_{i-1} \wedge b_i < B_{\min} & \text{(3b)} \\ q_i & \hat{r}_i < r_{i-1} \wedge b_i > B_{\max} & \text{(3c)} \end{cases}$$

The algorithm scans the available qualities Q and selects the representation with the highest average bit rate lower than or equal to the estimated throughput r modified by a factor λ, as presented in (3a). The algorithm will switch to the best achievable video bit rate if it differs from the current bit rate, excluding two scenarios: network throughput increases but the buffer level remains low (3b) or network throughput decreases but the buffer level remains high (3c). For ExoPlayer, the default values for $B_{\min}$ and $B_{\max}$ are 10 and 25 s.

The third algorithm, which we will refer to as *Tian2016*, avoids video bit rate variability resulting from throughput fluctuations and estimation errors. Additionally, when the available network bandwidth remains higher for a longer time than the current video bit rate, the algorithm increases the video bit rate steadily and decreases it more aggressively when the throughput falls. A proportional-integral-derivative (PID) controller is responsible for the player's decision regarding video bit rate selection. A PID implements a control loop with a feedback mechanism, which transfers the controller's output value to the system input. The system adjusts its output by computing the difference between a set point and a measured process variable. The authors claim that their algorithm quickly responds to congestion level shifts and maintains a stable video rate during throughput variations. We took the parameters for the algorithm from [23] from sections V and VI-D.

As was stated earlier, a DASH system fragments a video into segments. The segment length L is usually between 2 and 10 s; for example, the MSS algorithm uses 2–5 s [25]. In our work, we set the video segment length to 4 s.

3.3. Quality Measures

DASH systems strive to achieve the best possible QoE by selecting optimal video bit rates in given network conditions. This selection involves a few conflicting issues. Firstly, a user expects that the streaming video will have the highest possible bit rate available for the given network throughput. Secondly, the playback should be smooth, which a player achieves by avoiding frequent bit rate changes. Thirdly, a player should minimize the times when the video freezes on a user's screen due to the player's buffer depletion. Having identified the three main components of QoE, namely bit rate, bit rated variation and rebuffering time, we combine them into a single measure, where the QoE is calculated as a weighted sum of the parameters of each video segment in the range from 1 to N:

$$\text{QoE} = \sum_{i=1}^{N} q_i - \lambda \sum_{i=1}^{N-1} |q_{i+1} - q_i| - \mu \sum_{i=1}^{N} (K(b_i)),$$

where λ and μ are positive weights corresponding to video quality variations and rebuffering time. To count the buffer's starvation periods, we introduce the penalty function $K(b_i) = 1, b_i = 0$ and $K(b_i) = 0, b_i > 0$.

Video stalls disturb users more than bit rate switches do; thus, the value of μ is usually much higher than λ. A small λ indicates that bit rate switches are less nuisance for a user, while a large μ implies that the user is more concerned about rebuffering.

To obtain time- and video-independent measures with possibilities of comparison with different algorithms, we adapt from [26] a normalized QoE model

$$\text{nQoE} = \frac{\text{QoE}}{\text{QoE\_ref}}, \tag{4}$$

where QoE_ref is the QoE of another benchmark algorithm.

4. Prediction of Network Throughput

4.1. The Prediction Problem

We have measurements of the past throughput $\{r_t, r_{t-1}, \ldots, r_{t-n}\}$, and we are to predict the value of r_{t+m}, $m > 0$. Hence, we use a predictor $\hat{r}_{t+m}$, which takes the past throughput

$$\hat{r}_{t+m} = \phi(r_t, r_{t-1}, \ldots, r_{t-n}). \tag{5}$$

The problem is to choose ϕ, which minimizes the difference between $\hat{r}_{t+m}$ and r_{t+m}.

Typically, a DASH algorithm predicts network throughput for the nearest future r_{i+1}, using an average of past throughput measurements [27]

$$\hat{r}_{t+1} = \frac{1}{n} \sum_{k=1}^{n} r_{t+1-k}. \tag{6}$$

According to [28], DASH players apply the estimator (6) or its variation based on weighted, exponential or harmonic moving average. In our work, we improve the estimation by replacing the simple computation (6) with an LSTM ANN.

To measure the accuracy of the prediction algorithm, we chose the normalized root mean square error (NRMSE)

$$NRMSE = \frac{1}{\bar{r}} \sqrt{\frac{\sum_{t=1}^{N} (r_t - \hat{r}_t)^2}{N}}, \tag{7}$$

The NRMSE represents the squared difference between predictions $\hat{r}_t$ and observed values r_t divided by the number of observations N. An average of observed values $\bar{r}$ normalizes the calculation result. The lower the NRMSE value, the better the model describes empirical data. The normalization allows us to compare predictions among traces with different throughput levels.

4.2. Long Short-Term Memory Networks

An LTSM is a recurrent ANN, which can learn long-range dependencies in data. LSTM ANNs respond to the short-term memory problem of recurrent ANNs. Namely, a recurrent

ANN has difficulties retaining data from earlier computations because it misses information about long-memory dependencies. An LSTM ANN handles this problem by allowing its memory cells to select when to forget certain information, thus achieving the optimal time lags for time series problems. For this purpose, a network uses a memory block, which contains cells with self-connections, memorizing the temporary state and three adaptive, multiplicative gates (input, output and forget) to control information flow in the block—see Figure 3. The gates can learn to open and close, enabling LSTM memory cells to store data over long periods, which is desirable in time series prediction with long temporal dependency. Multiple memory blocks form a hidden layer. An LSTM ANN consists of an input layer, at least one hidden layer and an output layer.

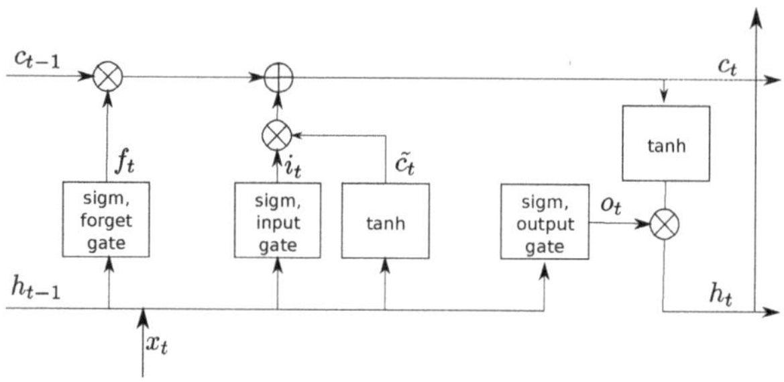

Figure 3. Structure of LSTM memory block: x_t is an input signal at the time t, c_{t-1} and c_t represent cell states at the previous $t-1$ and actual t times, h_{t-1} and h_t are output signals of the cell at $t-1$ and t times, b represents the bias. The symbol $\times$ represents the multiplication and $+$ summation of signals.

An LSTM's cell takes three inputs and generates two outputs:

$$(h_t, c_t) = \mathrm{CL}(h_{t-1}, c_{t-1}, x_t),$$

where x_t is an element of time series, c_{t-1} is a cell's memory signal from the previous time $t-1$, and h_{t-1} is a cell's output signal at the previous time $t-1$. At the time t, the cell generates two signals: c_t—a cell's state and h_t—a cell's output, which is passed to an upper ANN layer.

At every time step, the element x_t of a time sequence enters the cell. Both output signals h_t and c_t leave the cell at time t and return to that same cell at a time $t+1$. The cell can multiply or add signals, which are denoted by the symbols $\times$ and $+$. Cell gates also use two functions: *sigm* and *tanh*. The first represents the sigmoid function, which outputs numbers between zero and one ($\mathrm{sigm}: R \to [0,1]$), describing how much of each signal to let through. The value of zero transfers nothing, while a value of one transfers everything. The second function represents the hyperbolic tangent $\tanh : R \to [-1,1]$) and denotes which fraction of the candidate value $\tilde{c_t}$ updates the cell's state c_{t-1}.

An entering signal simultaneously reaches all three cell's gates. The *forget gate* decides which part of the memory from the previous state transfers to the current state. The gate multiplies the signal using its weight w_f and projects it into a unit interval using a sigmoid function

$$f_t = \mathrm{sigm}(w_f[h_{t-1}, x_t] + b_f).$$

The remaining two gates determine which new information to store in the cell. Firstly, the *input gate* decides which values to update, transforming them into a unit interval. Next, the *tanh* layer creates a new candidate value $\tilde{c_t}$ and adds it to the cell state later.

$$\begin{aligned} i_t &= \mathrm{sigm}(w_i[h_{t-1}, x_t] + b_i), \\ \tilde{c_t} &= \tanh(w_C[h_{t-1}, x_t] + b_c). \end{aligned}$$

Finally, the new state c_t replaces the old one c_{t-1}: the old state c_{t-1} multiplies the *forget gate* value f_t, and the candidate result $\tilde{c_t}$ multiplies the *input gate* value i_t

$$c_t = f_t \times c_{t-1} + i_t \times \tilde{c_t}. \tag{8}$$

Equation (8) controls the amount of information c_t, which will form the state of the cell at the time t.

In the cell's right-side, a multiplicative operation generates the output signal

$$h_t = o_t \times \tanh(c_t),$$

which is the product of the actual memory signal c_t computed in (8) and the signal

$$o_t = \rho(w_o[h_{t-1}, x_t] + b_o),$$

generated in the *output gate*. The output signal h_t describes how much information the cell transmits to other cells at the next time point.

For our study, we use two architectures: stacked and sequence to sequence. The differences between these architectures are described in [29].

4.2.1. Stacked Unidirectional LSTM

An ANN can consist of multiple hidden layers, which include LSTM cells. In such a case, the output of one hidden layer is fed as the input into the subsequent hidden layer. Such deep ANN architecture allows progressively building up a higher level of representations of sequence data and mapping more complex, non-linear dependencies, which a single-layered, shallow network does not discover.

A special case is stacked architecture, where a hidden layer is multiplied and stacked on top of each other, creating a grid of cells, as shown in Figure 4. Cells composing the grid share their connections' weights and biases. The outputs $\{h_t\}$ are forwarded as inputs to the next immediate layer above. The top layer produces $\hat{y_t}$, which corresponds to the expected unprocessed forecasts for the input time series. Each step of a training process calculates training error $e_t = f(y_t - \hat{y_t})$ and accumulates it until the end of the time series $E = \sum_{t=1}^{T} e_t$, where y_t is the actual value of the time series.

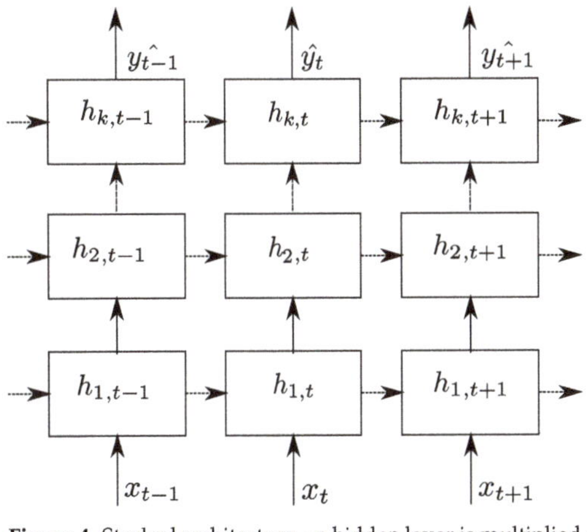

Figure 4. Stacked architecture—a hidden layer is multiplied and stacked on top of each other creating a grid of cells.

4.2.2. Stacked Bidirectional LSTM

A time series provided to an LSTM ANN is arranged chronologically: the information is passed in a positive direction from the time step $t-1$ to the time step t along the chain-like structure. However, it is also possible to process sequence data in both forward and backward directions. This idea is utilized in Bidirectional LSTM (BdLSTM) ANNs, which connect two hidden layers: a forward LSTM layer and a backward LSTM layer, to the same output layer, as illustrated in Figure 5. Thus, a BdLSTM ANN considers additionally backward dependencies, which pass information in a negative direction from the time step t to $t-1$. This allows a BdLSTM ANN to capture more patterns, which a unidirectional LSTM could filter. The BdLSTM ANN computes its forward layer output $\overrightarrow{h}$ iteratively from the positive sequence's inputs from time $t-n$ to $t-1$. Using the reversed inputs from time $t-1$ to $t-n$, the BdLSTM calculates the backward layer output $\overleftarrow{h}$.

The standard LSTM updating method based on the equations presented in Section 4.2 computes both forward and backward layer outputs. A BdLSTM layer generates an output h_t, calculated as

$$h_t = \rho(\overrightarrow{h}, \overleftarrow{h}),$$

where the ρ function is used to combine the two output sequences. Similar to the LSTM layer, the final output of a BdLSTM layer can be represented by $\hat{y}_T$, which is an expected unprocessed forecast.

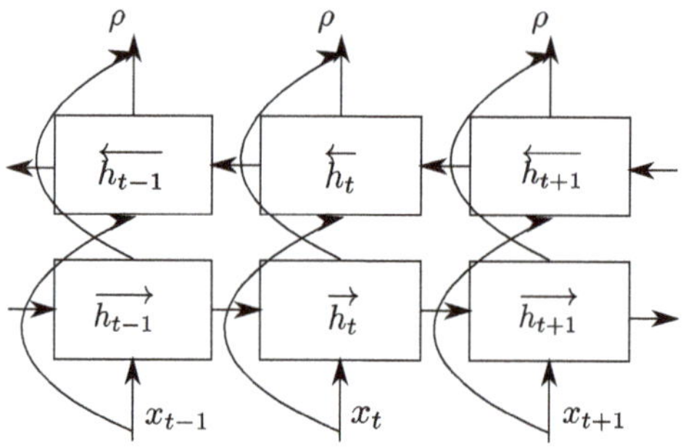

Figure 5. Bidirectional architecture of an LSTM network. BdLSTMs apply both forward and backward dependencies.

4.3. *Traffic Characteristics*

The traffic traces were collected in Ireland by a large telecommunication operator. They include static and mobility scenarios and total roughly 21 h. We divided them into four sets: 5G, 4G, static and mobile, as listed in Table 1. A more detailed description of the traffic generation, capture methodology and properties of the traces can be found in [30] for 4G and in [31] for the 5G traffic.

Table 1. Network capacity traces used in the experiments.

Trace	Activity	Length (min)	Average (MB/s)	Var. Range (MB/s)	Source
S4G	Static 4G	254	5.2	2.6-9.6	[30]
M4G	Mobile 4G	290	2.7	0.3-6.1	[30]
S5G	Static 5G	260	8.3	2.7-25	[31]
M5G	Mobile 5G	459	3.5	0.4-11	[31]

Table 1 shows that the average rates of the 5G traces are about 50% higher than 4G traces for the static scenario. Furthermore, the 5G traces achieve nearly three times higher variation rates, reaching about 25 MB/s, compared to 4G traces, which obtain 9.6 MB/s. In the mobile scenario, the differences between 5G and 4G are lower—the average throughput of 5G traces is only 27% higher than the 4G traces. The authors of [30] explain these differences by the lack of 5G base stations across the driving routes where the traffic was captured. Despite this restriction, the upper limit for the variation range is still almost two times higher for 5G than 4G.

We use r_i (1) to denote an average amount of traffic registered within a specified time. The length of a video segment is L, as stated in Section 3.1. A video player downloads segments at the rate $q_i \approx \tilde{q}_i$; hence, the player measures network throughput roughly every L seconds. In our model, we assumed $L = 4$ s.

4.4. *Traffic Prediction with an LSTM ANN*

For the time series prediction, an LSTM ANN uses a sliding window containing n most recent observations, as defined in (5). From every set of the traces listed in Table 1, we extracted ten overlapping sub-traces every 30 min. The models were trained on the first 10 min of each sub-trace and tested on its remaining part. We concatenated the obtained predictions, which allowed us to input their values into a video player and have uninterrupted playback.

For the prediction, we used a many-to-one architecture, where a throughput sequence $\{r_t\}$ is ANN's input, and a single throughput value r_{t+1} is a prediction. Firstly, we normalize the traffic flow sequence $\{r_t, r_{t-1}, \ldots, r_{t-n}\}$, i.e., remove trends and differentiate it obtaining $\{x_t, x_{t-1}, \ldots, x_{t-n}\}$, which is then delivered to the LSTM ANN. Both our LSTMs, unidirectional and bidirectional, consist of an input, two hidden and output layers. The top LSTM layer supplies its output to a conventional feed-forward ANN, which maps them into a single value that corresponds to an unprocessed prediction at time $t + 1$. Finally, we denormalize the output values from the feed-forward ANN, obtaining the prediction of network throughput $\hat{r}_{t+1}$ at $t + 1$. This process and the structure of the LSTM ANN used for predicting traffic flow are presented in Figure 6.

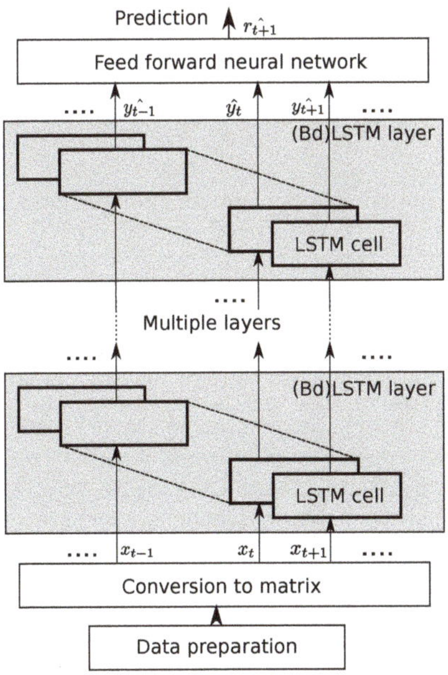

Figure 6. Architecture of the LSTM ANN used for network throughput prediction.

An LSTM ANN minimizes the NRMSE defined in (7) through training, which iteratively changes the weights and biases of the network to decrease the differences between the model output and observed values. For this purpose, we chose *Backpropagation Through Time* (BPTT) algorithm. The BPTT unfolds an LSTM ANN and converts it into a feed-forward ANN to apply the backpropagation algorithm. Then, the backpropagation algorithm finds the gradient of the cost taking network parameters as the input, which in the case of an LSTM are the same for every unfolded time step. The algorithm aggregates the errors from the unfolded time steps, updates the weights and folds the ANN back. Due to good performance, relative simplicity and implementation in popular software, the BPTT is popularly used for the RNN ANNs, including their LSTM variants. More information about the algorithm and its comparison with others can be found in [32].

4.5. Model Validation

The prediction is more accurate for traces generated by static terminals compared with the mobile ones; see Figure 7. Furthermore, the traffic from 4G systems obtained a lower NRMSE than the 5G traces. These differences may result from the positive correlation between the prediction errors and variance of the traces. The traces generated by mobile scenarios have higher variation than the static ones, see Table 1. Additionally, the 5G traces, due to higher network bandwidth, achieved higher peak values. The variation of the results is roughly on the same level, except for the 5G mobile scenario where the NRMSEs have a higher range. Compared to a simple moving average, the LSTM predictions achieved lower NRMSE. The difference is particularly notable for the mobile and 5G scenarios. In all scenarios, the best result was obtained by the BdLSTM.

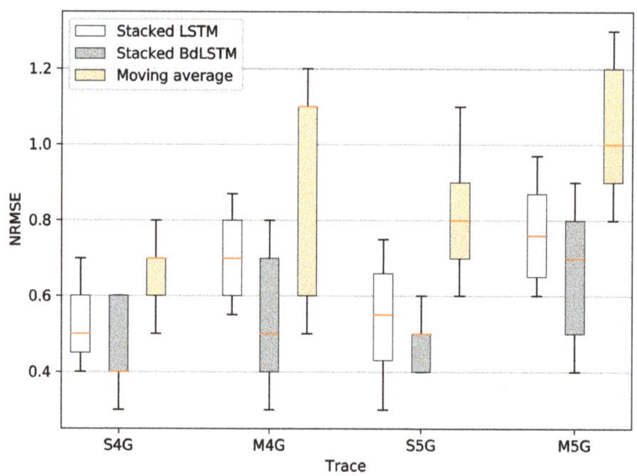

Figure 7. NRMSE of the analyzed prediction models for different traces. Every box on the plot represents predictions made on ten overlapping sub-traces every 30 minutes, which were extracted from traces listed in Table 1.

In Figure 8, we presented a fragment of the M5G trace to visualize the traffic characteristics and the prediction accuracy. The trace has an irregular structure without clearly identifiable patterns. The traffic frequently and abruptly changes its volume, producing oscillations between 2.5 and 5 MB/s. The prediction generally follows the traffic pattern relatively well; however, sometimes, it overestimates the scope of the oscillations, especially in the fragments where the traffic fluctuations have a higher frequency.

The prediction errors for the exemplary M5G trace have a normal distribution with a zero mean, as shown in Figure 9a. The errors are between −1.0 and 1.0 MB/s, and most do not exceed 0.5 MB/s. Compared with the case of the simple moving average shown in Figure 9b, its histogram is broader, and consequently, the errors have higher values than for the LSTM prediction.

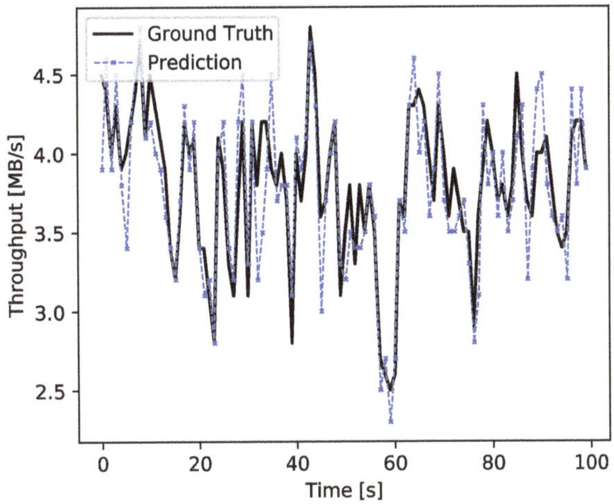

Figure 8. One-step prediction for the M5G trace. The prediction tends to overestimate the traffic oscillations, but the unidirectional LSTM ANN achieves better results than a naive predictor based on a simple moving average.

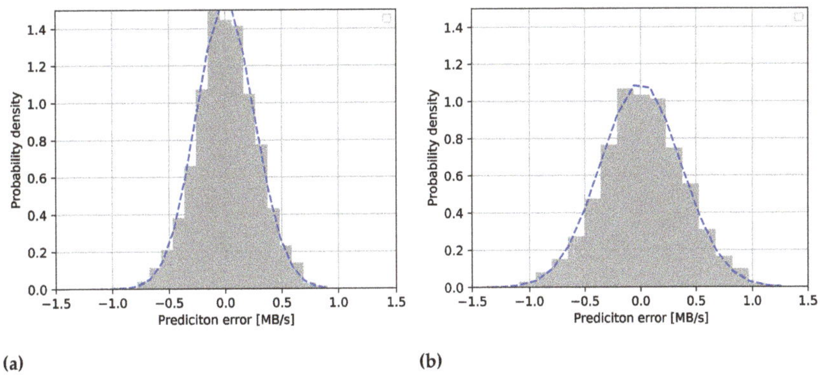

Figure 9. Distribution of prediction errors generated by (**a**) the unidirectional LSTM and (**b**) a simple moving average for the one-step prediction for M5G trace. The narrower prediction error interval for the LSTM ANN indicates fewer errors and better prediction accuracy.

4.6. Playback Algorithms Supported by Prediction

To assess the influence of the prediction on the quality of the video, we prepared a local media streaming system consisting of three Linux machines, as illustrated in Figure 10. The first machine ran a web server, the second one ran a traffic control facility *netem*, and the last one ran the *dash.js* player [24] in a Google Chrome browser. To emulate a mobile network, we used the *netem* module, which gave us control over bandwidth, packet delay and loss. For our experiments, we used the traffic from mobile environments examined in Section 4.3 as a network capacity trace, set a uniformly distributed delay from an interval [5 ms, 145 ms] and set a uniformly distributed packet loss from an interval [0.05%, 0.15%]. We placed six video files acquired from [33] at the web server and encoded them with 18 bit rates ranging from 0.1 to 20 MB/s. The server streamed the video to *dash.js* media player, which has an open-source code, which allows for the implementation and comparison of different playback algorithms.

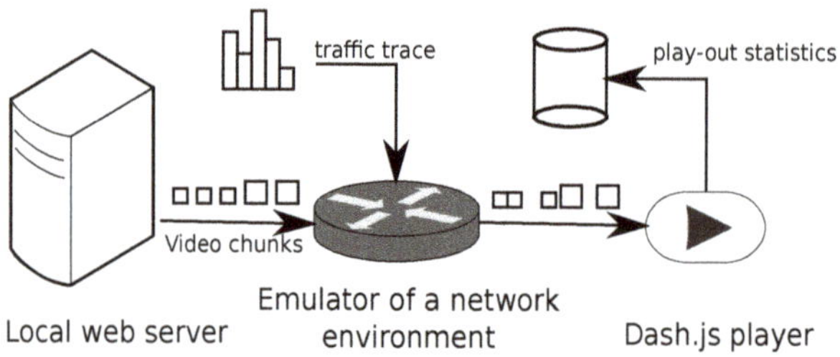

Figure 10. Experiment setup includes a web server, the *netem* module and the *dash.js* video player. The connection between the emulated network and the server has fixed parameters. The *netem* module shapes the traffic between the emulated network and the DASH player, providing variable transmission time of video segments.

The predicted throughput may integrate into the adaptation logic in two different ways. Firstly, it is possible to incorporate a prediction algorithm directly into a DASH player and replace its default throughput estimation algorithm. This approach requires translating the model into a native DASH player's code or wrapping the model into a package, which the player calls an external dependency. In summary, it requires deep integration into the player's implementation, which is challenging. In the second case, the prediction overwrites the measured throughput samples. The player acquires the information about throughput not directly from a network but from an output of the prediction model. This solution is easier to implement as it only requires minimal amendment of the player's code; hence, we replaced the default throughput estimation with the predicted values, see Figure 11. As we mentioned in Section 3.2, we also overwrote the default playback algorithm implemented in the plug-in using the simplified versions of MSS, ExoPlayer and Tian algorithms.

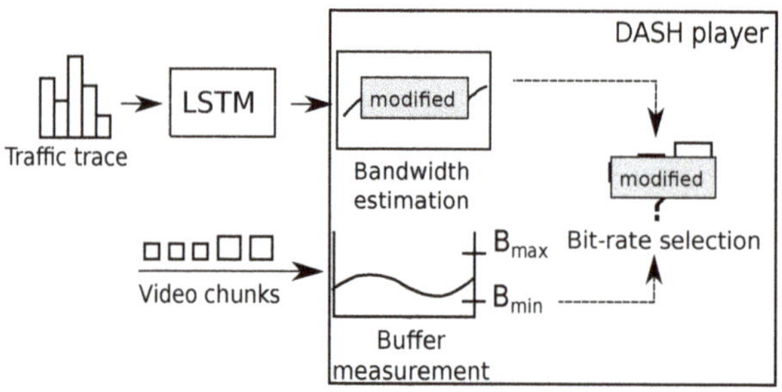

Figure 11. The architecture of the *dash.js* reference player. The modified estimation module acquires information about network throughput from the LSTM ANN. The MSS, ExoPlayer and Tian algorithms replaced the player's default playback logic.

Figure 12 shows a fragment of the performance of the MSS algorithm in its original version and the modified version with the support of the prediction made by a unidirectional LSTM. Both versions operate on the M5G trace from the set presented in Table 1. Compared with the original version, the prediction-supported algorithm switches its bit rates less often and more quickly recovers from bandwidth drops; nevertheless, its average bit rate is similar. Consequently, the less frequent switching of bit rates in the prediction-based version does not come at the cost of a lower bit rate.

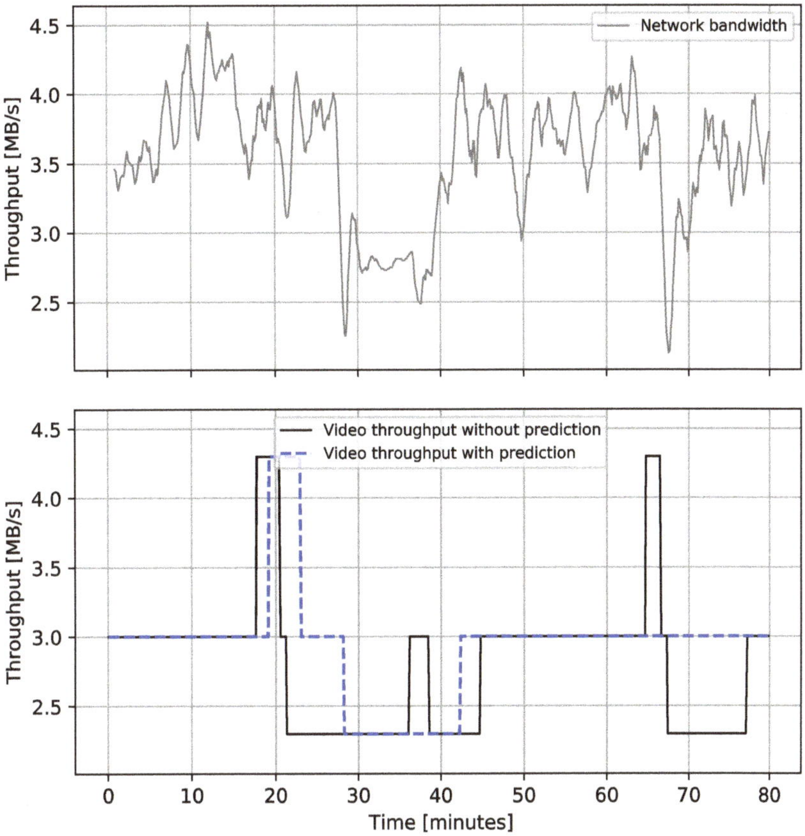

Figure 12. Performance of MSS algorithm without and with support of traffic prediction. Throughput prediction eliminates some of the video bit rate drops observed at the beginning, end and approximately the 20th min of the trace. However, as the prediction sometimes overestimates the throughput, the bit rate unnecessarily switches to a higher level at about the 20th min of the trace.

Figure 13 compares nQoE (4) between the LSTM prediction and simple moving average. The nQoE is higher from about 5% to over 20%, depending on a DASH player and streaming scenario. In the case of MSS and ExoPlayer, the average improvement is mostly between 10% and 15%. For static traces generated by the 4G system, this value is closer to 10%; for mobile scenarios, the improvement is about 17%. The mobile use case obtains better results on average, but the higher nQoE variation indicates that the refinement is not uniform but depends on network conditions. Throughput in mobile scenarios, especially involving the 5G network, has higher variability (see Table 1); therefore, some parts of the traces may be harder to predict. The poorer performance of the default prediction based on moving average can also contribute to better relative improvement. For Tian, the resulting pattern is similar, but the gain is a few percent lower than in the other two algorithms. The more complex algorithm based on a PID controller and less reliance on direct throughput estimation probably reduces the benefits of the prediction. Comparing the LSTM versions, the algorithms with support of the BdLSTM have slightly better scores for the MSS and ExoPlayer algorithms. In the case of Tian, the results are roughly the same.

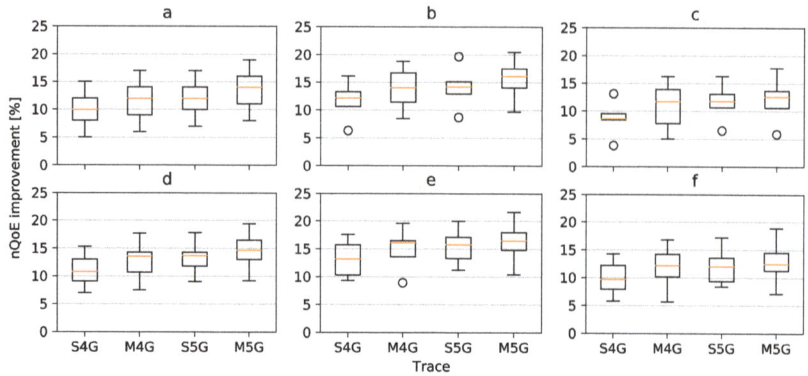

Figure 13. The relative improvement of the nQoE metric across different HAS algorithms and network traces with respect to the no-prediction: (**a**) MSS, LSTM, (**b**) ExoPlayer, LSTM, (**c**) Tian, LSTM, (**d**) MSS, BdLSTM, (**e**) ExoPlayer, BdLSTM, and (**f**) Tian, BdLSTM. The nQoE metrics are normalized to the no-prediction case set to 0% improvement. The scenarios with traces from the mobile environment obtain, on average, better results, but their variation is also higher.

4.7. Discussion of the Results

Our analysis shows that an LSTM ANN obtains better prediction accuracy compared with a naive predictor based on a moving average. The prediction achieved the best results for traffic registered by fixed terminals. The traffic acquired from mobile equipment, especially that operating on a 5G network, had a more irregular structure and was harder to predict for both the LSTM ANN and a moving average. However, for mobile traces, the accuracy of the prediction made by the moving average deteriorates even faster than the accuracy of an LTSM ANN. From the two LSTM versions we tested, the BdLSTM shows slightly better prediction accuracy compared to the unidirectional LSTM. Although we used an NRMSE (7), comparison with other research is not easy, as their NRMSE values differ significantly across different works. For instance, in [9], the NRMSE ranges between about 0.1 and 5.5 depending on an LSTM ANN and a data set used for the prediction, whereas in [12], the NRMSE is below 0.1 for all experiments. It is also unclear how the authors normalized their errors as they did not specify explicit mathematical formulas in their works. In other mentioned works in Section 2, e.g., [6–8,10], the authors provided non-normalized and non-comparative mean square error measures of prediction errors. Sometimes this measure is normalized as in [34] (normalized mean absolute error) or [13] (relative MSE) but is non-comparative with NRMSE without additional data.

Exemplary DASH algorithms supported by throughput prediction achieved better performance than their versions with default throughput estimators. Most gained the MSS and ExoPlayer playback strategies, while the results for Tian were lower. An explanation is the degree of complexity of the examined approaches. The first two algorithms rely heavily on rate estimation and use for this purpose moving averages, while the buffer measurements play a supportive role. In the case of Tian, the algorithm uses primarily the buffered video time as a feedback signal, while the expected throughput is estimated using support vector regression. This shift of focus results in a lower impact of prediction on the algorithm performance.

To our best knowledge, there is only a single work with which we can directly compare the results because every research uses its own set of measures. In our work, the quality improvement is between 5% and 25%, while for the perfect throughput prediction, Raca et al. [35] reached 55% using a similar nQoE (4) measure. The authors built a perfect estimator by looking explicitly into the future values of time series and applied it to three algorithms: ExoPlayer, Elastic [36] and Arbiter+ [37]. Similarly to our case, the simpler algorithm ExoPlayer profited the most from the prediction. The prediction gave a less impressive nQoE increase for the two other, more complex algorithms.

5. Conclusions

In this work, we improved the performance of adaptive streaming algorithms employing LSTM ANNs that predicted network throughput. As a result, the system provides higher video quality without longer buffering times or a higher frequency of bit rate changes. The improvement is more significant in 5G networks; nevertheless, advanced playback algorithms gain less from throughput prediction. The best results achieved relatively simple strategies based on throughput and buffer measurement. The improvement is not uniform across the usage scenario and depends on network conditions. The prediction gain was smaller for a more complex playback algorithm supported by the proportional-integral-derivative controller. From the two tested LSTM versions, the bidirectional LSTM gained a slightly better prediction score.

Our approach does not require modifications of the network infrastructure or the TCP stack. Unlike proprietary solutions, DASH has an open specification and leaves the implementation of the playback logic to third parties; thus, our solution can be integrated with adaptation algorithms. The prediction model relies only on past throughput measurement and works without additional network information, which can be hard to acquire. Therefore, the prediction accuracy and the quality improvement are less impressive than in some cited works, which support their prediction with additional data.

We identified several potential paths for further investigation. Firstly, the traffic models can be improved, as some authors report that the best results are usually obtained by hybrid techniques. It will be interesting to investigate how to leverage the additional information to increase the prediction accuracy, as more elaborate approaches may better adjust to network data and provide better input for adaptive algorithms. Secondly, there is room to extend our prediction beyond the current single video chunk horizon. Thirdly, one can investigate the optimal length of the training, test sets and a time horizon that provides the best prediction. Finally, we would like to move from the offline analysis to an online prediction made by a mobile device, which measures the throughput in real-time and provides information about battery, memory and processor usage.

Funding: The work was carried out within the statutory research project of the Department of Computer Networks and Systems.

Institutional Review Board Statement: Not applicable.

Informed Consent Statement: Not applicable.

Data Availability Statement: Publicly available datasets were analyzed in this study. The data can be found here https://www.ucc.ie/en/misl//research/datasets/ivid_4g_lte_dataset/ (accessed on 5 October 2022) and here https://github.com/uccmisl/5Gdataset (accessed on 5 October 2022).

Conflicts of Interest: The authors declare no conflict of interest.

References

1. Zou, X.K.; Erman, J.; Gopalakrishnan, V.; Halepovic, E.; Jana, R.; Jin, X.; Rexford, J.; Sinha, R.K. Can Accurate Predictions Improve Video Streaming in Cellular Networks? In Proceedings of the 16th International Workshop on Mobile Computing Systems and Applications, ACM, Santa Fe, NM, USA, 12–13 February 2015; pp. 57–62.
2. Katris, C.; Daskalaki, S. Comparing forecasting approaches for Internet traffic. *Expert Syst. Appl.* **2015**, *21*, 8172–8183. [CrossRef]
3. Liu, Y.; Lee, J.Y. An Empirical Study of Throughput Prediction in Mobile Data Networks. In Proceedings of the 2015 IEEE Global Communications Conference (GLOBECOM), San Diego, CA, USA, 6–10 December 2015; pp. 1–6.
4. Makridakis, S.; Spiliotis, E.; Assimakopoulos, V. Statistical and Machine Learning forecasting methods: Concerns and ways forward. *PLoS ONE* **2018**, *13*, e0194889. [CrossRef] [PubMed]
5. Oliveira, T.P.; Barbar, J.S.; Soares, A.S. Computer network traffic prediction: a comparison between traditional and deep learning neural networks. *Int. J. Big Data Intell.* **2016**, *3*, 28–37. [CrossRef]
6. Dalgkitsis, A.; Louta, M.; Karetsos, G.T. Traffic forecasting in cellular networks using the lstm rnn. In Proceedings of the 22nd Pan-Hellenic Conference on Informatics, Athens, Greece, 29 November–1 December 2018; pp. 28–33.

7. Zhuo, Q.; Li, Q.; Yan, H.; Qi, Y. Long short-term memory neural network for network traffic prediction. In Proceedings of the 2017 12th International Conference on Intelligent Systems and Knowledge Engineering (ISKE), Nanjing, China, 24–26 November 2017; pp. 1–6. [CrossRef]
8. Alsaade, F.W.; Hmoud Al-Adhaileh, M. Cellular Traffic Prediction Based on an Intelligent Model. *Mob. Inf. Syst.* **2021**, *2021*, 6050627. [CrossRef]
9. He, Q.; Moayyedi, A.; Dan, G.; Koudouridis, G.P.; Tengkvist, P. A Meta-Learning Scheme for Adaptive Short-Term Network Traffic Prediction. *IEEE J. Sel. Areas Commun.* **2020**, *38*, 2271–2283. [CrossRef]
10. Fu, R.; Zhang, Z.; Li, L. Using LSTM and GRU neural network methods for traffic flow prediction. In Proceedings of the 2016 31st Youth Academic Annual Conference of Chinese Association of Automation (YAC), Wuhan, China, 11–13 November 2016; pp. 324–328. [CrossRef]
11. Cao, S.; Liu, W. LSTM Network Based Traffic Flow Prediction for Cellular Networks. In *Simulation Tools and Techniques*; Song, H., Jiang, D., Eds.; Springer International Publishing: Cham, Switzerland, 2019; Volume 295, pp. 643–653. [CrossRef]
12. Trinh, H.D.; Giupponi, L.; Dini, P. Mobile traffic prediction from raw data using LSTM networks. In Proceedings of the 2018 IEEE 29th Annual International Symposium on Personal, Indoor and Mobile Radio Communications (PIMRC), Bologna, Italy, 9–12 September 2018; pp. 1827–1832.
13. Labonne, M.; López, J.; Poletti, C.; Munier, J.B. Short-Term Flow-Based Bandwidth Forecasting using Machine Learning. *arXiv* **2020**, arXiv: 2011.14421.
14. Azari, A.; Papapetrou, P.; Denic, S.; Peters, G. Cellular traffic prediction and classification: A comparative evaluation of LSTM and ARIMA. In Proceedings of the International Conference on Discovery Science, Split, Croatia, 28–30 October 2019; pp. 129–144.
15. Yue, C.; Jin, R.; Suh, K.; Qin, Y.; Wang, B.; Wei, W. LinkForecast: Cellular Link Bandwidth Prediction in LTE Networks. *IEEE Trans. Mob. Comput.* **2018**, *17*, 1582–1594. [CrossRef]
16. He, Q.; Koudouridis, G.P.; Dán, G. A comparison of machine and statistical time series learning for encrypted traffic prediction. In Proceedings of the 2020 International Conference on Computing, Networking and Communications (ICNC), Big Island, HI, USA, 17–20 February 2020; pp. 714–718.
17. Bentaleb, A.; Timmerer, C.; Begen, A.C.; Zimmermann, R. Performance Analysis of ACTE: A Bandwidth Prediction Method for Low-latency Chunked Streaming. *ACM Trans. Multimed. Comput. Commun. Appl.* **2020**, *16*, 1–24. [CrossRef]
18. Raca, D.; Zahran, A.H.; Sreenan, C.J.; Sinha, R.K.; Halepovic, E.; Jana, R.; Gopalakrishnan, V. On Leveraging Machine and Deep Learning for Throughput Prediction in Cellular Networks: Design, Performance, and Challenges. *IEEE Commun. Mag.* **2020**, *58*, 11–17. [CrossRef]
19. Santos, C.E.M.; da Silva, C.A.G.; Pedroso, C.M. Improving Perceived Quality of Live Adaptative Video Streaming. *Entropy* **2021**, *23*, 948. [CrossRef] [PubMed]
20. Biernacki, A. Traffic prediction methods for quality improvement of adaptive video. *Multimed. Syst.* **2018**, *24*, 531–547. [CrossRef]
21. Zambelli, A. *IIS Smooth Streaming Technical Overview*; Microsoft Corporation: Albuquerque, NM, USA, 2009.
22. ExoPlayer. 2022. Available online: https://github.com/google/ExoPlayer (accessed on 5 October 2022).
23. Tian, G.; Liu, Y. Towards Agile and Smooth Video Adaptation in HTTP Adaptive Streaming. *IEEE/ACM Trans. Netw.* **2016**, *24*, 2386–2399. [CrossRef]
24. (DASH-IF). JavaScript Reference Client. 2022. Available online: https://reference.dashif.org/dash.js/ (accessed on 5 October 2022).
25. Famaey, J.; Latré, S.; Bouten, N.; Van de Meerssche, W.; De Vleeschauwer, B.; Van Leekwijck, W.; De Turck, F. On the merits of SVC-based HTTP adaptive streaming. In Proceedings of the Integrated Network Management (IM 2013), Ghent, Belgium, 27–31 May 2013; pp. 419–426.
26. Belda, R.; de Fez, I.; Arce, P.; Guerri, J.C. Look ahead to improve QoE in DASH streaming. *Multimed. Tools Appl.* **2020**, *79*, 25143–25170. [CrossRef]
27. Li, Z.; Zhu, X.; Gahm, J.; Pan, R.; Hu, H.; Begen, A.C.; Oran, D. Probe and adapt: Rate adaptation for http video streaming at scale. *IEEE J. Sel. Areas Commun.* **2014**, *32*, 719–733. [CrossRef]
28. Bentaleb, A.; Taani, B.; Begen, A.C.; Timmerer, C.; Zimmermann, R. A Survey on Bitrate Adaptation Schemes for Streaming Media Over HTTP. *IEEE Commun. Surv. Tutorials* **2019**, *21*, 562–585. [CrossRef]
29. Hewamalage, H.; Bergmeir, C.; Bandara, K. Recurrent Neural Networks for Time Series Forecasting: Current Status and Future Directions. *Int. J. Forecast.* **2021**, *37*, 388–427. [CrossRef]
30. Raca, D.; Quinlan, J.J.; Zahran, A.H.; Sreenan, C.J. Beyond throughput: A 4G LTE dataset with channel and context metrics. In Proceedings of the 9th ACM Multimedia Systems Conference, Amsterdam, The Netherlands, 12–15 June 2018; pp. 460–465. [CrossRef]
31. Raca, D.; Leahy, D.; Sreenan, C.J.; Quinlan, J.J. Beyond throughput, the next generation: A 5G dataset with channel and context metrics. In Proceedings of the 11th ACM Multimedia Systems Conference, Istanbul, Turkey, 8–11 June 2020; pp. 303–308. [CrossRef]
32. Shrestha, A.; Mahmood, A. Review of Deep Learning Algorithms and Architectures. *IEEE Access* **2019**, *7*, 53040–53065. [CrossRef]
33. Zabrovskiy, A.; Feldmann, C.; Timmerer, C. Multi-codec DASH dataset. In Proceedings of the 9th ACM Multimedia Systems Conference, Amsterdam, The Netherlands, 12–15 June 2018; pp. 438–443. [CrossRef]
34. Kim, M. Network traffic prediction based on INGARCH model. *Wirel. Netw.* **2020**, *26*, 6189–6202. [CrossRef]

35. Raca, D.; Zahran, A.H.; Sreenan, C.J.; Sinha, R.K.; Halepovic, E.; Jana, R.; Gopalakrishnan, V.; Bathula, B.; Varvello, M. Incorporating Prediction into Adaptive Streaming Algorithms: A QoE Perspective. In Proceedings of the 28th ACM SIGMM Workshop on Network and Operating Systems Support for Digital Audio and Video, Amsterdam, The Netherlands, 12–15 June 2018; pp. 49–54. [CrossRef]
36. De Cicco, L.; Caldaralo, V.; Palmisano, V.; Mascolo, S. ELASTIC: A client-side controller for dynamic adaptive streaming over HTTP (DASH). In Proceedings of the 2013 20th International Packet Video Workshop, San Jose, CA, USA, 12–13 December 2013; pp. 1–8.
37. Zahran, A.H.; Raca, D.; Sreenan, C.J. Arbiter+: Adaptive rate-based intelligent http streaming algorithm for mobile networks. *IEEE Trans. Mob. Comput.* **2018**, *17*, 2716–2728. [CrossRef]

applied sciences

MDPI

Article

Quality Assessment of Dual-Parallel Edge Deblocking Filter Architecture for HEVC/H.265

Prayline Rajabai Christopher * and Sivanantham Sathasivam *

School of Electronics Engineering, Vellore Institute of Technology, Vellore 632014, Tamil Nadu, India
* Correspondence: prayline.c@vit.ac.in (P.R.C.); ssivanantham@vit.ac.in (S.S.)

Abstract: Preserving the visual quality is a major constraint for any algorithm in image and video processing applications. AVC and HEVC are the extensively used video coding standards for various video processing applications in recent days. These coding standards use filters to preserve the visual quality of the processed video. To retain the quality of the reconstructed video, AVC uses an in-loop filter, called the deblocking filter, while HEVC uses two in-loop filters, the sampling adaptive offset filter and the deblocking filter. These filters are implemented in hardware by adopting various optimization techniques such as reduction of power utilization, reduction of algorithm complexity, and consuming lesser area. The quality of the reconstructed video should not be impacted by these optimization measures. For the HEVC/H.265 coding standard, a parallel edge deblocking filter architecture is designed, and the effectiveness of the parallel edge filter architecture is evaluated using various quantization values for various resolutions. The quality of the parallel edge filter architecture is on par with the HEVC reference model.

Keywords: deblocking filter; quality assessment; parallel edge filter; HEVC/H.265; sample adaptive offset filter

Citation: Christopher, P.R.; Sathasivam, S. Quality Assessment of Dual-Parallel Edge Deblocking Filter Architecture for HEVC/H.265. *Appl. Sci.* **2022**, *12*, 12952. https://doi.org/10.3390/app122412952

Academic Editors: Zbigniew Lubniewski, Tadeus Uhl and Przemysław Falkowski-Gilski

Received: 14 October 2022
Accepted: 12 December 2022
Published: 16 December 2022

1. Introduction

Video compression is ineluctable in recent days owing to the rapid advancements in digital electronics. Several video coding techniques are widely used to compress video data, save storage space, and reduce channel bandwidth during transmission. HEVC is the last released video coding standard developed by the Joint Collaborative Team on Video Coding (JCT-VC) which has been used for more than half a decade in many multimedia applications. This video coding standard splits the raw input video frames into rectangular blocks which are transformed to the frequency domain and then predicted based on the former decoded video either by motion compensated prediction, interprediction, or intraprediction [1]. Due to this block-based transform coding followed by coarse quantization, there may not be uniformity in the intensity of the pixel transition within the two adjacent blocks. This nonuniformity in the transition of the intensity of the pixel values generates a visible discontinuity in the reconstructed video and, thus, degrades the quality of the decoded video.

Nonsmooth block boundaries commonly observed in earlier video coding standards performed at low and medium bit rates, such as visible block boundaries, color biases, and blurring effects [2], still exist in H.264 and H.265. Nonsmooth block boundaries are known as the blocking effect which is one of the most perceivable and objectionable artifacts of block-based compression methods [3,4]. Figure 1 shows the existence of blocking artifacts at the block boundary. It shows the variation of the pixel intensities between the edges of two 4 × 4 blocks. DBFs are used to reduce these blocking artifacts. Figure 2 shows the elimination of the blocking artifacts by smoothening the intensities of the pixels at the edges of a block. The computational complexity of the DBF algorithm is one-third of the H.264 video decoder [3] and one-fifth of the H.265 video decoder [5].

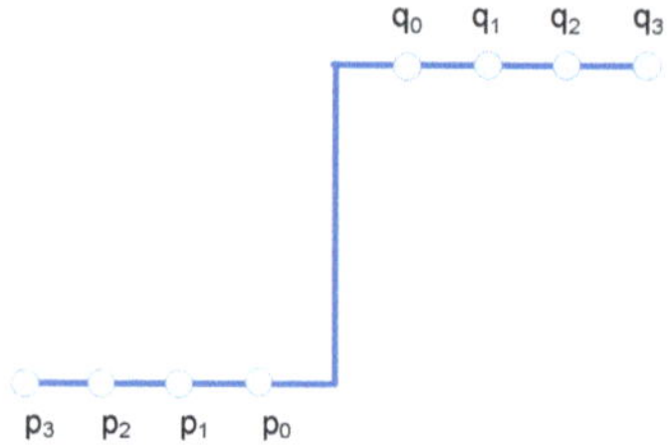

Figure 1. Variation of pixel intensities between block boundaries.

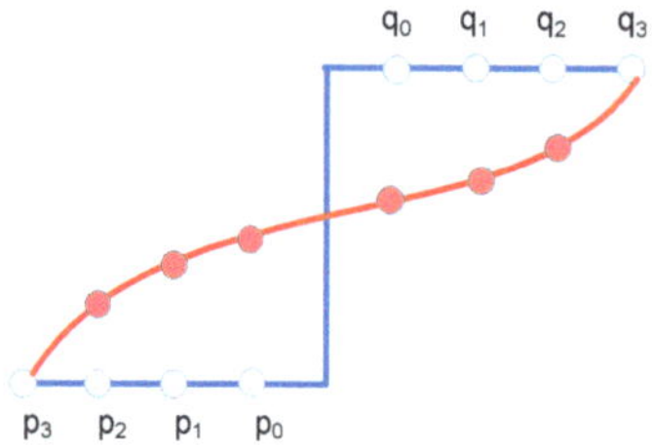

Figure 2. Smoothing of pixel intensities between block boundaries.

The discontinuities are perceivable for the human visual system (HVS) as blocking effects in the region with lower activity in the video frame [6]. Preserving the same visual quality as the original visual scene is a significant challenge. The AVC and HEVC coding standards utilize lossy compression algorithms such as DCT/DWT, and, hence, the originality of the pixel intensities are lost when the video data are reconstructed. However, measures are taken to upgrade the quality of the reconstructed video in the codec. AVC has an in-loop filter called the deblocking filter (DBF) to reduce visible discontinuities. HEVC standard reduces this visible discontinuity in the main profile by the use of two in-loop filters applied to the reconstructed video in succession. These two in-loop filters are the DBF and the sample adaptive offset (SAO) filter. Although these filters are employed to improve the reconstructed video's quality, these enhancement methods do not produce a perfect replica of the original image.

The DBF and SAO filters used in HEVC are optimized by modifying the filtering algorithms or by implementing the algorithms efficiently in hardware concerning cost and energy saving. Even though these optimization techniques enhance the quality of the video, we need to measure the quality to compare with different optimization techniques.

2. Related Work on Deblocking Filters

The deblocking filter and the sample adaptive offset (SAO) filter are the two filters used within the codec in the HEVC coding standard. To enhance the visual quality of the reconstructed frames, these filters are applied in two phases, the first stage applying the deblocking filter and the second stage applying the SAO filter. Blocking artifacts are eliminated by the deblocking filter, and the SAO filter enhances the visual quality by adding the offset values to the first-stage filtered pixel samples [7]. Edge offset or band offset are two possible offset values. The two filters for the HEVC coding standard are implemented in hardware adopting several architectural optimization techniques.

The SAO filter and deblocking filter are combined and implemented in hardware, or these filters are implemented separately as deblocking filter [2,8–14] and SAO filter [15]. Parallel and pipeline-based architectures are utilized to implement the area and throughput optimized deblocking filter. Few deblocking filter architectures use novel ordering for the filtering process to improve the performance. Several filter architectures were implemented by [8,16–23] to realize the H.265 deblocking filter. The deblocking filter architecture of H.264 is more complex compared to the deblocking filter architecture of H.265/HEVC deblocking filter [24]. SAO and deblocking filters were combined and are implemented by [17,23,25–27]. In [22,28], a graphics processing unit (GPU) was used to implement the in-loop filter using parallelism. A multicore coprocessor was used to implement the HEVC in-loop filtering in [19]. Convolutional neural network (CNN) was used by [29,30] to create an in-loop deblocking filter with coding unit categorization.

3. Quality Assessment Metrics

The HEVC standard is known to be advantageous for higher video resolutions such as HD and UHD videos with lower bit rates. The features of HEVC improve the compression ratio by 50% with an increase in the complexity by 150% compared to its former video coding standard AVC [31]. Research is ongoing to reduce the overall complexity of the coding standard without affecting the compression ratio. Overall complexity can be reduced if there is a reduction in the complexity of the various modules used in these coding standards, but reduction in computational complexity may deteriorate the quality of the encoded video stream.

Mean square error (*MSE*) and *PSNR* are the metrics used to evaluate the objective perceptional video quality [32,33]. Structural similarity (SSIM) is also employed to evaluate the video quality by [34,35], but none of the metrics correlate precisely with the perceptional quality of the HVS [36]. However, SSIM provides better results with respect to the perceptional quality of the HVS using the assessment of three components, viz., perceptional impact of changes in the luminance, contrast, and structure [37]. Despite the complexity of measuring SSIM, due to the assessment of three different components, it is more reliable compared to other measures [38]. Several block-edge impairment metrics were proposed and a generalized block-edge impairment metric (DBIM) was proposed by [39], which shows the difference in the perceptual quality. Equations (1)–(3) show the quality assessment of the reconstructed video using $PSNR$, MSE, and $SSIM$, respectively.

$$MSE(i,j) = \frac{(\sum_{i=1}^{M}\sum_{j=1}^{N}[f(i,j) - F(i,j)]^2)}{M.N} \tag{1}$$

$$PSNR = 20log_{10}\frac{255}{\sqrt{MSE}} \tag{2}$$

where $f(i,j)$ is the pixel value at ith row and jth column of the original video frame, $F(i,j)$ is the pixel value at ith row and jth column of the reconstructed/modified video frame, and M and N represent the width and the height of the video frame. The value of *PSNR* ranges from 30 dB to 40 dB as the quality of the modified video ranges from medium to high, respectively.

$$SSIM = \frac{(2\bar{x}\bar{y} + C_1)(2\sigma_{xy} + C_2)}{[\bar{x}^2 + \bar{y}^2 + C_1](\sigma_x^2 + \sigma_y^2 + C_2)} \tag{3}$$

where $\bar{x}$, $\bar{y}$ are the mean of x and y, respectively; σ_x ,σ_y, and σ_{xy} are the variance of x, the variance of y, and the covariance of x and y, respectively; and C_1 and C_2 are constants. $SSIM$ values range between -1 to 1, and the quality is said to be the best when the value is 1 [38].

Among the various metrics for video quality assessment, *PSNR* is the most desired quality assessment metric owing to its lucidity. Though *PSNR* is used to check the quality of the video to a great extent, it does not provide the actual perceptual quality as perceived by the HVS.

4. Parallel Edge Deblocking Filter Architecture (PEDBF)

The dual-parallel edge DBF architecture [40], employing five pipeline stages (V-DPEDBF) used to assess the quality, is shown in Figure 3. The (i) control unit, (ii) boundary Strength (BS) calculation unit, and (iii) filter unit are the three main modules in this architecture. The various operations of the filter architecture are administered by the control unit. An enable signal is used as a primary input from the external world to enable the control unit. The control unit oversees and coordinates a number of processes, including data fetch from the external storage, data fetch from the memory within PEDBF, data write to the PEDBF memory, data write to the external storage device, activation of the boundary strength calculator, and activation of the filter unit. The BS unit determines the values of the BS, which range from 0 to 2. The filter unit performs the filtering procedure in accordance with the computed BS value.

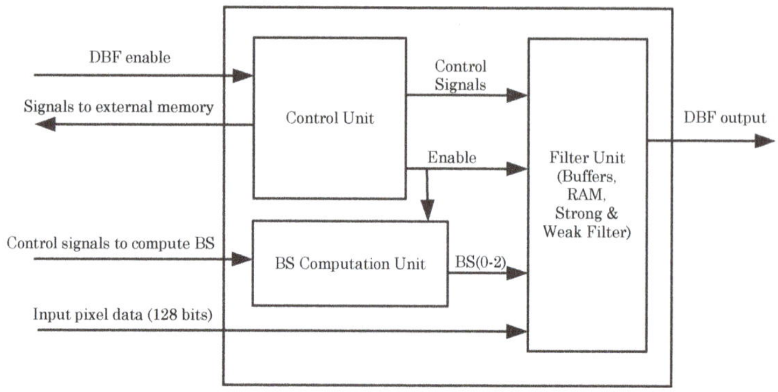

Figure 3. Architecture of V-DPEDBF for HEVC [40].

4.1. Control Unit

The control unit is activated by the control signal, which is one of the primary inputs to the filter architecture. The control unit of the HEVC DBF has a finite state machine to administer all the operations to handle the deblocking filtering process. When the control signal is not active, all the filtering process is deactivated and, thus, the modules are turned off. Hence, the DBF architecture consumes less power. When the enable signal is active, the control unit triggers the state machine and activates the filtering process in five stages. Five stages—memory read, parameter computation, filter determination, filtering, and memory write—are managed by the state machine during the filtering process. The pixel data are fetched from the external memory, which is outside the filter architecture, after the finite state machine is enabled. Then, 4 × 4 blocks (128 bits) of pixel data are fetched for each clock cycle from the external memory. Figure 4a depicts the sequencing of data fetch from the external memory for a largest coding unit (LCU), which is a 64 × 64 block, whereas Figure 4b,c depict the sequencing of data fetch from the external storage for a 16 × 16 block. The state machine activates the filter unit and the BS calculation unit by generating control signals once four 4 × 4 blocks of pixel data are in place.

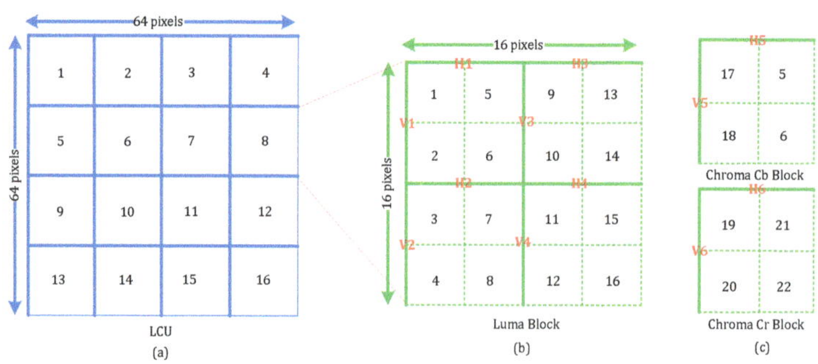

Figure 4. (**a**) Sequence of read for an LCU; (**b**,**c**) sequence of read for a 16 × 16 block.

The 16 × 16 luma block's vertical edges V1 and V2 are filtered in parallel in accordance with the determined BS value. Blocks 1 through 8 of the filtered pixel data are then transposed and stored in the internal memory. The filter unit is triggered once more to filter the edges V3 and V4 once the twelfth block of data is accessible. Blocks 9 through 16 of the vertically filtered data are eventually transposed and stored within the PEDBF architecture. After vertical filtering, the horizontal filter is applied by reading the pixel data that are stored within the PEDBF. The internal RAM receives the appropriate signals from the control unit to perform the required operation. H1 and H2 horizontal edges are parallel-filtered and then the filtered data are transposed. The frame memory, which is located external to the architecture, is subsequently written with the filtered data. The horizontal edges H3 and H4 are parallel filtered in a similar manner. After that, they are written to the external frame memory after being transposed. The Chroma Cb and Cr blocks filter the vertical edges (V5 and V6) first, followed by the horizontal edges (H5 and H6) using the same process. The filtered pixel data are subsequently saved in the external storage as 4 × 4 blocks, i.e., 128 bits for each clock cycle.

4.2. Boundary Strength Computation Unit

The boundary strength computation unit computes the BS value, as shown in Figure 5. The BS value range for the HEVC coding standard is between 0 and 2. The BS processing unit receives control signals regarding the pixel block received from the external buffer, such as whether the received pixel block is the left, right, top, or bottom edge of the frame, if it is inter/intracoded, and if its transform coefficients are not zero. The BS value is 0 if the block of pixel data read from the external buffer is a part of the left or top edge of a frame. In addition, the BS value is 0 if the data are not a part of the left or top edge of a frame, the two neighboring 8 × 8 blocks are not intracoded, if the block has nonzero transform coefficients, and if the motion vector is less than 4. The BS value is 1 if the transform coefficients of the block are nonzero and the adjacent blocks are not intracoded. The BS value is also 1 if the adjacent blocks are not intracoded and if the block does not have nonzero transform coefficients, and if the motion vector is higher than or equal to 4. The BS value is 2 if any of the above conditions are not met. DBF is triggered based on the calculated BS value. If the generated value of BS is 0, no filtering is performed; if the determined value is 1, a weak/normal filter is used; and if the stipulated value is 2, a strong filter is used.

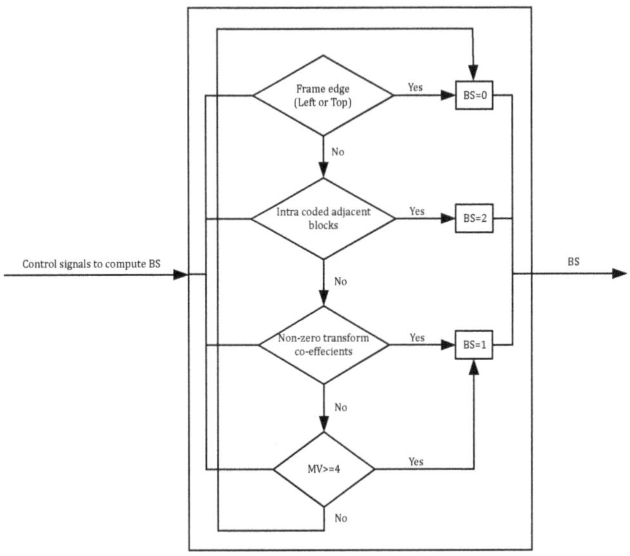

Figure 5. Boundary strength computation unit.

4.3. Filter Module

When the pixel data are ready to carry out the filtering operation, the control unit turns on the filter unit. This unit is the sophisticated computational unit. The architecture of the filter unit is shown in Figure 6. It has a parameter computation unit, buffers to store the pixel data, a filter decision block, internal memories, a strong filter, and a weak filter.

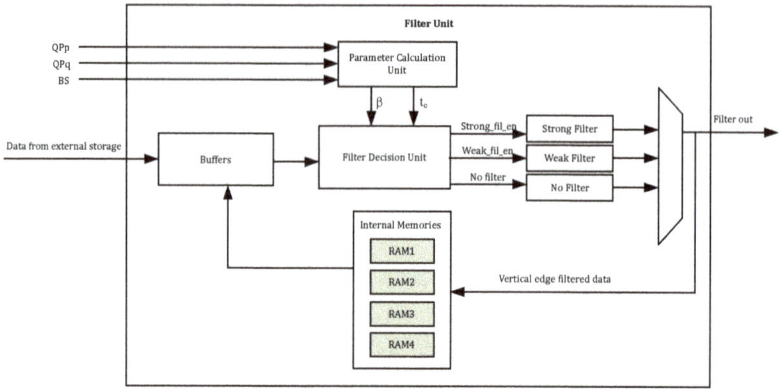

Figure 6. Filter unit of dual-edge deblocking filter architecture.

4.3.1. Parameter Calculation Unit

This unit calculates the filtering parameters such as β and t_c according to Tables 8–12 of [41]. Based on the quantization parameter values and the BS of the neighboring P block and Q block, referred to QP_p and QP_q, respectively, these filtering parameters are calculated. The LUT used to implement the parameter calculation unit has the outputs t_c and β. These output values are relative to the inputs BS, QP_p and QP_q.

4.3.2. Buffers

The filter unit of the dual-edge deblocking filter architecture uses eight buffers, as shown in Figure 7, each of which can store a 4 × 4 block of pixels (128 bits). These buffers are initialized with all zeroes. Before the control unit begins the filtering process, the block of pixels 1–4 indicated in Figure 4b from the external memory are stored into the corresponding buffers Q1 BUF, Q2 BUF, Q3 BUF, and Q4 BUF. The filtering operation is carried out along the vertical edges V1 and V2; at the same time, the block of pixels 5–8 is sent to the buffers P1 BUF, P2 BUF, P3 BUF, and P4 BUF, respectively. Figure 8 indicates the relative buffer for both the luma and chroma blocks, along with the outline of each 4 × 4 pixel block. The filtered data are saved to the internal RAM after the edges V1 and V2 have been filtered. Similar techniques are used to filter V3 and V4 vertical edges. For horizontal filtering, the pixel data kept in the internal memory were subsequently transferred to these storage units. The same process is utilized for horizontal filtering as for vertical filtering.

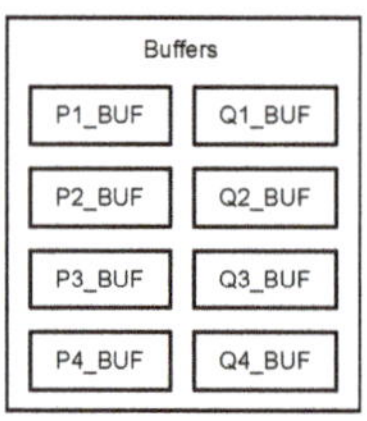

Figure 7. Buffers to store the pixel block.

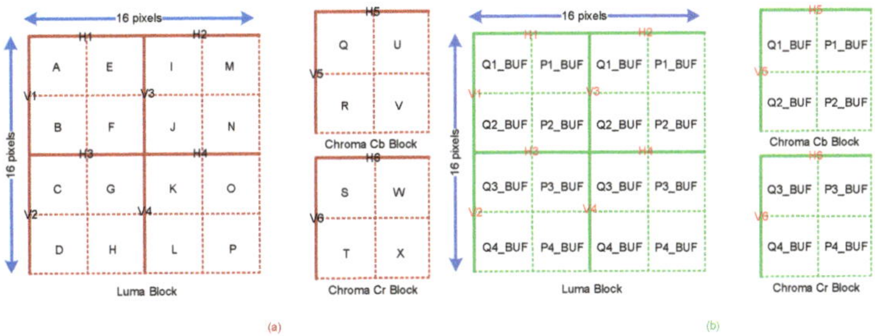

Figure 8. Mapping of pixel blocks to buffers: (**a**) pixel blocks; (**b**) buffers.

4.3.3. Filter Decision Unit

Based on the values of the parameters β and t_c, the strength of the filtering procedure to be used for a 4 × 4 pixel block is determined. The pixel threshold values for the two neighboring blocks are decided by the filter decision unit.

4.3.4. Internal Memories

The memories utilized within the filter architecture employ four dual-port RAM to store the pixel data filtered vertically. A 4 × 4 block of pixel data can be stored in each of the four segments of the 64-byte RAM (16 bytes or 128 bits). Figures 9 and 10 depict the transfer of the pixel data for the luma and the chroma Cb, Cr blocks from the RAM to the buffers. The amount of external memory access cycles is reduced by this method of data storage. It also avoids the utilization of transpose buffers. The first two memory regions are used exclusively during the chroma block filtering procedure, leaving the other regions unused.

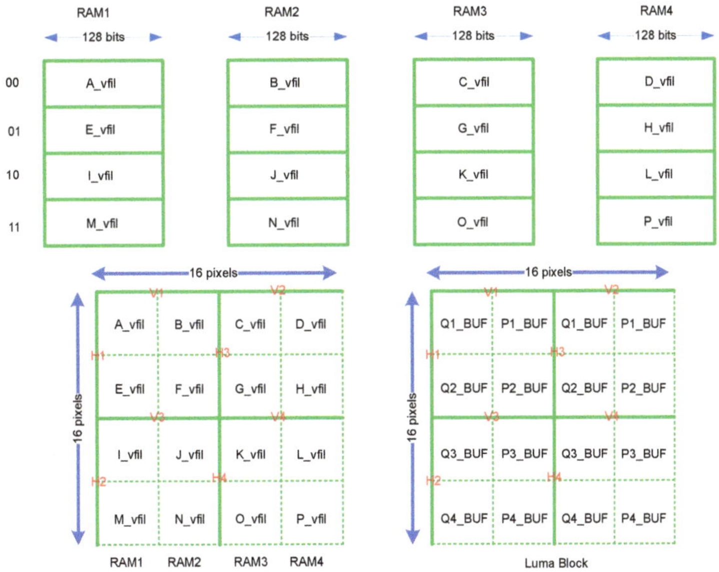

Figure 9. Mapping of internal memory to buffer before horizontal filtering for luma block.

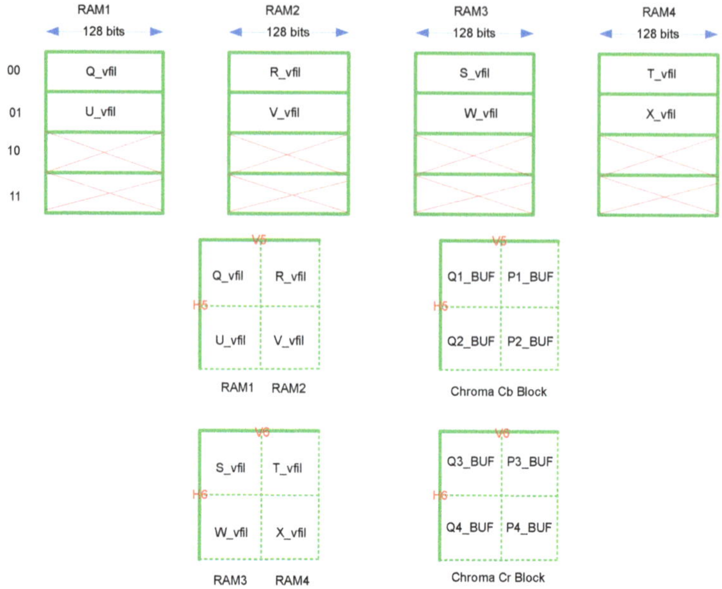

Figure 10. Mapping of internal memory to buffer before horizontal filtering for chroma block.

4.3.5. Filter Modules

The architecture includes filter modules that make use of both strong and weak/normal filters. Any of these filters are activated according to the filtering decisions determined by the filter decision unit. Therefore, any one of the weak filter or the strong filter is activated to execute the filtering process. The filtering process is omitted if the filtering decision unit determines for no filtering. The weak or strong filtering is executed according to the filtering equations specified in [41]. The similarities in these equations allow for the creation and implementation of a resource-sharing architecture for the filter module, which optimizes the area. The data from the horizontally filtered pixel block are then saved in the external memory, while the data from the vertically filtered pixel block are stored in the internal memory.

4.4. Resource Sharing Architecture

Depending on the deblocking filtering technique used in [9], which follows the HEVC standard, the weak and strong filter units are implemented. The filter architecture is constructed in such a way that it shares the common resources that are used in these equations by taking advantage of the similarity in the filtering equations. This technique thenceforward brings down the area and the power utilization. The pixels of the two neighboring blocks are updated for the strong filter using the third parameter of the clip3 function, which is an equation involving two or more of the pixel values of the adjacent P and Q blocks. Adders and shifters are utilized to implement these equations in the hardware. It is identified that $p_0 + q_0$ is employed in all the expressions and $p_0 + q_0 + 2$ is used in most of the expressions. In addition, $p_1 + 2$ and $q_1 + 2$ are used twice. Hence, to add $p_0 + q_0$, an eight-bit binary adder is employed and the output is fanned out to all the equations. This adder's output is also used as an input for another binary adder, whose other input is 2, producing the result $p_0 + q_0 + 2$, which is then utilized in the appropriate expressions. The same mechanism is used to implement the addition operations $p_1 + 2$ and $q_1 + 2$, and the output of these adders is reused wherever necessary. As a result, the shared resources are distributed over several equations, and an area-optimized filter is used. Figure 11 depicts the resource sharing architecture used for the third argument of the clip3 function for the strong filter.

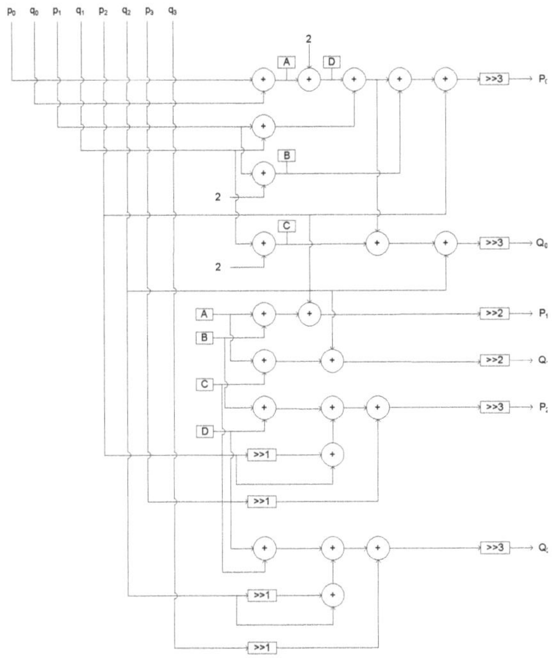

Figure 11. Resource sharing architecture of strong filter—partial view.

5. Quality of PEDBF for HEVC/H.265

The optimization techniques employed in the hardware architectures of the DBF filter should not deteriorate the quality of the reconstructed video for any reason. Degradation in the quality of the filtered video will lead to poor performance and coding inefficiency. The DBF algorithm is highly adaptive and data-dependent, as the current edge to be filtered depends on the previously filtered pixel blocks. Since different filter ordering is followed, these changes in the ordering should not affect the quality of the filtered video. Hence, the quality assessment of the dual-edge DBF for H.265 coding standard is measured using *PSNR*.

Quality Assessment Procedure

To measure the quality of the V-DPEDBF architecture, the following steps are executed sequentially.

1. Original raw video is given as input to the HEVC Test Model to obtain the reconstructed video by disabling the in-loop filter
2. Reconstructed video data from the HM is given as input to the dual-parallel edge deblocking filter (PEDBF) architecture.
3. The input video to PEDBF architecture is split into luminance (Y), chrominance Cb (U) and, chrominance Cr (V) components.
4. Each video component is segregated into image frames and stored into frame buffers.
5. Each image frame of the Y, U, and V components is fetched from the frame buffers.
6. Each image frame is split into uniform pixel blocks of size 16 × 16.
7. Each 16 × 16 block is again split into four 8 × 8 blocks.
8. DBF is applied at the boundary of two 8 × 8 blocks simultaneously for vertical filtering.
9. The vertically filtered output data are transposed and moved into pixel buffers to perform horizontal edge filtering.

10. Horizontal edge filtering is performed after the vertical edge filtering is performed for the entire 16 × 16 block.
11. The horizontally filtered output is transposed and stored into the frame buffer.
12. The filtered video of PEDBF is reconstructed from the frame buffer.
13. The *PSNR* of the filtered video is computed.
14. Original raw video is given as input to the HEVC Test Model to obtain the reconstructed video by enabling the in-loop filter.
15. The *PSNR* of the filtered video from the HEVC reference software is computed and the results are compared.

The flow diagram to perform the quality assessment is shown in Figure 12.

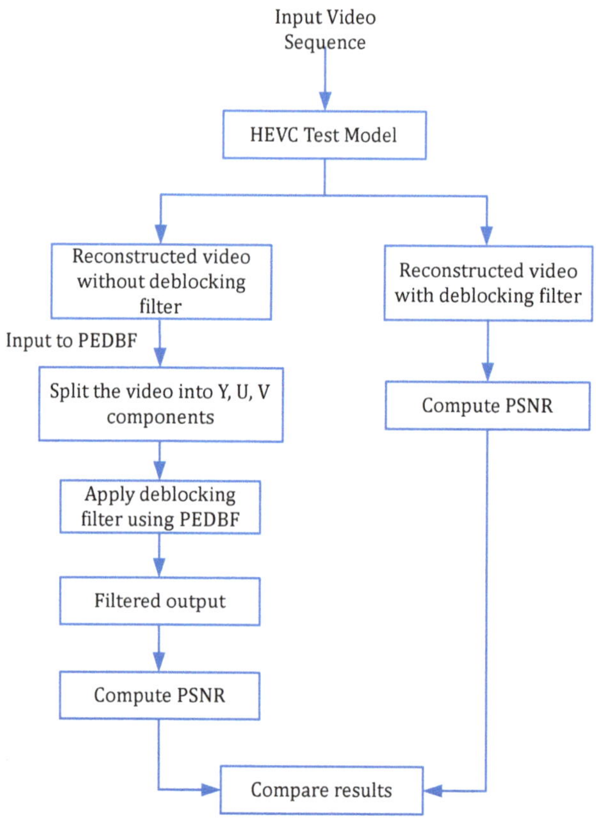

Figure 12. Flow diagram to perform quality assessment.

6. Filter Selection

The filter architecture uses four resource sharing edge filters that operate simultaneously to filter the edges of two 8 × 8 blocks of pixels. Each 8 × 8 block's filtering decision is made in accordance with the β, t_c parameters and in accordance with the threshold values of the two adjacent pixel blocks. The following criteria are employed for selecting the type of filter:

1. If the BS value is zero, then no filtering is performed.
2. If the BS value is nonzero, then Equations (4)–(8) are computed; and if Equation (8) is not satisfied, then no filtering is performed.

$$dp0 = abs(p_{2,0} - 2p_{1,0} + p_{0,0}) \tag{4}$$

$$dp3 = abs(p_{2,3} - 2p_{1,3} + p_{0,3}) \tag{5}$$

$$dq0 = abs(q_{2,0} - 2q_{1,0} + q_{0,0}) \tag{6}$$

$$dq3 = abs(q_{2,3} - 2q_{1,3} + q_{0,3}) \tag{7}$$

$$dp0 + dp3 + dq0 + dq3 < \beta \tag{8}$$

3. If Equation (8) is satisfied then Equations (9)–(16) are computed to decide a strong or weak filter. If Equations (11)–(16) are satisfied, then a strong filter is used.

$$dpq0 = dp0 + dq0 \tag{9}$$

$$dpq3 = dp3 + dq3 \tag{10}$$

$$dpq0 < \beta/8 \tag{11}$$

$$dpq3 < \beta/8 \tag{12}$$

$$abs(p_{3,0} - p_{0,0}) + abs(q_{0,0} - q_{3,0}) < \beta/8 \tag{13}$$

$$abs(p_{3,3} - p_{0,3}) + abs(q_{0,3} - q_{3,3}) < \beta/8 \tag{14}$$

$$abs(p_{0,0} - q_{0,0}) < 2.5t_c \tag{15}$$

$$abs(p_{0,3} - q_{0,3}) < 2.5t_c \tag{16}$$

4. If any one of Equations (11)–(16) are not satisfied, then Equations (17)–(20) are computed and if Equation (20) is satisfied, a weak filter is used.

$$dp0 + dp3 < \beta/16 \tag{17}$$

$$dq0 + dq3 < \beta/16 \tag{18}$$

$$\delta_0 = 9(q_{0,0} - p_{0,0}) - 3(q_{0,1} - p_{0,1}) \gg 4 \tag{19}$$

$$abs(\delta_0) < 10t_c \tag{20}$$

5. If Equation (20) is not satisfied then the filtering process is skipped (no filter).

7. Filtering Operation

The deblocking filtering operations are performed on the edges of the luma and the chroma Cb and Cr pixel blocks based on the computed boundary strength value. Boundary strength is computed based on the coding information given by the codec.

7.1. Luma Block

A strong filter is used, and up to three pixels on either side of the block edges are adjusted as in Equation (21) to Equation (26) for a luma block if the computed BS value is 2.

$$p_0 = Clip3(p_0 - 2t_c, p_0 + 2t_c, (p_2 + 2p_1 + 2p_0 + 2q_0 + q_1 + 4) \gg 3) \tag{21}$$

$$p_1 = Clip3(p_1 - 2t_c, p_1 + 2t_c, (p_2 + p_1 + p_0 + q_0 + 2) \gg 2) \tag{22}$$

$$p_2 = Clip3(p_2 - 2t_c, p_2 + 2t_c, (2p_3 + 3p_2 + p_1 + p_0 + q_0 + 4) \gg 3) \tag{23}$$

$$q_0 = Clip3(q_0 - 2t_c, q_0 + 2t_c, (p_1 + 2p_0 + 2q_0 + 2q_1 + q_2 + 4) \gg 3) \tag{24}$$

$$q_1 = Clip3(q_1 - 2t_c,\ q_1 + 2t_c\ ,\ (q_2 + q_1 + q_0 + p_0 + 2) \gg 2) \tag{25}$$

$$q_2 = Clip3(q_2 - 2t_c,\ q_2 + 2t_c\ ,\ (2q_3 + 3q_2 + q_1 + q_0 + p_0 + 4) \gg 3) \tag{26}$$

If BS = 1, then a weak filter is applied and the pixels on either side of the block edges are modified as in Equations (27)–(32).

$$\Delta = 9(q_0 - p_0) - 3(q_1 - p_1) + 8 \gg 4 \tag{27}$$

If Abs(Δ)<10t_c then

$$\Delta = Clip3(-t_c,\ t_c, \Delta) \tag{28}$$

$$p_0 = Clip1_Y(p_0 + \Delta) \tag{29}$$

$$q_0 = Clip1_Y(q_0 - \Delta) \tag{30}$$

when dEp = 1, then

$$p_1 = Clip1_Y(p_1 + \Delta_p) \tag{31}$$

where $\Delta_p = Clip3(-\frac{t_c}{2}, \frac{t_c}{2}, (((p_2 + p_0 + 1) \gg 1) - p_1 +\ \Delta) \gg 1)$.

When dEq = 1, then

$$q_1 = Clip1_Y(q_1 + \Delta_q) \tag{32}$$

where $\Delta_q = Clip3(-\frac{t_c}{2}, \frac{t_c}{2}, (((q_2 + q_0 + 1) \gg 1) - q_1 +\ \Delta) \gg 1)$

7.2. Chroma Block—Cb and Cr

Chroma blocks are filtered only if BS = 2, and no filtering is performed when BS = 1 or 0. When BS = 2, the strong filter is applied, and the pixels p_0 and q_0 alone on either sides of the block edges are modified as in Equations (33)–(35).

$$\Delta_c = Clip3(-t_c, t_c, (((p_0 - q_0) \ll 2) + p_1 - q_1 + 4) \gg 3) \tag{33}$$

$$p_0 = ClipY(p_0 + \Delta_c) \tag{34}$$

$$q_0 = ClipY(q_0 - \Delta_c) \tag{35}$$

8. Results and Discussion

The quality assessment of the five-stage pipelined dual-edge deblocking filter architecture implemented for the HEVC standard [40] is performed using Matlab. Different test video sequences are used to obtain the *PSNR* values of the PEDBF architecture using different QP values. The typical QP value is 32. Hence, the quality assessment is performed with the typical QP value (32), the QP value lesser than the typical value (27), and QP value greater than the typical QP (37). Figure 13 shows an image frame of the original video sequence. This original/raw video sequence is given as input to the HM software to obtain two different types of encoded video. One type of encoded video is obtained from the HM software with the deblocking filter turned off (without DBF), and another type of encoded video is obtained from the HM software with the deblocking filter turned on (with DBF). Figures 14 and 15 show a sample frame of the output video sequence from the HM software without DBF and with DBF, respectively. The encoded output video from the HM software without DBF is used to perform the quality analysis of the PEDBF architecture. This video encoded without DBF is given as the input video sequence to the PEDBF implemented in Matlab to undergo deblocking filtering. Figure 16 shows a sample frame of the output video obtained from the PEDBF. The *PSNR* value of the output video obtained from the PEBDF is computed.

Figure 13. Frame of the original input video.

Figure 14. Frame of output video from HM with DBF disabled.

Figure 15. Frame of output video from HM with DBF enabled.

Figure 16. Frame of output video from dual PEDBF.

The quality of this architecture is assessed by comparing the *PSNR* value of the output video of the PEDBF architecture and the *PSNR* value of the video obtained using the HM software with DBF turned on for different QP values. The results for QCIF video sequences are tabulated in Table 1 and the comparison graph is shown in Figure 17 with QP = 32. The results for CIF video sequences are tabulated in Table 2 and the comparison graph is shown in Figure 18 with QP = 32. It is noted that the implemented architecture shows a slight improvement in quality concerning the *PSNR* metrics. The execution time for QCIF and CIF video sequences are tabulated in Tables 3 and 4, respectively, and the comparison graphs are shown in Figures 19 and 20. It is seen that, as the two edges are filtered in parallel, the filtering time is reduced by 50%, and thus the throughput of the architecture is improved by 50% with the increase in quality.

Table 1. *PSNR* of QCIF video sequences with QP = 32.

Video Sequence	HM (dB)	PEDBF (dB)
Akiyo	37.3833	37.59186
Coastguard	34.4138	36.40767
Foreman	35.4274	36.58643
Mobile Calendar	32.2366	32.85071
Hall	36.4334	37.235
Carphone	36.1972	36.4323
Miss America	38.8972	39.12775

Table 2. *PSNR* of CIF video sequences with QP = 32.

Video Sequence	HM (dB)	PEDBF (dB)
Akiyo	38.307737	39.2812
Coastguard	35.03723	34.9852
Foreman	35.8951	35.99
Mobile Calendar	33.04411	33.0432
Hall	36.93208	37.4513
Tempete	34.09227	34.224

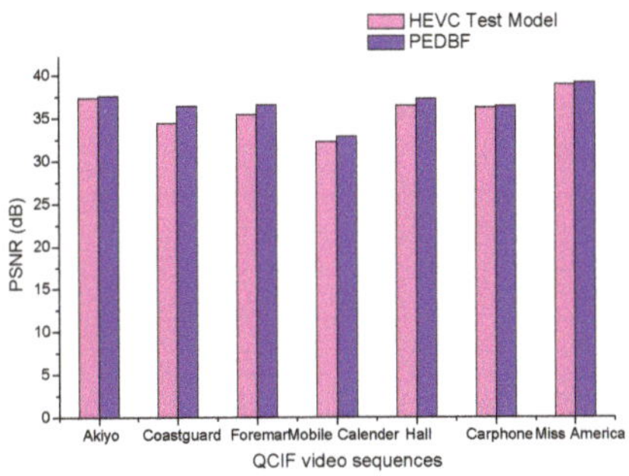

Figure 17. Comparison of *PSNR* with HM for QCIF video sequences using QP = 32.

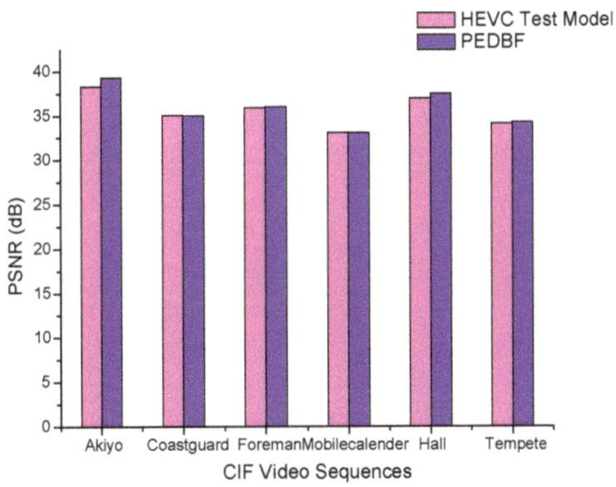

Figure 18. Comparison of *PSNR* with HM for CIF video sequences using QP = 32.

Table 3. Processing time of QCIF video sequences with QP = 32.

Video Sequence	HM (Sec)	PEDBF (Sec)
Akiyo	64.678	39.4553
Coastguard	85.103	44.81924
Foreman	82.368	45.6632
Mobile Calendar	85.673	43.21693
Hall	64.064	37.28175
Carphone	73.358	43.57872
Miss America	23.811	13.27997

Table 4. Processing time of CIF video sequences with QP = 32.

Video Sequence	HM (Sec)	PEDBF (Sec)
Akiyo	264.989	165.49741
Coastguard	376.373	196.2231
Foreman	278.388	176.485
Mobile Calendar	312.898	193.3375
Hall	219.416	112.2503
Tempete	229.776	152.46557

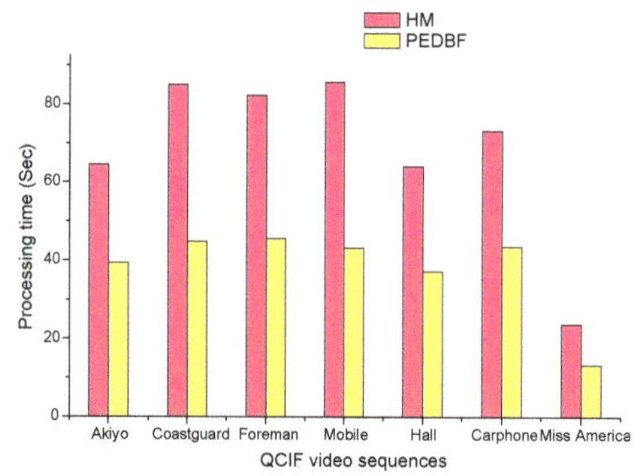

Figure 19. Comparison of processing time with HM for QCIF video sequences using QP = 32.

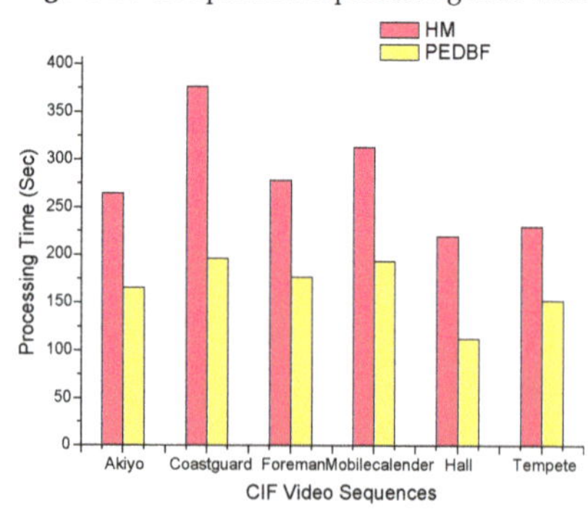

Figure 20. Comparison of processing time with HM for CIF video sequences using QP = 32.

Tables 5 and 6 show the results of QCIF and CIF video sequences, respectively, filtered with QP = 37, and the comparison graphs are shown in Figures 21 and 22. The execution time for QCIF and CIF video sequences are tabulated in Tables 7 and 8, and the comparison graphs are shown in Figures 23 and 24.

Table 5. *PSNR* of QCIF video sequences with QP = 37.

Video Sequence	HM (dB)	PEDBF (dB)
Akiyo	34.0182	34.893362
Coastguard	31.1392	33.457211
Foreman	32.1997	33.862149
Mobile Calendar	28.3189	29.568311
Hall	32.971	34.430916
Carphone	32.9865	34.303814
Miss America	36.4185	37.267113

Table 6. *PSNR* of CIF video sequences with QP = 37.

Video Sequence	HM (dB)	PEDBF (dB)
Akiyo	36.2985	36.716387
Coastguard	31.8043	33.970701
Foreman	33.0051	34.390624
Mobile Calendar	29.2222	30.454212
Hall	34.5437	35.553274
Tempete	30.6828	32.131876

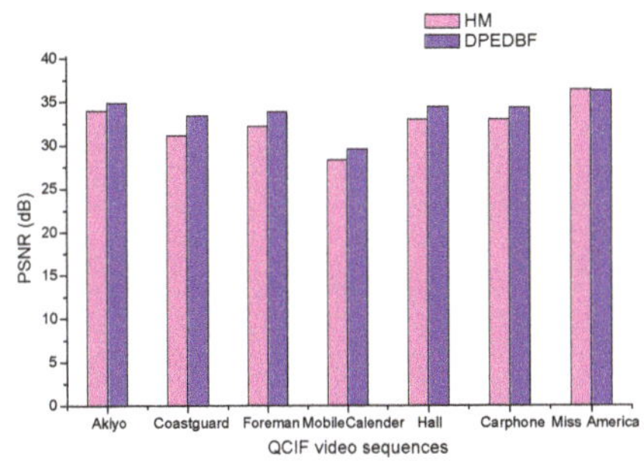

Figure 21. Comparison of *PSNR* with HM for QCIF video sequences using QP = 37.

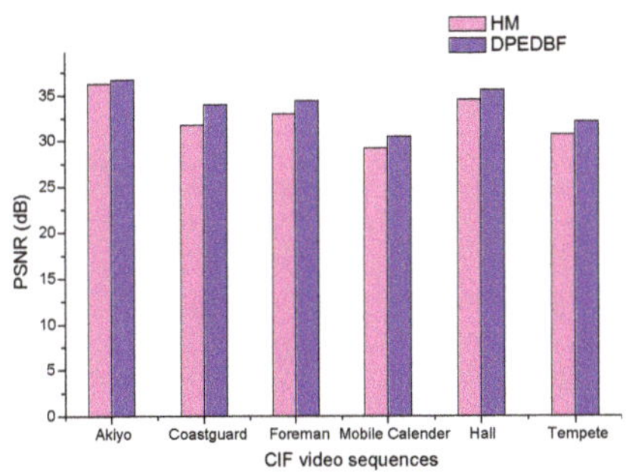

Figure 22. Comparison of *PSNR* with HM for CIF video sequences using QP = 37.

Table 7. Processing time of QCIF video sequences with QP = 37.

Video Sequence	HM (Sec)	PEDBF (Sec)
Akiyo	46.354	23.904816
Coastguard	52.02	27.976072
Foreman	48.861	28.335635
Mobile Calendar	70.905	38.544973
Hall	50.138	28.943229
Carphone	63.938	33.216266
Miss America	23.778	15.006227

Table 8. Processing time of CIF video sequences with QP = 37.

Video Sequence	HM (Sec)	PEDBF (Sec)
Akiyo	361.664	139.577545
Coastguard	323.229	141.569951
Foreman	293.987	161.030763
Mobile Calendar	461.949	225.603688
Hall	280.68	142.62714
Tempete	200.476	112.878238

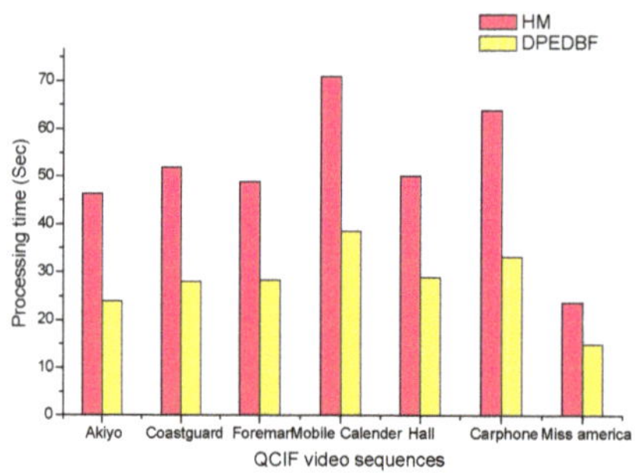

Figure 23. Comparison of processing time with HM for QCIF video sequences using QP = 37.

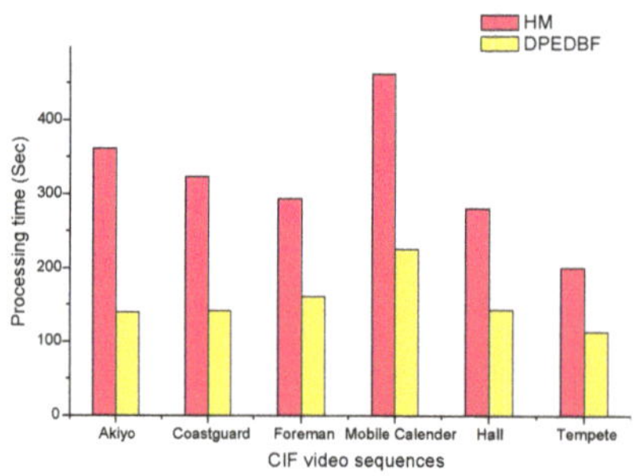

Figure 24. Comparison of processing time with HM for CIF video sequences using QP = 37.

Tables 9 and 10 show the results of QCIF and CIF video sequences, respectively, filtered with QP = 27, and the comparison graphs are shown in Figures 25 and 26. The execution time for QCIF and CIF video sequences are tabulated in Tables 11 and 12, and the comparison graphs are shown in Figures 27 and 28.

Table 9. *PSNR* of QCIF video sequences with QP = 27.

Video Sequence	HM (dB)	PEDBF (dB)
Akiyo	40.8999	40.891213
Coastguard	38.1673	39.927605
Foreman	38.9294	39.793245
Mobile Calendar	36.6292	36.744202
Hall	39.8733	40.164219
Carphone	39.6617	39.917189
Miss America	41.5719	41.766507

Table 10. *PSNR* of CIF video sequences with QP = 27.

Video Sequence	HM (dB)	PEDBF (dB)
Akiyo	42.2861	42.332459
Coastguard	38.6598	40.411503
Foreman	39.2529	40.130872
Mobile Calendar	37.1378	37.29197
Hall	40.1375	40.525791
Tempete	38.0572	38.495716

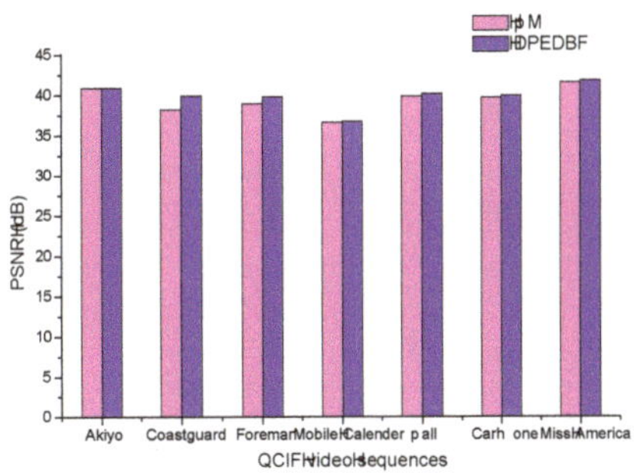

Figure 25. Comparison of *PSNR* with HM for QCIF video sequences using QP = 27.

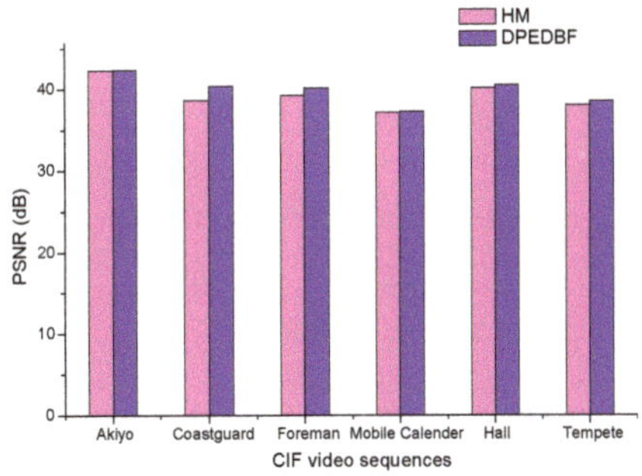

Figure 26. Comparison of *PSNR* with HM for CIF video sequences using QP = 27.

Table 11. Processing time of QCIF video sequences with QP = 27.

Video Sequence	HM (Sec)	PEDBF (Sec)
Akiyo	66.61	32.722607
Coastguard	72.627	39.74846
Foreman	81.681	41.068033
Mobile Calendar	120.716	58.555709
Hall	67.687	35.016408
Carphone	95.34	55.400211
Miss America	24.741	13.174376

Table 12. Processing time of CIF video sequences with QP = 27.

Video Sequence	HM (Sec)	PEDBF (Sec)
Akiyo	205.041	100.410937
Coastguard	277.788	150.710927
Foreman	250.539	126.669947
Mobile Calendar	352.151	208.336164
Hall	228.818	117.526938
Tempete	263.123	156.781587

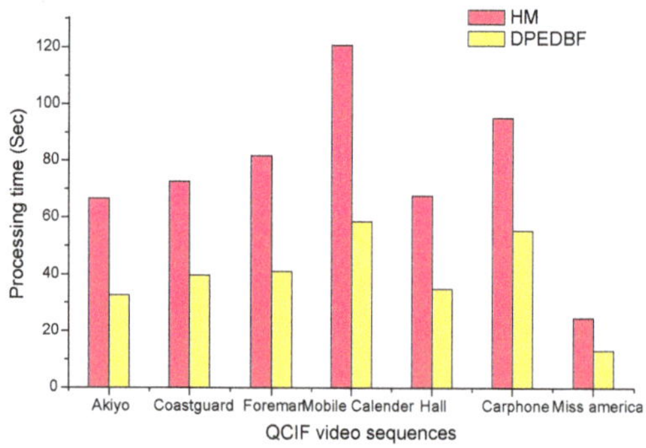

Figure 27. Comparison of processing time with HM for QCIF video sequences using QP = 27.

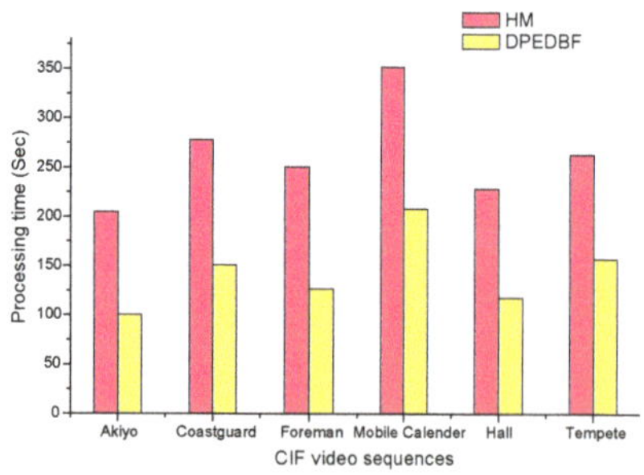

Figure 28. Comparison of processing time with HM for CIF video sequences using QP = 27.

Based on the experiments carried out with different video sequences of different resolutions for different QP values, it is identified that as the QP value increases, the quality decreases and results in lower *PSNR* values. In addition, the execution time decreases as the QP value increases. This behavior is as expected with any video coding algorithm. Hence, the PEDBF architecture complies with the coding standards. The execution time of the PEDBF architecture is reduced to almost half of the execution time of the HM, owing to the fact of filtering two edges in parallel.

9. Conclusions

The quality assessment of any algorithm implemented in hardware is strongly required as the optimization techniques do not affect the quality of the reconstructed video data in any of the processing steps. Among the various quality assessment metrics, we use the metric *PSNR* to assess the quality of the video owing to its simplicity. Raw video data of two different resolutions (QCIF and CIF) are taken and the quality of the filtered video is checked with the quality of the video obtained from the HEVC Test Model. It is noted that the quality of the parallel edge filter architecture does not affect the quality of the reconstructed video, and it shows slight improvements compared to the HEVC Test Model. In one second, ten to twelve separate images can be processed by the human visual system by perceiving each image discretely. One image is held in the visual cortex for around one of fifteen parts in a second. Therefore, as the frame rate is higher, perception of the moving picture will be smooth. Hence, the processing time decreases in the dual-parallel edge DBF, the frame rate will increase, and, thus, the perceptual quality is increased.

Author Contributions: Conceptualization, P.R.C. and S.S.; methodology, P.R.C. and S.S.; validation, P.R.C.; formal analysis, P.R.C.; investigation, P.R.C. and S.S.; resources, P.R.C.; data curation, P.R.C.; writing—original draft preparation, P.R.C.; writing—review and editing, P.R.C. and S.S.; supervision, S.S. All authors have read and agreed to the published version of the manuscript.

Funding: The APC is funded by Vellore Institute of India, Vellore, Tamil Nadu, India.

Institutional Review Board Statement: Not applicable.

Informed Consent Statement: Not applicable.

Data Availability Statement: The reference software for HEVC called HM (HEVC Test Model) is from HM software repository https://hevc.hhi.fraunhofer.de/svn/svn_HEVCSoftware (accessed on 1 October 2022), and the test video sequences used for quality assessment are from the Video Trace Library of Arizona State University http://trace.eas.asu.edu/yuv/index.html (accessed on 1 October 2022) and Video Information Processing Lab of National Chiao Tung University http://vip.cs.nctu.edu.tw/resource_seq.html (accessed on 1 October 2022).

Acknowledgments: The authors would like to acknowledge Vellore Institute of Technology, Vellore, Tamil Nadu, India, for providing all the necessary facilities for the research.

Conflicts of Interest: The authors declare no conflict of interest.

References

1. Wedi, T.; Musmann, H.G. Motion-and aliasing-compensated prediction for hybrid video coding. *IEEE Trans. Circuits Syst. Video Technol.* **2003**, *13*, 577–586. [CrossRef]
2. Zhou, W.; Zhang, J.; Zhou, X.; Liu, Z.; Liu, X. A high-throughput and multi-parallel VLSI architecture for HEVC deblocking filter. *IEEE Trans. Multimed.* **2016**, *18*, 1034–1047. [CrossRef]
3. Pourazad, M.T.; Doutre, C.; Azimi, M.; Nasiopoulos, P. HEVC: The new gold standard for video compression: How does HEVC compare with H. 264/AVC? *IEEE Consum. Electron. Mag.* **2012**, *1*, 36–46. [CrossRef]
4. Singh, J.; Singh, S.; Singh, D.; Uddin, M. Blocking artefact detection in block-based DCT compressed images. *Int. J. Signal Imaging Syst. Eng.* **2011**, *4*, 181–188. [CrossRef]
5. Vanne, J.; Viitanen, M.; Hamalainen, T.D.; Hallapuro, A. Comparative rate-distortion-complexity analysis of HEVC and AVC video codecs. *IEEE Trans. Circuits Syst. Video Technol.* **2012**, *22*, 1885–1898. [CrossRef]
6. Tai, S.C.; Chen, Y.Y.; Sheu, S.F. Deblocking filter for low bit rate MPEG-4 video. *IEEE Trans. Circuits Syst. Video Technol.* **2005**, *15*, 733–741.
7. Wang, Y.; Guo, X.; Lu, Y.; Fan, X.; Zhao, D. GPU-based optimization for sample adaptive offset in HEVC. In Proceedings of the 2016 IEEE International Conference on Image Processing (ICIP), Phoenix, AZ, USA, 25–28 September 2016; pp. 829–833.

8. Li, M.; Zhou, J.; Zhou, D.; Peng, X.; Goto, S. De-blocking Filter Design for HEVC and H. 264/AVC. In *Lecture Notes in Computer Science, Proceedings of the 13th Pacific-Rim Conference on Multimedia, Singapore, Singapore, 4–6 December 2012*; Springer: Berlin/Heidelberg, Germany, 2012; pp. 273–284.
9. Hsu, P.K.; Shen, C.A. The VLSI Architecture of a Highly Efficient Deblocking Filter for HEVC Systems. *IEEE Trans. Circuits Syst. Video Technol.* **2017**, *27*, 1091–1103. [CrossRef]
10. Ye, X.; Ding, D.; Yu, L. A cost-efficient hardware architecture of deblocking filter in HEVC. In Proceedings of the Visual Communications and Image Processing Conference, Valletta, Malta, 7–10 December 2014; pp. 209–212.
11. Bae, J. Register array-based VLSI architecture of H. 265/HEVC loop filter. *IEICE Electron. Express* **2013**, *10*, 20130161. [CrossRef]
12. Tang, G.; Zeng, X.; Fan, Y. An SRAM-free HEVC Deblocking Filter VLSI Architecture for 8K Application. In Proceedings of the 2018 14th IEEE International Conference on Solid-State and Integrated Circuit Technology (ICSICT), Qingdao, China, 31 October–3 November 2018; pp. 1–3. [CrossRef]
13. Peesapati, R.; Das, S.; Baldev, S.; Ahamed, S.R. Design of streaming deblocking filter for HEVC decoder. *IEEE Trans. Consum. Electron.* **2017**, *63*, 1–9. [CrossRef]
14. Jiang, L.; Yang, Q.; Zhu, Y.; Deng, J. A parallel implementation of deblocking filter based on video array architecture for HEVC. In Proceedings of the 2016 Seventh International Green and Sustainable Computing Conference (IGSC), Hangzhou, China, 7–9 November 2016; pp. 1–7. [CrossRef]
15. Park, S.; Ryoo, K. Hardware design of HEVC in-loop filter for ultra-HD video encoding. *Lect. Notes Electr. Eng.* **2019**, *518*, 405–409. [CrossRef]
16. Shen, W.W.; Shang, Q.; Shen, S.; Fan, Y.; Zeng, X. A high-throughput VLSI architecture for deblocking filter in HEVC. In Proceedings of the IEEE International Symposium on Circuits and Systems, Beijing, China, 19–23 May 2013; pp. 673–676.
17. Shen, S.; Shen, W.; Fan, Y.; Zeng, X. A pipelined VLSI architecture for Sample Adaptive Offset (SAO) filter and deblocking filter of HEVC. *IEICE Electron. Express* **2013**, *10*, 20130272. [CrossRef]
18. Ozcan, E.; Adibelli, Y.; Hamzaoglu, I. A high performance deblocking filter hardware for high efficiency video coding. *IEEE Trans. Consum. Electron.* **2013**, *59*, 714–720. [CrossRef]
19. Hautala, I.; Boutellier, J.; Hannuksela, J.; Silvén, O. Programmable low-power multicore coprocessor architecture for HEVC/H. 265 in-loop filtering. *IEEE Trans. Circuits Syst. Video Technol.* **2014**, *25*, 1217–1230. [CrossRef]
20. Yan, C.; Zhang, Y.; Dai, F.; Wang, X.; Li, L.; Dai, Q. Parallel deblocking filter for HEVC on many-core processor. *Electron. Lett.* **2014**, *50*, 367–368. [CrossRef]
21. Mody, M.; Nandan, N.; Hideo, T. High throughput VLSI architecture supporting HEVC loop filter for Ultra HDTV. In Proceedings of the 2013 IEEE Third International Conference on Consumer Electronics—Berlin (ICCE-Berlin), Berlin, Germany, 9–11 September 2013; pp. 54–57.
22. Wang, Y.; Guo, X.; Fan, X.; Lu, Y.; Zhao, D.; Gao, W. Parallel In-Loop Filtering in HEVC Encoder on GPU. *IEEE Trans. Consum. Electron.* **2018**, *64*, 276–284. [CrossRef]
23. Kim, H.; Ko, J.; Park, S. An Efficient Architecture of In-Loop Filters for Multicore Scalable HEVC Hardware Decoders. *IEEE Trans. Multimed.* **2017**, *20*, 810–824. [CrossRef]
24. Kotra, A.M.; Raulet, M.; Deforges, O. Comparison of different parallel implementations for deblocking filter of HEVC. In Proceedings of the 2013 IEEE International Conference on Acoustics, Speech and Signal Processing, Vancouver, BC, Canada, 26–31 May 2013; pp. 2721–2725.
25. Liu, L.; Chen, Y.; Deng, C.; Yin, S.; Wei, S. Implementation of in-loop filter for HEVC decoder on reconfigurable processor. *IET Image Process.* **2017**, *11*, 685–692. [CrossRef]
26. Shen, W.; Fan, Y.; Bai, Y.; Huang, L.; Shang, Q.; Liu, C.; Zeng, X. A Combined Deblocking Filter and SAO Hardware Architecture for HEVC. *IEEE Trans. Multimed.* **2016**, *18*, 1022–1033. [CrossRef]
27. Baldev, S.; Shukla, K.; Gogoi, S.; Rathore, P.K.; Peesapati, R. Design and Implementation of Efficient Streaming Deblocking and SAO Filter for HEVC Decoder. *IEEE Trans. Consum. Electron.* **2018**, *64*, 127–135. [CrossRef]
28. Jiang, W.; Mei, H.; Lu, F.; Jin, H.; Yang, L.T.; Luo, B.; Chi, Y. A novel parallel deblocking filtering strategy for HEVC/H.265 based on GPU. *Concurr. Comput. Pract. Exp.* **2016**, *28*, 4264–4276. [CrossRef]
29. Dai, Y.; Liu, D.; Zha, Z.J.; Wu, F. A CNN-Based In-Loop Filter with CU Classification for HEVC. In Proceedings of the 2018 IEEE Visual Communications and Image Processing (VCIP), Taichung, Taiwan, 9–12 December 2018; pp. 1–4.
30. Park, W.S.; Kim, M. CNN-based in-loop filtering for coding efficiency improvement. In Proceedings of the 2016 IEEE 12th Image, Video, and Multidimensional Signal Processing Workshop (IVMSP), Bordeaux, France, 11–12 July 2016; pp. 1–5. [CrossRef]
31. Zuo-Cheng, Z.; Ke-Bin, J. Key technologies and new developments of next generation video coding standard HEVC. In Proceedings of the 2013 Ninth International Conference on Intelligent Information Hiding and Multimedia Signal Processing, Beijing, China, 16–18 October 2013; pp. 125–128.
32. Na, T.; Kim, M. A novel no-reference PSNR estimation method with regard to deblocking filtering effect in H. 264/AVC bitstreams. *IEEE Trans. Circuits Syst. Video Technol.* **2013**, *24*, 320–330.
33. Tanchenko, A. Visual-PSNR measure of image quality. *J. Vis. Commun. Image Represent.* **2014**, *25*, 874–878. [CrossRef]
34. Ou, T.S.; Huang, Y.H.; Chen, H.H. SSIM-based perceptual rate control for video coding. *IEEE Trans. Circuits Syst. Video Technol.* **2011**, *21*, 682–691.

35. Wang, S.; Rehman, A.; Wang, Z.; Ma, S.; Gao, W. Perceptual video coding based on SSIM-inspired divisive normalization. *IEEE Trans. Image Process.* **2012**, *22*, 1418–1429. [CrossRef] [PubMed]
36. Winkler, S. *Digital Video Quality: Vision Models and Metrics*; John Wiley and Sons: Hoboken, NJ, USA, 2005.
37. Dosselmann, R.; Yang, X.D. A comprehensive assessment of the structural similarity index. *Signal Image Video Process.* **2011**, *5*, 81–91. [CrossRef]
38. Wang, Y. Survey of Objective Video Quality Measurements. 2006. Available online: https://digitalcommons.wpi.edu/computerscience-pubs/42 (accessed on 10 September 2019).
39. Wei, W.Y. *Deblocking Algorithms in Video and Image Compression Coding*; National Taiwan University: Taipei, Taiwan, 2009.
40. Christopher, P.R.; Sathasivam, S. Five-stage pipelined dual-edge deblocking filter architecture for H.265 video codec. *IEICE Electron. Express* **2019**, *16*, 20190500. [CrossRef]
41. *H.265-ITU-T*; SERIES H: Audiovisual and Multimedia Systems—Infrastructure of Audiovisual Services—Coding of Moving Video, High Efficiency Video Coding. Telecommunication Standardization Sector of ITU: Paris, France, 2018; Recommendation ITU-T H.265.

applied sciences

MDPI

Article

Assessment of the Quality of Video Sequences Performed by Viewers at Home and in the Laboratory

Janusz Klink [1,*], Stefan Brachmański [2] and Michał Łuczyński [2]

[1] Department of Telecommunications and Teleinformatics, Faculty of Information and Telecommunication Technology, Wrocław University of Science and Technology, Wybrzeże Wyspiańskiego 27, 50-370 Wrocław, Poland

[2] Photonics and Microsystems Department of Acoustics, Multimedia and Signal Processing, Faculty of Electronic, Wrocław University of Science and Technology, Wybrzeże Wyspiańskiego 27, 50-370 Wrocław, Poland

* Correspondence: janusz.klink@pwr.edu.pl

Featured Application: The results of the research may be helpful in the setup of video quality assessment procedures in order to achieve results as close as possible to the quality experienced by the end users of the video streaming services.

Abstract: The paper presents the results of subjective and objective quality assessments of H.264-, H.265-, and VP9-encoded video. Most of the literature is devoted to subjective quality assessment in well-defined laboratory circumstances. However, the end users usually watch the films in their home environments, which may be different from the conditions recommended for laboratory measurements. This may cause significant differences in the quality assessment scores. Thus, the aim of the research is to show the impact of environmental conditions on the video quality perceived by the user. The subjective assessment was made in two different environments: in the laboratory and in users' homes, where people often watch movies on their laptops. The video signal was assessed by young viewers who were not experts in the field of quality assessment. The tests were performed taking into account different image resolutions and different bit rates. The research showed strong correlations between the obtained results and the coding bit rates used, and revealed a significant difference between the quality scores obtained in the laboratory and at home. As a conclusion, it must be underlined that the laboratory tests are necessary for comparative purposes, while the assessment of the video quality experienced by end users should be performed under circumstances that are as close as possible to the user's home environment.

Keywords: video quality; objective quality assessment methods; subjective quality assessment methods; user experience; user perception; QoE; video codecs; H.264 (AVC); H.265 (HEVC); VP9

Citation: Klink, J.; Brachmański, S.; Łuczyński, M. Assessment of the Quality of Video Sequences Performed by Viewers at Home and in the Laboratory. *Appl. Sci.* **2023**, *13*, 5025. https://doi.org/10.3390/app13085025

Academic Editor: Yutaka Ishibashi

Received: 27 February 2023
Revised: 4 April 2023
Accepted: 14 April 2023
Published: 17 April 2023

1. Introduction

For many years, in everyday life, television played the role of the most important medium. Significant changes are currently being observed, especially among the young generation. Today's youth are increasingly willing to watch TV broadcasts, including movies, via the Internet using mobile devices or laptops. An important issue is the quality of the video delivered to the users. There are two general approaches to quality assessment, namely, subjective and objective. The recommendations of the International Telecommunication Union (ITU) [1] specify, in detail, the conditions for performing measurements related to the subjective assessment of video quality. In general, the assessment should be conducted under laboratory conditions, possibly simulating home conditions. The universal measurement room should be able to meet the requirements of an idealized room as well as a domestic room. Recommendation BT.500 [1] defines the general viewing conditions for subjective assessments in a laboratory and in the home environment. However, it should be taken into account that the general viewing conditions for the home environment do not guarantee that they will suit the specific home conditions of each user. An Internet user usually watches a video transmission under conditions that do not always meet the requirements of the ITU BT.500 recommendation. Consequently, the evaluation of video quality performed under real home conditions and under home conditions emulated in the laboratory will not necessarily be the same. Moreover, conducting the research in real users' locations allows us to spread the test environment across a much wider population of service customers. In general, the viewing conditions can significantly impact the results obtained. The purpose of the evaluation and the audience to whom the evaluation is devoted may determine the acceptable circumstances of the test. The next issue of such a subjective evaluation is its high cost, because many factors must be included in the experimental design and many subjects (human testers) must be involved. Therefore, many studies try to find a replacement for these methods by modeling, simulating the real world in an artificial environment, or using objective approaches to quality assessment [2]. In the next step, different approaches to quality modeling may be applied, taking into account the different points of view of various stakeholders in the media streaming process [3]. However, the results obtained from objective methods may not always correlate with subjective users' scores, especially when the circumstances of subjective quality assessment are changing. The authors anticipate that the video quality assessment performed in the artificial environment (even if the conditions emulate an 'average' home) may give different results from the scores obtained under real home conditions. Examining this may provide an answer as to whether the assessment of video quality experienced by users may (or may not) always be replaced by laboratory tests. The next issue occurs when we talk about comparisons between the subjective and objective results of the quality assessment. This is not a trivial task, taking into account that there are plenty of objective methods and quality metrics. There is a long list of literature that describes their characteristics. Some studies discuss their strengths and weaknesses, as well as their usability to predict video quality assessed by end users, especially when talking about metrics such as mean squared error (MSE) and peak signal-to-noise ratio (PSNR) [4–6]. Others show that the correspondence between objective and subjective scores also depends on the video content and show that some metrics, such as the structural similarity (SSIM) index, present characteristics similar to the human visual system (HVS) and their results are closer to users' subjective scores [7,8]. Finally, there are papers that give a comprehensive view of the different factors that influence the degradation of the video content delivered to the user, present a broad review of objective video quality assessment methods, their classification, and performance comparison [9], and give a survey of the evolution of these methods, analyzing their characteristics, advantages, and drawbacks [10,11]. Most of them are good enough for comparison and benchmarking purposes [12–15], but some give results that present stronger correlations with quality of experience (QoE) scores, given by users during subjective quality assessment, than others. Mapping the quality of service (QoS) onto QoE allows us to build proper QoE models. However, finding general relationships between QoS and QoE is not an easy task. Sometimes, the content of the video may influence the perceptual-based quality assessment in specific circumstances [16–19]. This is why big content providers and streaming platforms, which use Dynamic Adaptive Streaming over HTTP (DASH) mechanisms to provide their content via the Internet, use different coding bit rate ladders according to the video content provided [20,21]. Furthermore, the bit rate coding ladder for specific video content may depend on the video codec [22]. The authors chose three objective quality metrics, namely, the PSNR [23], SSIM [24], and the video multimethod assessment fusion (VMAF) [25], from the long list of metrics proposed in the literature. The PSNR metric is often used because it has a clear physical meaning and is simple to calculate. It presents good results when assessing the influence of some degradation factors on the quality of specific video footage, e.g., before and after compression. However, it may not always be sufficiently correlated with subjective quality assessment scores. PSNR is memoryless, which means that it is calculated pixel by pixel, independently, for each pair of corresponding frames of the two compared videos and assumes that the video quality is independent of the spatial and temporal relationships between the samples of the source footage. Reordering the pixels in the reference and examined videos, in the same way, does not change the PSNR values, although the subjective quality may change. Moreover, it can be found in the literature that video signals are highly structured and the ordering of pixels carries important perceptual structural information about the contents of the visual scene [4]. This led to also taking into account other video quality metrics, such as SSIM and VMAF, which take into account the fact that natural image signals are highly structured and may possibly better correlate with subjective quality assessment scores [11,24–26]. When there is no possibility to access the original footage, then no-reference (NR) image or video quality assessment methods can be used to evaluate the quality of the material delivered to the end user. The original footage may be distorted at any stage of the media delivery chain, that is, during acquisition, processing, compression, transmission, decoding, or presentation at the receiver's site. Therefore, it is important to use quality assessment methods that are based on a good representation of different types of distortions and may use them for a proper evaluation. Early NR quality assessment methods usually took into account specific distortion types, such as blur [27], blocking [28], and ringing artifacts [29]. In real situations, the distortion types are usually not known in advance; thus, recently, more attention has been paid to general-purpose NR methods. These metrics attempt to learn the knowledge when evaluating the quality of images and characterize the general rules of image distortions. On the basis of this knowledge, image quality prediction models can be established and adapted to unknown distortions [30]. There are many approaches based on deep convolutional neural networks (DCNN) NR image quality assessment (IQA) [31–33]. They emphasize a good distortion representation, which is crucial for the performance of NR-IQA or blind image quality assessment (BIQA). In [34], the relationship between different distortion levels and their types is analyzed. The authors proposed a new approach, named 'GraphIQA', which presents a distortion graph representation-based deep learning BIQA. General-purpose BIQA models suffer from catastrophic forgetting, which refers to the tendency of a neural network to 'forget'. A solution to this problem may be the lifelong blind image quality assessment (LIQA) approach, which not only learns new distortions, but can also mitigate the catastrophic forgetting of identified distortions [35]. The main purpose of our work was to assess the influence of the environment on the video quality experienced by the user and to find correlations with the results of the objective quality assessment. The objective evaluation was based on the full reference (FR) method, where not only the distorted video, but also the reference footage was available.

The goals of the research were to:

1. Conduct a comparative analysis of the video quality assessment results obtained under laboratory and real home (not lab-emulated) conditions;
2. Find correlations between objective results and subjective assessment scores, taking into account the influence of the test environment.

The results of the research should answer the question of whether laboratory tests can replace the video quality assessment conducted in users' homes and reduce testing costs. Furthermore, the research should show which type of subjective quality assessment is more closely correlated with objective quality assessment methods and which metric is worth using.

The video quality assessment was made taking into account:

1. H.264, H.265, and VP9 encodings [36–38];
2. The bit rate (from 300 kbps to 6000 kbps);
3. Resolutions (640 × 360—ninth high definition (nHD), 858 × 480—standard definition (SD), 1280 × 720—high definition (HD), and 1920 × 1080—full high definition (Full HD)).

The paper is organized as follows. After the introduction, Section 2 describes the video test sample preparation procedure and the methods used in the research. In the next section, the results of the subjective and objective quality assessment are presented and discussed. At the end, the results are summarized and the conclusions drawn.

2. Materials and Methods

The first step of the research consisted of using the subjective video quality assessment. From many different video quality assessment methods [1,39–42], the comparative method Double Stimulus Impairment Scale Method (DSSM) was used in the study. The DSSM is recommended by the International Telecommunication Union (ITU), and the measurement technique is described in the BT.500 recommendation [1]. The evaluation consists of comparing the reference video sequence (reference signal) with the evaluated sequence. The reference signal was presented first and assessed second. The task of the observer (viewer) was to assess the degree of deterioration of the second signal in relation to the first signal. The rating was given on a five-point mean opinion score (MOS) scale, where 5 means invisible quality deterioration, 4—noticeable but not annoying, 3—slightly annoying, 2—annoying, and 1—very annoying [39]. The video sequences were presented to the observers in single pairs (pattern-evaluated sequence). Each pair was assessed separately. The reference and evaluated video sequences were separated by a gray screen presented to the observers for about 2 s.

Measurements were made for two cases:

1. Evaluation in the laboratory;
2. Ratings at the viewer's home.

The evaluation of video quality for condition 1, that is, in the laboratory, was carried out in a room adapted for the evaluation of video signals, equipped with a 60-inch TV screen. The laboratory room met the requirements of the recommendations of the International Telecommunication Union [1,39,43,44]. Its additional advantage was the fact that all participants in the research knew about it, so it did not affect the distraction of the students related to the adaptation to the location of the research. In turn, the video quality assessment for Case 2 was made in home conditions, i.e., not ideal, but still ensuring the quality assessment by the consumer. All participants in the measurements evaluated the video sequence on high-definition television (HDTV) monitors with a resolution of 1920 × 1080 [45]. The standard test material was a 20 s video sequence (without sound) with a resolution of 1920 × 1080 pixels in AVI format. The length of the video footage was twice as long as the (minimum) value proposed in [1]. This decision does not negatively affect the results of the subjective assessment, but, in the case of objective evaluation, allows one to calculate quality metrics based on a larger dataset, which, in the case of films with varied dynamics and content of the presented scenes, may positively influence the proper calculation of objective quality metrics, which will be more representative and more correlated with the subjective assessment. However, there are studies that take into account longer video samples. This case was described in [46], where the authors considered 180 s samples for the evaluation of QoE in adaptive video streaming over wireless networks. Longer samples allow for a better evaluation of the quality perceived by users, especially when transmission disturbances may occur irregularly and at relatively longer intervals. The test footage included horse racing start scenes (see Figure 1) [47].

The original sequence was encoded in H.264, H.265, and VP09 with different resolutions and different bit rates. Four resolutions were taken into account in the research: 640 × 360 (360p), 858 × 480 (480p), 1280 × 720 (720p), and 1920 × 1080 (1080p). For the coding techniques and a specific resolution, various transmission conditions were simulated with 18 bit rates: 300, 400, 500, 600, 700, 800, 900, 1000, 1500, 2000, 2500, 3000, 3500, 4000, 4500, 5000, 5500, and 6000 kbps. The test material was presented to viewers divided into encoding technique and resolution. The test videos with different transmission conditions (bit rate) were randomly presented to the viewers. Each group of viewers evaluated the video signal subjected to one encoding technique for all resolutions and bit rates. In both cases, the team of observers consisted of second-year electronics students at the Faculty of Electronics, Photonics, and Microsystems of the Wrocław University of Science and Technology, aged 20–21, with normal visual acuity and correct color discrimination. As recommended by the International Telecommunication Union BT. 500 [1], the minimum number of observers should be 15. In the presented studies, three groups were created for home measurements and three groups for laboratory measurements. Each group evaluated a different type of coding. The number of individual test groups examining individual codecs was as follows:

1. H.264—25 people under home conditions and 45 people under laboratory conditions;
2. H.265—35 people under home conditions and 35 people under laboratory conditions;
3. VP09—30 people under home conditions and 40 people under laboratory conditions.

Figure 1. An example frame from the original video.

The different size of the groups was, among other reasons, the result of a different number of individuals willing to participate in a given measurement session, the effect of a statistical analysis of observer ratings, and the elimination of those observers who were characterized by low participation (regardless of the bit rate or resolution, they gave the same quality rating). Before beginning the measurements, the participants were familiarized with the assessment method and had one training session. During the training, the observers became acquainted with the technique of presenting the test material and how to assess the changes in video quality. After the training, the actual measurements began. After viewing the original and encoded sequences, each study participant recorded their assessment of quality deterioration in a special form. In the second part of the research, the authors performed video quality assessment using the objective double stimulus method, which relies on a comparison of the encoded video samples with the reference original video (see Figure 2).

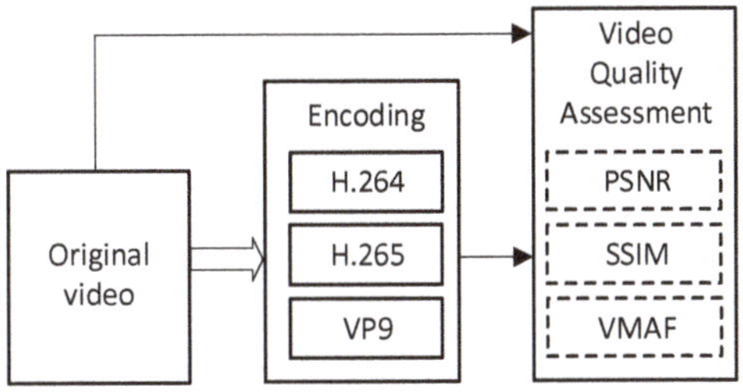

Figure 2. Video quality assessment using double stimulus method.

The original video footage was encoded using the FFmpeg [48] tool with implemented H.264, H.265, and VP9 video codecs. Four spatial resolutions and coding bit rates in the range from 300 to 6000 kbps, mentioned above, were taken into account. There were 216 video samples (3 codecs × 4 spatial resolutions × 18 coding bit rates) prepared in total.

Each set of video samples for specific spatial resolution should be compared with source video footage of the same resolution. This way, the quality of each set of videos is objectively assessed independently of the other sets of videos. When it comes to subjective quality assessment, each set of videos should be presented on a display with proper resolution that should be fitted to the resolution of the assessed video. This may be difficult to achieve when the quality assessment is performed by many different users in their home environments, where a specific display resolution may be used by default. Thus, the authors assumed that the objective assessment should be conducted using one display resolution. The most popular spatial resolution of the displays used by end users was 1920 × 1080 pixels (FHD). Therefore, the sample preparation process was a bit more complicated than just encoding. It also included upsizing all of the videos of smaller spatial resolution, i.e., 640 × 360, 858 × 480 and 1280 × 720, to FHD (Figure 3). This way, the authors wanted to achieve the same effect observed on the end-user equipment, which usually resizes smaller resolution videos to the maximum display size, with FHD resolution set by default.

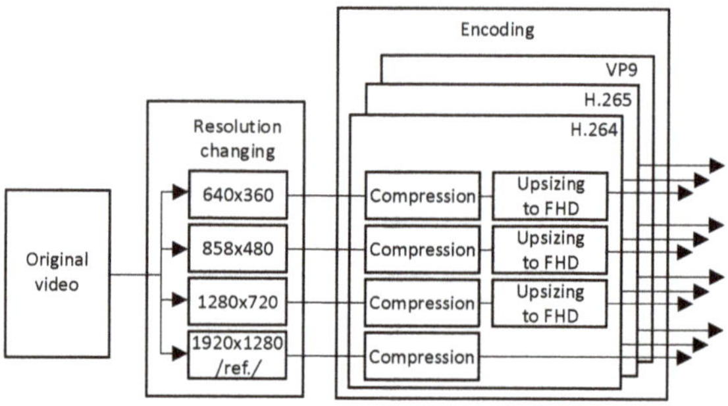

Figure 3. Video sample preparation procedure.

After this preparation, the tested video samples were objectively assessed, by comparison with the reference video (denoted in Figure 3 as '1920 × 1280/ref./'), using three metrics, i.e., PSNR, SSIM, and VMAF. Finally, these results could be compared with the subjective user scores. A detailed description of the methodology in the form of a flow diagram of the work is presented in Figure 4.

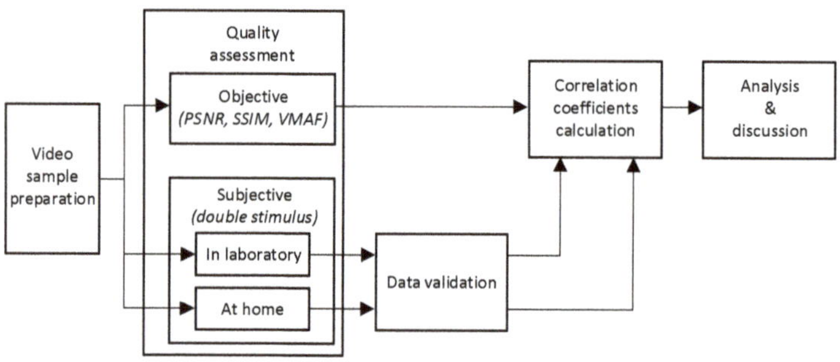

Figure 4. Flow diagram of the work.

3. Results and Discussion

The results of the subjective assessment of the quality of the video were entered into a spreadsheet and subjected to statistical analysis according to the procedure described in the ITU-R BT.500 recommendation [1]. In accordance with this recommendation, a 95% confidence interval was adopted. The mean value of the MOS score in the group of observers was calculated separately for each encoding technique, screen resolution, and bit rate.

3.1. Subjective Quality Assessment of Video Encoded Using H.264 Standard

The H.264 standard [36], also known as MPEG-4 Part 10 or AVC (advanced video coding), was introduced in 2003 as a result of cooperation between the ITU-T Q.6/SG16 Video Coding Experts Group (VCEG) and ISO/IEC Moving Picture Experts Group (MPEG). The team formed in this way is known as the Joint Video Team (JVT). The H.264 standard uses differential compression, in which the current image is created based on one or more previous images, taking into account the differences that occurred between them at that time. In relation to earlier solutions, the H.264 standard uses a number of improvements, on the one hand, allowing a reduction in the bit rate with unchanged image quality, and on the other hand, significantly increasing the demand for computing power during encoding. The degradation of the quality of the video signal encoded in the H.264 standard was assessed in a group of 25 people at home and 45 people in a laboratory. The obtained results of the measurements made under home conditions are presented in Table 1 and graphically in Figure 5, with the laboratory conditions presented in Table 2 and Figure 6. In addition to the MOS mean value, the tables also include the standard deviation values (S), as well as the values of the confidence interval coefficient (δ) calculated according to the ITU BT.500 recommendation [1].

Table 1. Mean value of the video quality assessment (MOS) for H.264 codec, standard deviation (S), and confidence interval coefficient (δ) for four resolutions—measurements at home [49].

Bit Rate	640 × 360			858 × 480			1280 × 720			1920 × 1080		
(kbps)	MOS	S	δ	MOS	S	δ	MOS	S	δ	MOS	S	δ
300	1.00	0.00	0.00	1.00	0.00	0.00	1.00	0.00	0.00	1.05	0.23	0.10
400	1.11	0.32	0.15	1.21	0.42	0.19	1.26	0.45	0.20	1.26	0.45	0.20
500	1.50	0.51	0.24	1.42	0.69	0.31	1.58	0.69	0.31	1.42	0.51	0.23
600	1.74	0.73	0.33	1.72	0.57	0.27	1.84	0.60	0.27	1.68	0.48	0.21
700	1.94	0.73	0.34	2.11	0.58	0.27	2.11	0.57	0.26	1.89	0.32	0.14
800	2.33	0.49	0.22	2.26	0.65	0.29	2.39	0.70	0.32	2.21	0.54	0.24
900	2.42	0.51	0.23	2.47	0.62	0.30	2.61	0.70	0.32	2.58	0.51	0.23
1000	2.71	0.92	0.44	2.79	0.54	0.24	2.83	0.62	0.29	2.84	0.50	0.23
1500	2.89	0.88	0.39	3.00	0.49	0.22	3.16	0.76	0.34	3.37	0.50	0.22
2000	2.95	0.71	0.32	3.11	0.68	0.31	3.50	0.51	0.24	3.79	0.63	0.28
2500	3.00	0.67	0.30	3.24	0.75	0.36	3.67	0.59	0.27	4.06	0.73	0.34
3000	3.05	0.62	0.28	3.33	0.77	0.35	3.88	0.78	0.37	4.28	0.67	0.31
3500	3.17	0.62	0.29	3.42	0.77	0.35	4.06	0.68	0.33	4.41	0.51	0.24
4000	3.22	0.55	0.25	3.58	0.84	0.38	4.11	0.58	0.27	4.56	0.51	0.24
4500	3.33	0.59	0.27	3.68	0.89	0.40	4.19	0.66	0.32	4.67	0.49	0.22
5000	3.38	0.72	0.35	3.84	0.76	0.34	4.25	0.58	0.28	4.78	0.43	0.20
5500	3.44	0.62	0.28	3.95	0.71	0.32	4.32	0.58	0.26	4.83	0.38	0.18
6000	3.57	0.65	0.34	4.06	0.73	0.34	4.37	0.50	0.22	4.94	0.24	0.11

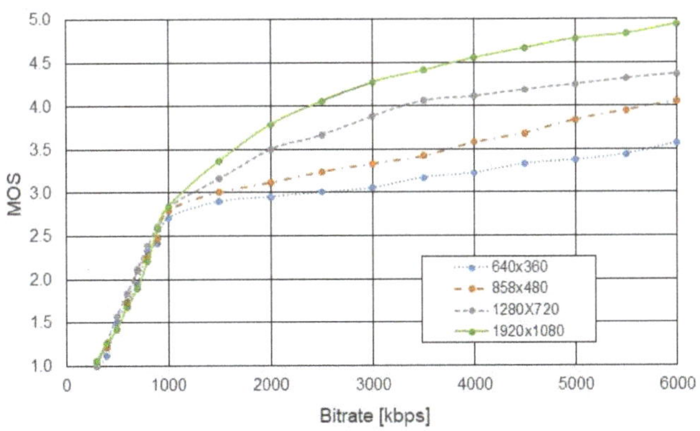

Figure 5. Results of the subjective quality assessment (MOS) for the H.264-encoded video as a function of bit rate for different spatial resolutions—measurements at home [49].

Table 2. Mean value of the video quality assessment (MOS) for H.264 codec, standard deviation (S), and confidence interval coefficient (δ) for four resolutions—measurements in laboratory.

Bit Rate	640 × 360			858 × 480			1280 × 720			1920 × 1080		
(kbps)	MOS	S	δ	MOS	S	δ	MOS	S	δ	MOS	S	δ
300	1.33	0.48	0.15	1.21	0.42	0.13	1.16	0.43	0.13	1.05	0.32	0.10
400	1.54	0.51	0.16	1.51	0.55	0.16	1.33	0.57	0.17	1.19	0.45	0.13
500	1.85	0.71	0.22	1.95	0.49	0.15	1.81	0.55	0.16	1.35	0.57	0.17
600	2.18	0.79	0.25	2.26	0.62	0.19	2.24	0.66	0.20	1.58	0.59	0.18
700	2.69	0.52	0.16	2.63	0.66	0.20	2.71	0.56	0.17	1.98	0.64	0.19
800	2.90	0.55	0.17	2.95	0.49	0.15	2.86	0.64	0.19	2.40	0.54	0.16
900	2.97	0.49	0.15	3.00	0.62	0.19	3.12	0.54	0.16	2.81	0.55	0.16
1000	3.08	0.62	0.20	3.16	0.69	0.21	3.26	0.54	0.16	3.16	0.57	0.17
1500	3.38	0.49	0.15	3.60	0.69	0.21	3.74	0.66	0.20	3.86	0.47	0.14
2000	3.49	0.64	0.20	3.81	0.70	0.21	4.07	0.70	0.21	4.26	0.49	0.15
2500	3.54	0.79	0.25	3.91	0.53	0.16	4.19	0.59	0.18	4.40	0.54	0.16
3000	3.59	0.64	0.20	4.05	0.58	0.17	4.28	0.55	0.16	4.49	0.51	0.15
3500	3.61	0.72	0.23	4.07	0.74	0.22	4.33	0.47	0.14	4.54	0.55	0.17
4000	3.68	0.66	0.21	4.14	0.74	0.22	4.40	0.49	0.15	4.63	0.49	0.15
4500	3.74	0.60	0.19	4.21	0.60	0.18	4.49	0.51	0.15	4.70	0.46	0.14
5000	3.79	0.77	0.24	4.26	0.62	0.19	4.56	0.50	0.15	4.79	0.41	0.12
5500	3.82	0.51	0.16	4.36	0.48	0.15	4.65	0.48	0.14	4.84	0.37	0.11
6000	3.85	0.49	0.15	4.40	0.49	0.15	4.72	0.45	0.14	4.91	0.29	0.09

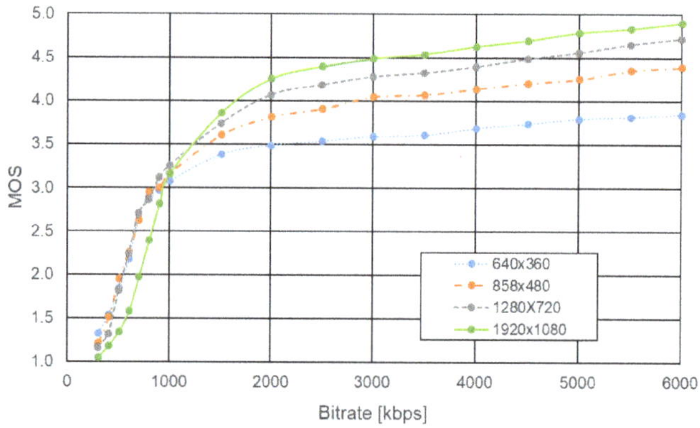

Figure 6. Results of the subjective quality assessment (MOS) for the H.264-encoded video as a function of bit rate for different spatial resolutions—measurements in the laboratory.

The statistical analysis of the results showed that up to a bit rate of 1000 kbps, the resolution does not affect the assessment of the video image quality made at home, while the laboratory measurements show a slightly lower assessment for the resolution of 1920 × 1080. Above this bit rate, the video quality depends on the resolution, and as expected, the video signal with a resolution of 1920 × 1080 is the highest rated. For this resolution, an MOS rating of at least 4 was achieved in home conditions at a bit rate starting from 2500 kbps and in laboratory conditions from approximately 1500 kbps. On the other hand, for a resolution of 1280 × 720, the MOS value of 4 was achieved at home at 3500 kbps and in laboratory conditions at 2000 kbps. Comparing the results of the MOS evaluation obtained in the laboratory and home conditions, it can be seen that viewers rated the video image presented under laboratory conditions more highly; only for the highest resolution at bit rates up to 600 kbps was the opposite MOS result found. Table 3 and Figure 7 show the difference of ΔMOS in the evaluation of video quality obtained for measurements made in the laboratory and home conditions according to Formula (1):

$$\Delta MOS = MOS_L - MOS_H, \tag{1}$$

where MOS_L is the evaluation obtained under laboratory conditions and MOS_H is the evaluation obtained under home conditions.

Table 3. Value of the difference in ΔMOS (for H.264 codec) between the results obtained under laboratory and home conditions.

Bit Rate	ΔMOS			
(kbps)	640 × 360	858 × 480	1280 × 720	1920 × 1080
300	0.33	0.21	0.16	0.00
400	0.43	0.30	0.06	−0.08
500	0.35	0.53	0.24	−0.07
600	0.44	0.53	0.40	−0.10
700	0.75	0.52	0.60	0.08
800	0.56	0.69	0.47	0.18
900	0.55	0.53	0.51	0.24
1000	0.37	0.37	0.42	0.32
1500	0.49	0.60	0.58	0.49
2000	0.54	0.70	0.57	0.47
2500	0.54	0.67	0.52	0.34
3000	0.54	0.71	0.40	0.21
3500	0.44	0.65	0.26	0.13
4000	0.46	0.56	0.28	0.07
4500	0.40	0.53	0.30	0.03
5000	0.42	0.41	0.31	0.01
5500	0.37	0.41	0.34	0.00
6000	0.27	0.34	0.35	−0.04

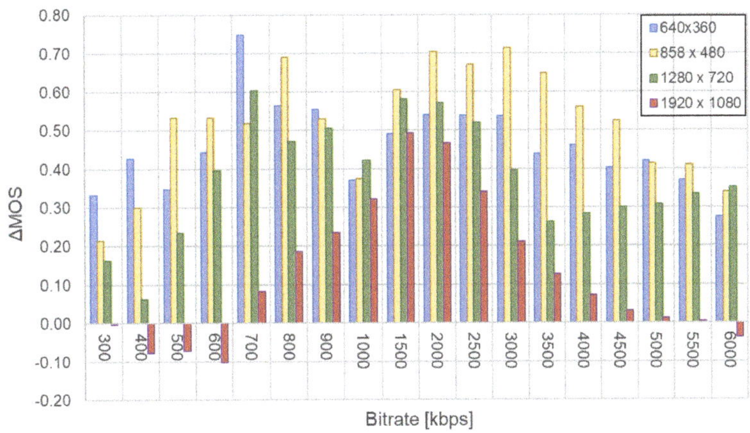

Figure 7. Difference between the results of the subjective evaluation of the H.264-encoded video (ΔMOS), conducted in the laboratory and at home, as a function of bit rate for different resolutions.

The statistical analysis of the results obtained using the *t*-test showed that at the confidence level of $\alpha = 0.05$, there is no basis to accept the hypothesis of the identity of the results obtained under laboratory and home conditions. The *t*-test values obtained using the Statistica tool for each resolution are as follows:

- 640 × 360; $t = 17.6 > t\ \alpha = 2.1$, at $\alpha = 0.05$;
- 858 × 480; $t = 14.9 > t\ \alpha = 2.1$, at $\alpha = 0.05$;
- 1280 × 720; $t = 10.7 > t\ \alpha = 2.1$, at $\alpha = 0.05$;
- 1920 × 1080; $t = 2.9 > t\ \alpha = 2.1$, with $\alpha = 0.05$.

It can be concluded that the differences between the MOS values obtained under the laboratory and home conditions are significant.

3.2. Subjective Quality Assessment of Video Encoded Using H.265 Standard

The H.265 standard [37], also known as high-efficiency video coding (HEVC), was originally published on 13 April 2013 and is currently the most recent and most efficient video coding system. This standard was created in cooperation between the Video Coding Experts Group (VCEG) and the Moving Picture Experts Group. The H.265 standard ensures the compression of videos in very high resolution (2 K, 4 K, 8 K, etc.) and also allows the use of images with increasingly higher resolutions on mobile devices. The H.265 standard offers up to twice the compression compared to H.264. Video compression is based on motion prediction; that is, when there are no changes to a pixel, the codec references that pixel instead of reproducing it. The motion prediction and compensation procedure have also been improved. Another improvement is the enlargement of the macroblock from 16 × 16 pixels (H.264) to 64 × 64 pixels, which is especially important in high-definition movies. The quality degradation of the video signal encoded in the H265 standard was evaluated in a group of 35 people, under both home and laboratory conditions. The results of the measurements taken at home are presented in Table 4 and graphically represented in Figure 8, with the laboratory conditions presented in Table 5 and Figure 9. In addition to the mean MOS value, the tables also include the standard deviation (S) and the values of the confidence interval coefficient (δ) calculated in accordance with the ITU BT.500 recommendation [1]. The statistical analysis of the results showed that up to a bit rate of 600 kbps, the resolution does not affect the evaluation of video image quality. In turn, by comparing the quality ratings for 1280 × 720 and 1920 × 1080 resolutions, it can be observed that there is no significant difference in the image quality rating for bit rates up to 900 kbps for home measurements and up to 1000 kbps under laboratory conditions.

Table 4. Mean value of the video quality assessment (MOS) for H.265 codec, standard deviation (S), and confidence interval coefficient (δ) for four resolutions—measurements at home.

Bit Rate	640 × 360			858 × 480			1280 × 720			1920 × 1080		
(kbps)	MOS	S	δ	MOS	S	δ	MOS	S	δ	MOS	S	δ
300	1.09	0.34	0.16	1.09	0.33	0.11	1.09	0.28	0.10	1.12	0.38	0.13
400	1.33	0.56	0.22	1.35	0.50	0.17	1.48	0.59	0.20	1.42	0.65	0.22
500	1.73	0.79	0.33	1.74	0.85	0.29	1.76	0.74	0.25	1.79	0.73	0.24
600	1.97	0.73	0.26	1.97	0.71	0.24	2.16	0.62	0.21	2.18	0.71	0.24
700	2.09	0.72	0.26	2.23	0.77	0.27	2.42	0.67	0.24	2.50	0.76	0.26
800	2.33	0.66	0.22	2.55	0.82	0.28	2.69	0.65	0.22	2.82	0.72	0.25
900	2.47	0.56	0.22	2.76	0.74	0.25	2.94	0.55	0.19	3.06	0.55	0.19
1000	2.56	0.59	0.27	2.94	0.73	0.25	3.13	0.29	0.10	3.35	0.57	0.19
1500	2.75	0.72	0.23	3.18	0.58	0.19	3.44	0.51	0.18	3.79	0.57	0.19
2000	2.93	0.84	0.22	3.35	0.56	0.19	3.64	0.49	0.17	4.03	0.57	0.19
2500	3.06	0.86	0.33	3.58	0.58	0.20	3.85	0.53	0.18	4.26	0.68	0.23
3000	3.24	0.81	0.21	3.68	0.60	0.20	4.00	0.60	0.21	4.44	0.65	0.22
3500	3.33	0.75	0.16	3.82	0.85	0.29	4.18	0.54	0.18	4.53	0.58	0.19
4000	3.44	0.72	0.15	3.88	0.61	0.21	4.25	0.58	0.20	4.65	0.46	0.15
4500	3.52	0.78	0.00	4.03	0.62	0.21	4.31	0.60	0.21	4.71	0.41	0.14
5000	3.63	0.65	0.20	4.09	0.63	0.22	4.39	0.65	0.23	4.79	0.41	0.14
5500	3.69	0.63	0.16	4.18	0.76	0.26	4.53	0.58	0.20	4.85	0.37	0.13
6000	3.84	0.80	0.19	4.21	0.78	0.27	4.59	0.51	0.18	4.88	0.37	0.13

Above these bit rates, the video quality is clearly resolution dependent. For a video signal with a resolution of 1920 × 1080, the MOS rating exceeds the value of 4.0 for the bit rate starting from 2000 kbps for home measurements and approximately 1500 kbps for laboratory measurements. On the other hand, for the resolution of 1280 × 720, the MOS value = 4.0 is reached for a bit rate of 3000 kbps for home measurements, and for laboratory measurements for a bit rate of 2000 kbps. Level 4.0 was also exceeded for a resolution of 858 × 480 with a bit rate of at least 4500 kbps for the home measurements and 2500 kbps for laboratory measurements. A video signal with a resolution of 640 × 360 under home conditions does not reach MOS = 4.0, while under laboratory conditions, starting at 4500 kbps, the MOS reaches a value of 4.0.

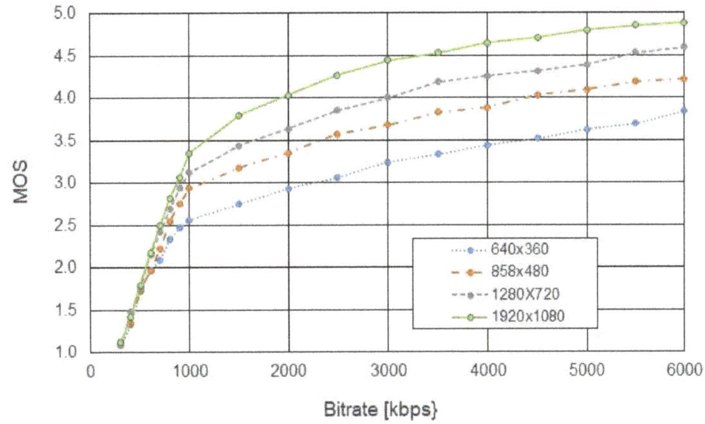

Figure 8. Results of the subjective quality assessment (MOS) for the H.265-encoded video as a function of bit rate for different spatial resolutions—measurements at home.

Table 5. Mean value of the video quality assessment (MOS) for H.265 codec, standard deviation (S), and confidence interval coefficient (δ) for four resolutions—measurements in laboratory.

Bit Rate	640 × 360			858 × 480			1280 × 720			1920 × 1080		
(kbps)	MOS	S	δ	MOS	S	δ	MOS	S	δ	MOS	S	δ
300	1.15	0.37	0.16	1.22	0.42	0.17	1.09	0.29	0.12	1.13	0.34	0.14
400	1.55	0.51	0.22	1.43	0.51	0.22	1.61	0.50	0.20	1.65	0.49	0.20
500	1.85	0.75	0.33	1.96	0.71	0.29	2.09	0.42	0.17	2.04	0.64	0.26
600	2.15	0.59	0.26	2.26	0.54	0.22	2.48	0.59	0.24	2.48	0.51	0.21
700	2.35	0.59	0.26	2.57	0.66	0.27	2.87	0.34	0.14	2.83	0.58	0.24
800	2.50	0.51	0.22	2.87	0.55	0.22	3.09	0.42	0.17	3.13	0.76	0.31
900	2.60	0.50	0.22	3.05	0.49	0.20	3.26	0.45	0.18	3.35	0.49	0.20
1000	2.80	0.62	0.27	3.17	0.49	0.20	3.43	0.51	0.21	3.48	0.59	0.24
1500	3.20	0.52	0.23	3.57	0.51	0.21	3.78	0.42	0.17	3.91	0.51	0.21
2000	3.40	0.50	0.22	3.91	0.60	0.24	4.09	0.67	0.27	4.22	0.52	0.21
2500	3.55	0.76	0.33	4.04	0.47	0.19	4.30	0.56	0.23	4.43	0.51	0.21
3000	3.70	0.47	0.21	4.17	0.39	0.16	4.43	0.59	0.24	4.57	0.51	0.21
3500	3.80	0.41	0.18	4.26	0.45	0.18	4.48	0.59	0.24	4.70	0.47	0.19
4000	3.89	0.32	0.14	4.30	0.47	0.19	4.57	0.51	0.21	4.78	0.42	0.17
4500	4.00	0.00	0.00	4.35	0.49	0.20	4.61	0.50	0.20	4.83	0.39	0.16
5000	4.05	0.39	0.17	4.39	0.50	0.20	4.70	0.47	0.19	4.87	0.34	0.14
5500	4.10	0.31	0.13	4.39	0.50	0.20	4.74	0.45	0.18	4.91	0.29	0.12
6000	4.15	0.37	0.16	4.41	0.50	0.21	4.78	0.42	0.17	4.91	0.29	0.12

Compared to the H.264-encoding standard, much higher MOS rating values are observed for the H.265 standard. Comparing the results of the MOS evaluation obtained under laboratory and home conditions, it can be seen that the viewers rated the video image presented under laboratory conditions more highly; Table 6 and Figure 10 show the difference of ΔMOS in the evaluation of video quality obtained for measurements made under laboratory and home conditions according to Formula (1).

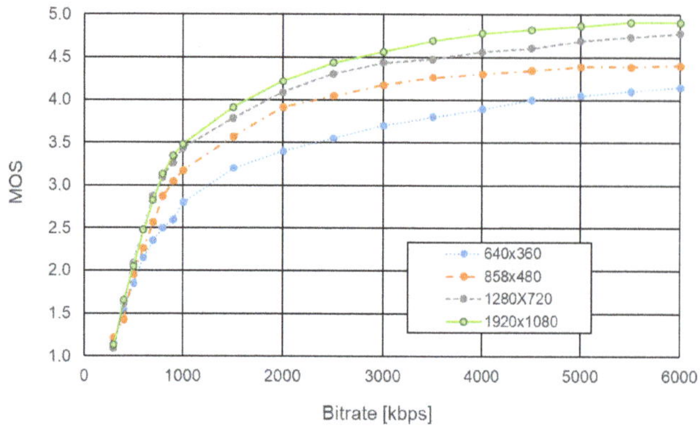

Figure 9. Results of the subjective quality assessment (MOS) for the H.265-encoded video as a function of bit rate for different spatial resolutions—measurements in the laboratory.

Table 6. Value of the difference in ΔMOS (for H.265 codec) between the results obtained under laboratory and home conditions.

Bit Rate	ΔMOS			
(kbps)	640 × 360	858 × 480	1280 × 720	1920 × 1080
300	0.06	0.13	0.00	0.01
400	0.22	0.08	0.12	0.23
500	0.12	0.22	0.33	0.25
600	0.18	0.29	0.32	0.30
700	0.26	0.34	0.45	0.33
800	0.17	0.32	0.40	0.31
900	0.13	0.29	0.32	0.29
1000	0.24	0.23	0.31	0.13
1500	0.45	0.39	0.35	0.12
2000	0.47	0.56	0.45	0.19
2500	0.49	0.47	0.46	0.17
3000	0.46	0.50	0.43	0.12
3500	0.47	0.44	0.30	0.17
4000	0.46	0.43	0.32	0.14
4500	0.48	0.32	0.30	0.12
5000	0.43	0.30	0.31	0.08
5500	0.41	0.21	0.21	0.06
6000	0.31	0.20	0.19	0.03

The statistical analysis of the results obtained using the *t*-test showed that at the confidence level of $\alpha = 0.05$, there is no basis to accept the hypothesis of the identity of the results obtained under laboratory and home conditions. The *t*-test values obtained with the Statistica tool for each resolution are as follows:

- 640 × 360; t = 9.1 > t α = 2.1, at $\alpha = 0.05$;
- 858 × 480; t = 10.4 > t α = 2.1, at $\alpha = 0.05$;
- 1280 × 720; t = 10.9 > t α = 2.1, at $\alpha = 0.05$;
- 1920 × 1080; t = 7.3 > t α = 2.1, at $\alpha = 0.05$.

It can be concluded that the differences between the MOS values obtained under the laboratory and home conditions are significant.

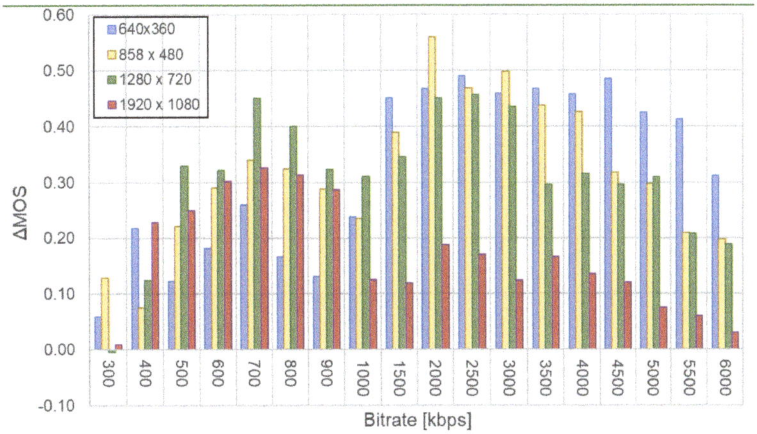

Figure 10. Difference between the results of the subjective evaluation of the H.265-encoded video (ΔMOS), conducted in the laboratory and at home, as a function of bit rate for different resolutions.

3.3. Subjective Quality Assessment of Video Encoded Using VP9 Standard

The VP9 standard, developed by Google, was the last evaluated coding technique. The VP9 codec is used, among others, on YouTube. The VP9 codec is based on an open-source license, uses the Webm container, and is basically MKV (the H.264 and H.265 codecs use the MP4 container) [40]. The degradation of the quality of the video signal encoded in the VP9 standard was assessed in a group of 30 people at home and 40 people in a laboratory. The results of the measurements made under home conditions are presented in Table 7 and graphically in Figure 11, with the results of the laboratory conditions in Table 8 and Figure 12. In addition to the MOS mean value, the tables also include the standard deviation values (S), as well as the values of the confidence interval coefficient (δ) calculated according to the ITU BT.500 recommendation [1]. The statistical analysis of the results showed that up to a bit rate of 2000 kbps, there is no difference in the assessment of image quality with resolutions of 1920 × 1080 and 1280 × 720 in home measurements; for higher speeds, slight differences can be observed in favor of the image with higher resolution. However, the differences in the quality assessment are within the designated confidence interval. For both resolutions, the MOS value of 4.0 is exceeded at 3000 kbps. The quality assessment made at home for other resolutions is comparable to the H.265 standard, which is probably related to the young people's habits, as this standard is very popular, among others, on YouTube. In turn, the statistical analysis of the results obtained in the laboratory measurements showed that up to a bit rate of 1000 kbps, there is no difference in the assessment of image quality for all of the resolutions assessed. Above this bit rate, the video quality slightly depends on the resolution and, as expected, the video signal with a resolution of 1920 × 1080 is rated the highest, for which MOS = 4 was already achieved at a bit rate of 2000 kbps. An MOS value of 4.0 was obtained for 1280 × 720 at a bit rate of 2500 kbps and for 858 × 480 at a bit rate of 3000 kbps. The smallest resolution, i.e., 640 × 360, achieves the worst MOS values, but starting from 4500 kbps, the quality rating reaches the level of 4.0, just like in home measurements.

Table 7. Mean value of the video quality assessment (MOS) for VP9 codec, standard deviation (S), and confidence interval coefficient (δ) for four resolutions—measurements at home.

Bit Rate	640 × 360			858 × 480			1280 × 720			1920 × 1080		
(kbps)	MOS	S	δ	MOS	S	δ	MOS	S	δ	MOS	S	δ
300	1.12	0.35	0.14	1.24	0.46	0.18	1.26	0.45	0.13	1.29	0.48	0.19
400	1.48	0.51	0.20	1.52	0.51	0.20	1.69	0.56	0.17	1.62	0.76	0.29
500	1.76	0.43	0.17	1.80	0.71	0.28	2.07	0.64	0.19	2.00	0.86	0.32
600	1.92	0.56	0.22	1.96	0.71	0.28	2.41	0.55	0.17	2.25	0.75	0.28
700	2.09	0.57	0.23	2.04	0.74	0.30	2.71	0.55	0.17	2.43	0.92	0.34
800	2.30	0.62	0.24	2.29	0.58	0.23	2.93	0.71	0.22	2.64	0.85	0.32
900	2.38	0.66	0.25	2.46	0.60	0.24	3.12	0.50	0.15	2.85	0.80	0.30
1000	2.48	0.51	0.20	2.60	0.50	0.20	3.26	0.45	0.13	3.00	0.76	0.28
1500	2.81	0.74	0.28	2.96	0.71	0.27	3.64	0.48	0.15	3.37	0.81	0.31
2000	2.96	0.85	0.33	3.19	0.72	0.28	3.90	0.43	0.13	3.71	0.56	0.21
2500	3.07	0.72	0.27	3.35	0.70	0.27	4.07	0.51	0.16	3.93	0.57	0.21
3000	3.19	0.72	0.27	3.50	0.59	0.23	4.29	0.60	0.18	4.14	0.60	0.22
3500	3.30	0.64	0.24	3.64	0.66	0.26	4.43	0.55	0.17	4.36	0.56	0.21
4000	3.44	0.59	0.22	3.77	0.62	0.24	4.52	0.55	0.17	4.46	0.51	0.19
4500	3.52	0.51	0.19	3.85	0.58	0.22	4.57	0.55	0.17	4.61	0.51	0.19
5000	3.63	0.65	0.24	3.96	0.71	0.27	4.64	0.48	0.15	4.67	0.49	0.19
5500	3.70	0.62	0.24	4.04	0.71	0.27	4.69	0.47	0.14	4.70	0.48	0.18
6000	3.74	0.61	0.23	4.15	0.63	0.24	4.74	0.45	0.13	4.75	0.46	0.17

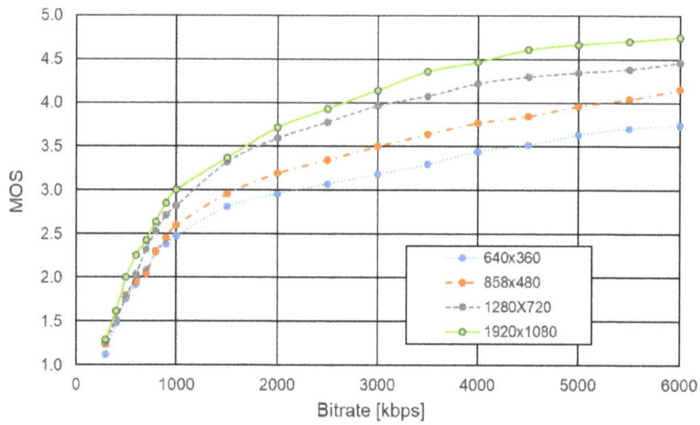

Figure 11. Results of the subjective quality assessment (MOS) for the VP9-encoded video as a function of bit rate for different spatial resolutions—measurements at home.

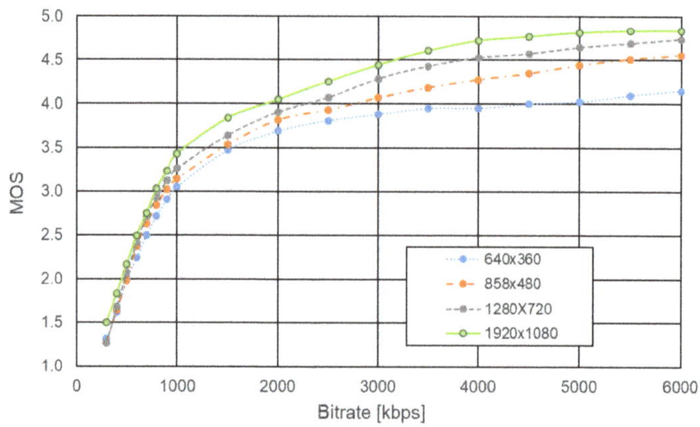

Figure 12. Results of the subjective quality assessment (MOS) for the VP9-encoded video as a function of bit rate for different spatial resolutions—measurements in laboratory.

Comparing the results of the MOS evaluation obtained in the laboratory and home conditions, it can be seen that the viewers rated the video image presented under laboratory conditions more highly; Table 9 and Figure 13 show the difference in ΔMOS in the evaluation of video quality obtained for measurements made under laboratory and home conditions according to Formula (1).

Table 8. Mean value of the video quality assessment (MOS) for VP9 codec, standard deviation (S), and confidence interval coefficient (δ) for four resolutions—measurements in laboratory.

Bit Rate	640 × 360			858 × 480			1280 × 720			1920 × 1080		
(kbps)	MOS	S	δ	MOS	S	δ	MOS	S	δ	MOS	S	δ
300	1.32	0.47	0.14	1.27	0.45	0.14	1.26	0.45	0.13	1.50	0.55	0.17
400	1.62	0.58	0.18	1.64	0.58	0.17	1.69	0.56	0.17	1.83	0.66	0.20
500	2.00	0.44	0.13	1.98	0.51	0.15	2.07	0.64	0.19	2.17	0.61	0.20
600	2.24	0.48	0.15	2.37	0.62	0.18	2.41	0.55	0.17	2.49	0.63	0.19
700	2.50	0.51	0.15	2.63	0.49	0.15	2.71	0.55	0.17	2.74	0.76	0.23
800	2.71	0.46	0.14	2.84	0.43	0.13	2.93	0.71	0.22	3.02	0.47	0.14
900	2.90	0.30	0.09	3.02	0.34	0.10	3.12	0.50	0.15	3.23	0.43	0.13
1000	3.05	0.44	0.13	3.14	0.41	0.12	3.26	0.45	0.13	3.43	0.50	0.15
1500	3.48	0.51	0.15	3.53	0.50	0.15	3.64	0.48	0.15	3.84	0.48	0.14
2000	3.69	0.56	0.17	3.81	0.45	0.13	3.90	0.43	0.13	4.05	0.43	0.13
2500	3.81	0.45	0.14	3.93	0.40	0.12	4.07	0.51	0.16	4.26	0.44	0.13
3000	3.88	0.33	0.10	4.07	0.40	0.12	4.29	0.60	0.18	4.44	0.50	0.15
3500	3.95	0.38	0.11	4.19	0.50	0.15	4.43	0.55	0.17	4.60	0.49	0.15
4000	3.95	0.44	0.13	4.28	0.50	0.15	4.52	0.55	0.17	4.72	0.45	0.14
4500	4.00	0.58	0.18	4.35	0.53	0.16	4.57	0.55	0.17	4.77	0.43	0.13
5000	4.05	0.44	0.14	4.44	0.50	0.15	4.64	0.48	0.15	4.81	0.39	0.12
5500	4.10	0.43	0.13	4.51	0.51	0.15	4.69	0.47	0.14	4.83	0.38	0.11
6000	4.15	0.48	0.15	4.56	0.50	0.15	4.74	0.45	0.13	4.84	0.37	0.11

Table 9. Value of the difference in ΔMOS (for VP9 codec) between the results obtained under laboratory and home conditions.

Bit Rate	ΔMOS			
(kbps)	640 × 360	858 × 480	1280 × 720	1920 × 1080
300	0.20	0.03	−0.01	0.21
400	0.14	0.12	0.17	0.22
500	0.24	0.18	0.29	0.17
600	0.32	0.41	0.38	0.24
700	0.41	0.59	0.39	0.32
800	0.42	0.55	0.39	0.38
900	0.52	0.56	0.40	0.38
1000	0.57	0.54	0.44	0.43
1500	0.66	0.57	0.32	0.47
2000	0.73	0.62	0.31	0.33
2500	0.74	0.58	0.29	0.33
3000	0.70	0.57	0.32	0.30
3500	0.66	0.55	0.35	0.25
4000	0.51	0.51	0.30	0.26
4500	0.48	0.50	0.28	0.16
5000	0.42	0.48	0.30	0.15
5500	0.39	0.47	0.31	0.13
6000	0.41	0.40	0.28	0.09

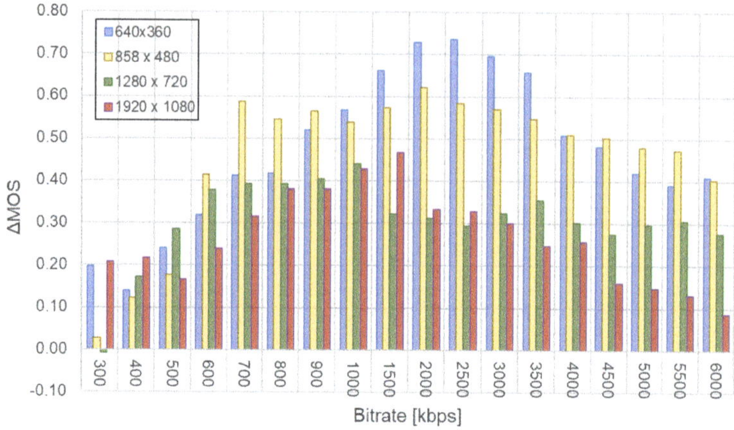

Figure 13. Difference between the results of the subjective evaluation of the VP9-encoded video (ΔMOS), conducted in the laboratory and at home, as a function of bit rate for different resolutions.

The statistical analysis of the results obtained using the *t*-test showed that at the confidence level of $\alpha = 0.05$, there is no basis to accept the hypothesis of the identity of the results obtained under laboratory and home conditions. The *t*-test values obtained with the Statistica tool for each resolution are as follows:

- 640×360; $t = 11.1 > t\ \alpha = 2.1$, at $\alpha = 0.05$;
- 858×480; $t = 11.3 > t\ \alpha = 2.1$, at $\alpha = 0.05$;
- 1280×720; $t = 13.0 > t\ \alpha = 2.1$, at $\alpha = 0.05$;
- 1920×1080; $t = 10.5 > t\ \alpha = 2.1$, at $\alpha = 0.05$.

It can be concluded that the differences between the MOS values obtained under the laboratory and home conditions are significant.

3.4. Objective Quality Assessment of Video Encoded Using H.264, H.265, and VP9 Standards

The results of the objective video quality assessment are presented using three metrics: PSNR, SSIM, and VMAF (see Figures 14–16).

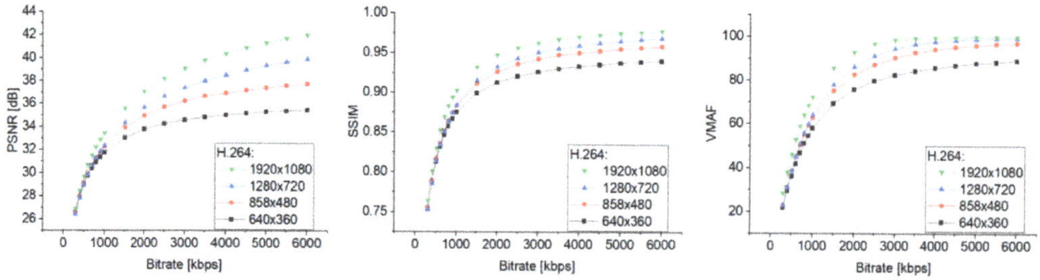

Figure 14. Relationship of the objective assessment of video quality encoded in the H.264 standard vs. bit rate for the resolutions 640×360, 858×480, 1280×720, and 1920×1080.

It can be noted that, just like in the case of subjective quality assessment, the objective video quality results are directly proportional to the used coding bit rate, which is valid for all presented metrics and video codecs. The most significant changes in quality are observed for low bit rates, while for higher bit rates, the quality changes seem to be very small or imperceptible. The results are also consistent with those presented in the literature, where the H.265 and VP9 codecs are more efficient than the H.264 codec. A very important issue here is the problem of the spatial resolutions of the examined videos. Here, each set of video samples of a specific resolution was compared (double-stimulus method) with a reference footage of proper resolution, i.e., 360p reference with 360p test sample, 480p reference with 480p test sample, etc. This resulted in higher quality values for videos with higher spatial resolution, which was consistent with the results of the subjective assessment. The correlation coefficients between these objective results and subjective quality assessment scores in the laboratory and in users' homes for each codec and video spatial resolution were determined and are presented in Tables 10–12.

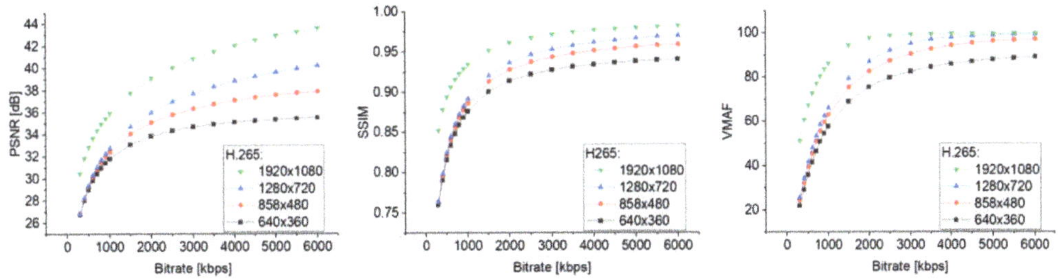

Figure 15. Relationship of the objective assessment of video quality encoded in the H.265 standard vs. bit rate for the resolutions 640 × 360, 858 × 480, 1280 × 720, and 1920 × 1080.

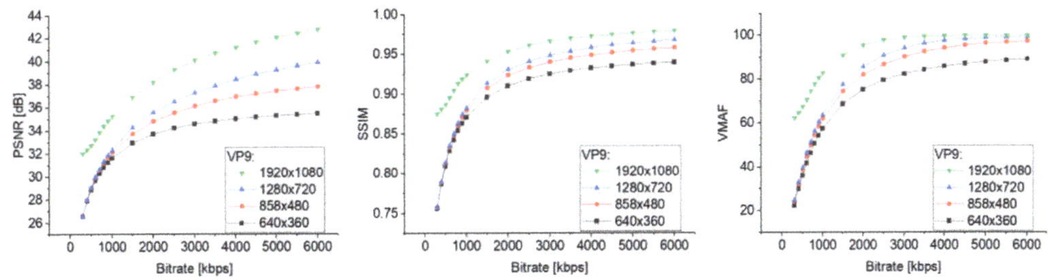

Figure 16. Relationship of the objective assessment of video quality encoded in the VP9 standard vs. bit rate for the resolutions 640 × 360, 858 × 480, 1280 × 720, and 1920 × 1080.

Table 10. Correlations between QoS and QoE values (in the lab and home) for H.264 encoded video.

		QoS vs. QoE Correlations							
		Lab				Home			
		360p	480p	720p	1080p	360p	480p	720p	1080p
PSNR	360p	0.981	0.996	0.995	0.986	0.986	0.988	0.995	0.985
	480p	0.968	0.989	0.990	0.987	0.980	0.989	0.998	0.994
	720p	0.951	0.978	0.981	0.982	0.970	0.986	0.995	0.996
	1080p	0.949	0.976	0.979	0.981	0.968	0.985	0.994	0.996
SSIM	360p	0.989	0.997	0.995	0.977	0.988	0.983	0.987	0.970
	480p	0.987	0.997	0.996	0.981	0.988	0.985	0.990	0.975
	720p	0.985	0.997	0.996	0.984	0.988	0.987	0.993	0.980
	1080p	0.989	0.997	0.995	0.978	0.988	0.982	0.987	0.970
VMAF	360p	0.974	0.992	0.994	0.992	0.983	0.988	0.998	0.992
	480p	0.973	0.992	0.993	0.992	0.982	0.987	0.998	0.993
	720p	0.971	0.990	0.992	0.993	0.979	0.984	0.996	0.992
	1080p	0.981	0.994	0.994	0.990	0.982	0.979	0.989	0.980

Table 11. Correlations between QoS and QoE values (in the lab and home) for H.265 encoded video.

		QoS vs. QoE Correlations							
		Lab				Home			
		360p	**480p**	**720p**	**1080p**	**360p**	**480p**	**720p**	**1080p**
PSNR	360p	0.998	0.997	0.997	0.998	0.991	0.995	0.996	0.997
	480p	0.999	0.991	0.990	0.993	0.993	0.994	0.995	0.996
	720p	0.995	0.981	0.980	0.983	0.991	0.989	0.991	0.989
	1080p	0.991	0.975	0.973	0.977	0.989	0.986	0.987	0.985
SSIM	360p	0.993	0.997	0.999	0.998	0.986	0.991	0.992	0.993
	480p	0.995	0.998	0.999	0.999	0.988	0.993	0.994	0.995
	720p	0.995	0.998	0.999	0.999	0.988	0.993	0.994	0.995
	1080p	0.994	0.996	0.998	0.998	0.987	0.991	0.993	0.993
VMAF	360p	0.998	0.995	0.992	0.995	0.989	0.994	0.995	0.998
	480p	0.997	0.995	0.992	0.995	0.988	0.993	0.995	0.997
	720p	0.995	0.996	0.992	0.995	0.985	0.991	0.993	0.996
	1080p	0.976	0.989	0.991	0.989	0.965	0.976	0.977	0.981

Table 12. Correlations between QoS and QoE values (in the lab and home) for VP9 encoded video.

		QoS vs. QoE Correlations							
		Lab				Home			
		360p	**480p**	**720p**	**1080p**	**360p**	**480p**	**720p**	**1080p**
PSNR	360p	0.997	0.998	0.999	0.998	0.994	0.990	0.996	0.993
	480p	0.990	0.996	0.997	0.997	0.997	0.997	0.999	0.999
	720p	0.980	0.989	0.990	0.991	0.995	0.999	0.997	0.999
	1080p	0.951	0.966	0.968	0.972	0.978	0.989	0.984	0.987
SSIM	360p	0.998	0.997	0.997	0.995	0.988	0.981	0.989	0.985
	480p	0.998	0.998	0.998	0.996	0.991	0.984	0.991	0.988
	720p	0.998	0.998	0.998	0.997	0.991	0.986	0.993	0.990
	1080p	0.988	0.991	0.992	0.995	0.990	0.994	0.999	0.995
VMAF	360p	0.995	0.996	0.997	0.998	0.993	0.993	0.998	0.995
	480p	0.994	0.995	0.996	0.998	0.992	0.993	0.998	0.996
	720p	0.994	0.994	0.995	0.997	0.989	0.990	0.997	0.993
	1080p	0.993	0.988	0.987	0.991	0.978	0.980	0.991	0.983

All tests and calculated correlations were conducted for the selected video resolutions and a limited number of coding bit rates (i.e., 18 coding bit rates for each video sample of a specific resolution). To validate these results and check how much of the whole population of different cases is well described by this research, a determination coefficient (R^2) was calculated for each previously determined correlation.

Taking into account each codec, it can be stated that the determination coefficients fluctuated as follows:

1. For the H.264 codec: from 0.9 to 0.996;
2. For the H.265 codec: from 0.931 to 0.998;
3. For the VP9 codec: from 0.905 to 0.998.

This means that the obtained results of the correlations well describe 90 to 99 percent of the whole population. This leads to the conclusion that the obtained correlations are very strong and that they are representative for a population that can be much wider than the video set that was used during the research.

4. Conclusions

The authors presented the problem of subjective quality assessment conducted in different environments. Most papers and formal regulations recommend performing such tests in a laboratory under special circumstances. It is understandable that test conditions must be strictly determined, especially when the procedure must be repeatable and should give representative results that are comparable with those of other laboratories. However, in the case of watching the video at home, the environment may not meet the laboratory conditions described in formal recommendations. This may cause the quality experienced by the home user to differ from the quality measured in the laboratory. The results of our investigations confirmed these assumptions and showed statistically significant differences. This implies the need to separate these two types of environments and to conduct the tests in both depending on their purpose. The second part of the research was devoted to objective video quality evaluation and identifying the relationships between their results and the results of the subjective assessment conducted in different environments. The authors observed very high correlations between all three sets of results, i.e., objective, subjective in the laboratory, and subjective at home. The very high determination coefficients imply that the results obtained from testing a limited number of video samples may produce conclusions that can be generalized to the entire population. Obviously, here, the QoS/QoE models can be made, but their parameters must be determined separately for laboratory and home environments. Searching for better quality models for other environments (different from laboratory) may help to better fit the video delivered to its recipients.

Author Contributions: Conceptualization and methodology, J.K. and S.B.; objective quality assessment, J.K.; subjective quality assessment, S.B. with M.Ł.'s support; writing—original draft preparation, J.K. and S.B.; visualization, J.K. All authors have read and agreed to the published version of the manuscript.

Funding: This research received no external funding.

Informed Consent Statement: Informed consent was obtained from all subjects involved in the study.

Data Availability Statement: Data sharing is not applicable to this article.

Acknowledgments: The paper presents the results of the statutory research carried out at Wroclaw University of Science and Technology. The authors would like to thank Wroclaw Centre for Networking and Supercomputing for providing the computing resources that were used for the digital processing of the tested video samples.

Conflicts of Interest: The authors declare no conflict of interest.

References

1. *ITU-R BT 500-14*; Methodologies for the Subjective Assessment of the Quality of Television Images. ITU: Geneva, Switzerland, 2020.
2. Fela, R.F.; Zacharov, N.; Forchhammer, S. Comparison of Full Factorial and Optimal Experimental Design for Perceptual Evaluation of Audiovisual Quality. *J. Audio Eng. Soc.* **2023**, *71*, 4–19. [CrossRef]
3. Barman, N.; Martini, M.G. QoE Modeling for HTTP Adaptive Video Streaming–A Survey and Open Challenges. *IEEE Access* **2019**, *7*, 30831–30859. [CrossRef]
4. Wang, Z.; Bovik, A.C. Mean squared error: Love it or leave it? A new look at Signal Fidelity Measures. *IEEE Signal Process. Mag.* **2009**, *26*, 98–117. [CrossRef]
5. Huynh-Thu, Q.; Ghanbari, M. The accuracy of PSNR in predicting video quality for different video scenes and frame rates. *Telecommun. Syst.* **2012**, *49*, 35–48. [CrossRef]
6. Klink, J.; Uhl, T. Video Quality Assessment: Some Remarks on Selected Objective Metrics. In Proceedings of the 2020 28th International Conference Software, Telecommun. Comput. Networks, SoftCOM, Split, Croatia, 17–19 September 2020.
7. Vranjes, M.; Rimac-Drlje, S.; Zagar, D. Objective video quality metrics. In Proceedings of the ELMAR 2007, Zadar, Croatia, 12–14 September 200; pp. 45–49. [CrossRef]
8. Kotevski, Z.; Mitrevski, P. Performance Assessment of Metrics for Video Quality Estimation. In Proceedings of the International Scientific Conference on Information, Communication and Energy Systems and Technologies, Macedonia, Greece, 23–26 June 2010; pp. 693–696.
9. Chikkerur, S.; Sundaram, V.; Reisslein, M.; Karam, L.J. Objective video quality assessment methods: A classification, review, and performance comparison. *IEEE Trans. Broadcast.* **2011**, *57*, 165–182. [CrossRef]

10. Akramullah, S.; Akramullah, S. Video quality metrics. In *Digital Video Concepts, Methods, and Metrics*; Apress: New York, NY, USA, 2014; pp. 101–160.
11. Chen, Y.; Wu, K.; Zhang, Q. From QoS to QoE: A Tutorial on Video Quality Assessment. *IEEE Commun. Surv. Tutor.* **2015**, *17*, 1126–1165. [CrossRef]
12. Hanhart, P.; Korshunov, P.; Ebrahimi, T. Benchmarking of quality metrics on ultra-high definition video sequences. In Proceedings of the 2013 18th International Conference on Digital Signal Processing (DSP), Santorini, Greece, 1–3 July 2013; pp. 1–8.
13. Hanhart, P.; Bernardo, M.V.; Pereira, M.; Pinheiro, A.M.G.; Ebrahimi, T. Benchmarking of objective quality metrics for HDR image quality assessment. *EURASIP J. Image Video Process.* **2015**, *2015*, 39. [CrossRef]
14. Klink, J. A Method of Codec Comparison and Selection for Good Quality Video Transmission Over Limited-Bandwidth Networks. *Sensors* **2021**, *21*, 4589. [CrossRef]
15. Barman, N.; Martini, M.G. H. 264/MPEG-AVC, H. 265/MPEG-HEVC and VP9 codec comparison for live gaming video streaming. In Proceedings of the 2017 Ninth International Conference on Quality of Multimedia Experience (QoMEX), Erfurt, Germany, 31 May–2 June 2017; pp. 1–6.
16. You, J.; Reiter, U.; Hannuksela, M.M.; Gabbouj, M.; Perkis, A. Perceptual-based quality assessment for audio–visual services: A survey. *Signal Process. Image Commun.* **2010**, *25*, 482–501. [CrossRef]
17. Akhtar, Z.; Siddique, K.; Rattani, A.; Lutfi, S.L.; Falk, T.H. Why is Multimedia Quality of Experience Assessment a Challenging Problem? *IEEE Access* **2017**, *7*, 117897–117915. [CrossRef]
18. Rassool, R. VMAF reproducibility: Validating a perceptual practical video quality metric. In Proceedings of the 2017 IEEE International Symposium on Broadband Multimedia Systems and Broadcasting (BMSB), Cagliari, Italy, 7–9 June 2017; pp. 1–2.
19. Moldovan, A.-N.; Ghergulescu, I.; Muntean, C.H. VQAMap: A Novel Mechanism for Mapping Objective Video Quality Metrics to Subjective MOS Scale. *IEEE Trans. Broadcast.* **2016**, *62*, 610–627. [CrossRef]
20. Bentaleb, A.; Taani, B.; Begen, A.C.; Timmerer, C.; Zimmermann, R. A Survey on Bitrate Adaptation Schemes for Streaming Media Over HTTP. *IEEE Commun. Surv. Tutor.* **2018**, *21*, 562–585. [CrossRef]
21. Sani, Y.; Mauthe, A.; Edwards, C. Adaptive Bitrate Selection: A Survey. *IEEE Commun. Surv. Tutor.* **2017**, *19*, 2985–3014. [CrossRef]
22. Zabrovskiy, A.; Feldmann, C.; Timmerer, C. Multi-codec DASH dataset. In Proceedings of the 9th ACM Multimedia Systems Conference, Amsterdam, The Netherlands, 12–15 June 2018; pp. 438–443.
23. Tanchenko, A. Visual-PSNR measure of image quality. *J. Vis. Commun. Image Represent.* **2014**, *25*, 874–878. [CrossRef]
24. Wang, Z.; Bovik, A.C.; Sheikh, H.R.; Simoncelli, E.P. Image Quality Assessment: From Error Visibility to Structural Similarity. *IEEE Trans. Image Process.* **2004**, *13*, 600–612. [CrossRef]
25. Bampis, C.G.; Li, Z.; Bovik, A.C. Spatiotemporal Feature Integration and Model Fusion for Full Reference Video Quality Assessment. *IEEE Trans. Circuits Syst. Video Technol.* **2018**, *29*, 2256–2270. [CrossRef]
26. Wang, Z.; Bovik, A.C. A universal image quality index. *IEEE Signal Process. Lett.* **2002**, *9*, 81–84. [CrossRef]
27. Li, L.; Lin, W.; Wang, X.; Yang, G.; Bahrami, K.; Kot, A.C. No-Reference Image Blur Assessment Based on Discrete Orthogonal Moments. *IEEE Trans. Cybern.* **2015**, *46*, 39–50. [CrossRef]
28. Li, L.; Zhu, H.; Yang, G.; Qian, J. Referenceless Measure of Blocking Artifacts by Tchebichef Kernel Analysis. *IEEE Signal Process. Lett.* **2013**, *21*, 122–125. [CrossRef]
29. Liu, H.; Klomp, N.; Heynderickx, I. A No-Reference Metric for Perceived Ringing Artifacts in Images. *IEEE Trans. Circuits Syst. Video Technol.* **2009**, *20*, 529–539. [CrossRef]
30. Zhu, H.; Li, L.; Wu, J.; Dong, W.; Shi, G. MetaIQA: Deep meta-learning for no-reference image quality assessment. In Proceedings of the IEEE/CVF Conference on Computer Vision and Pattern Recognition, Seattle, WA, USA, 14 June 2020–19 June 2020; pp. 14143–14152.
31. Kang, L.; Ye, P.; Li, Y.; Doermann, D. Convolutional neural networks for no-reference image quality assessment. In Proceedings of the 2014 IEEE Computer Vision and Pattern Recognition, Columbus, OH, USA, 23–28 June 2014; pp. 1733–1740.
32. Bosse, S.; Maniry, D.; Müller, K.R.; Wiegand, T.; Samek, W. Deep neural networks for no-reference and full-reference image quality assessment. *IEEE Trans. Image Process.* **2017**, *27*, 206–219. [CrossRef] [PubMed]
33. Zhang, W.; Ma, K.; Yan, J.; Deng, D.; Wang, Z. Blind Image Quality Assessment Using a Deep Bilinear Convolutional Neural Network. *IEEE Trans. Circuits Syst. Video Technol.* **2018**, *30*, 36–47. [CrossRef]
34. Sun, S.; Yu, T.; Xu, J.; Zhou, W.; Chen, Z. GraphIQA: Learning Distortion Graph Representations for Blind Image Quality Assessment. *IEEE Trans. Multimedia* **2022**, *14*, 1–14. [CrossRef]
35. Liu, J.; Zhou, W.; Li, X.; Xu, J.; Chen, Z. LIQA: Lifelong Blind Image Quality Assessment. *IEEE Trans. Multimedia* **2022**, *14*, 1–13. [CrossRef]
36. *ITU-T Rec*; H.264. Audiovisual and Multimedia Systems: Infrastructure of Audiovisual Services-Coding of Moving Video, Advanced Video Coding for Generic Audiovisual Services. International Telecommunication Union: Geneva, Switzerland, 2021.
37. *ITU-T Rec*; H.265. Infrastructure of Audiovisual Services—Coding of Moving Video. High Efficiency Video Coding. International Telecommunication Union: Geneva, Switzerland, 2021.
38. Grange, A.; De Rivaz, P.; Hunt, J. VP9 Bitstream Decoding Process Specification. *WebM Project*. 2016. Available online: http://downloads.webmproject.org.storage.googleapis.com/docs/vp9/vp9-bitstream-specification-v0.6-20160331-draft.pdf (accessed on 25 February 2023).

39. *ITU-T Rec*; P.910. Subjective Video Quality Assessment Methods for Multimedia Applications. International Telecommunication Union: Geneva, Switzerland, 2021.
40. Mukherjee, D.; Bankoski, J.; Grange, A.; Han, J.; Koleszar, J.; Wilkins, P.; Xu, Y.; Bultje, R. The latest open-source video codec VP9—An overview and preliminary results. In Proceedings of the 2013 Picture Coding Symposium, San Jose, CA, USA, 8–11 December 2013; pp. 390–393. [CrossRef]
41. Winkler, S. Video quality measurement standards—Current status and trends. In Proceedings of the 2009 7th International Conference on Information, Communications and Signal Processing (ICICS), Macau, China, 8–10 December 2009; pp. 1–5.
42. Winkler, S. On the properties of subjective ratings in video quality experiments. In Proceedings of the 2009 International Workshop on Quality of Multimedia Experience, Lippstadt, Germany, 5–7 September 2022; pp. 139–144. [CrossRef]
43. *ITU-T Recommendation P.913*; Methods for the Subjective Assessment of Video Quality, Audio Quality and Audiovisual Quality of Internet Video and Distribution Quality Television in Any Environment. ITU: Geneva, Switzerland, 2021.
44. Harysandi, D.K.; Oktaviani, R.; Meylani, L.; Vonnisa, M.; Hashiguchi, H.; Shimomai, T.; Aris, N.A.M. International Telecommunication Union-Radiocommunication Sector P. 837-6 and P. 837-7 performance to estimate Indonesian rainfall. *Telkomnika* **2020**, *18*, 2292–2303.
45. *ITU-R BT 709-6*; Parameter Values for the HDTV Standards for Production and International Programme Exchange BT Series Broadcasting Service. ITU: Geneva, Switzerland, 2015.
46. Taha, M.; Ali, A.; Lloret, J.; Gondim, P.R.L.; Canovas, A. An automated model for the assessment of QoE of adaptive video streaming over wireless networks. *Multimedia Tools Appl.* **2021**, *80*, 26833–26854. [CrossRef]
47. Mercat, A.; Viitanen, M.; Vanne, J. UVG dataset: 50/120fps 4K sequences for video codec analysis and development. In Proceedings of the 11th ACM Multimedia Systems Conference, Istanbul, Turkey, 8– 11 June 2020; pp. 297–302.
48. FFmpeg: A Complete, Cross -Platform Solution to Record, Convert and Stream Audio and Video. Available online: https://ffmpeg.org/ (accessed on 25 February 2023).
49. Brachmański, S.; Klink, J. Subjective Assessment of the Quality of Video Sequences by the Young Viewers. In Proceedings of the 30 th International Conference on Software, Telecommunications and Computer Networks (SoftCOM 2022), Split, Croatia, 22–24 September 2022; pp. 1–6.

applied sciences

MDPI

Article

Application of Transfer Learning and Convolutional Neural Networks for Autonomous Oil Sheen Monitoring

Jialin Dong [1], Katherine Sitler [2], Joseph Scalia [2], Yunhao Ge [3], Paul Bireta [4], Natasha Sihota [4], Thomas P. Hoelen [4] and Gregory V. Lowry [1,*]

1 Department of Civil and Environmental Engineering, Carnegie Mellon University, Pittsburgh, PA 15213, USA
2 Department of Civil and Environmental Engineering, Colorado State University, Fort Collins, CO 80523, USA
3 Department of Computer Science, University of Southern California, Los Angeles, CA 90007, USA
4 Chevron Technical Center, San Ramon, CA 94583, USA
* Correspondence: glowry@cmu.edu

Abstract: Oil sheen on the water surface can indicate a source of hydrocarbon in underlying subaquatic sediments. Here, we develop and test the accuracy of an algorithm for automated real-time visual monitoring of the water surface for detecting oil sheen. This detection system is part of an automated oil sheen screening system (OS-SS) that disturbs subaquatic sediments and monitors for the formation of sheen. We first created a new near-surface oil sheen image dataset. We then used this dataset to develop an image-based Oil Sheen Prediction Neural Network (OS-Net), a classification machine learning model based on a convolutional neural network (CNN), to predict the existence of oil sheen on the water surface from images. We explored the effectiveness of different strategies of transfer learning to improve the model accuracy. The performance of OS-Net and the oil detection accuracy reached up to 99% on a test dataset. Because the OS-SS uses video to monitor for sheen, we also created a real-time video-based oil sheen prediction algorithm (VOS-Net) to deploy in the OS-SS to autonomously map the spatial distribution of sheening potential of hydrocarbon-impacted subaquatic sediments.

Keywords: oil sheen; oil pollution monitoring; convolutional neural network; transfer learning

Citation: Dong, J.; Sitler, K.; Scalia, J.; Ge, Y.; Bireta, P.; Sihota, N.; Hoelen, T.P.; Lowry, G.V. Application of Transfer Learning and Convolutional Neural Networks for Autonomous Oil Sheen Monitoring. *Appl. Sci.* **2022**, *12*, 8865. https://doi.org/10.3390/app12178865

Academic Editors: Zbigniew Lubniewski, Tadeus Uhl and Przemysław Falkowski-Gilski

Received: 26 July 2022
Accepted: 30 August 2022
Published: 3 September 2022

1. Introduction

Oil sheens are common, and their source and type need to be adequately characterized. An oil sheen is an iridescent appearance that forms on the water's surface when oil spreads on water [1]. Oil sheen can be produced from the release of small amounts of hydrocarbons that are entrapped in subaquatic sediments after they are disturbed or can occur from natural microbial activity, forming a film on the water surface. The hydrocarbons in the sediments can come from anthropogenic activities (e.g., oil and gas extraction, transportation, and petroleum exploration) or natural processes (e.g., natural oil seeps, erosion of sedimentary rocks, and organic matter from the soil) [2,3]. Sheens are unsightly and hydrocarbon sheen can potentially be detrimental to wildlife [4], economic development [5], and the environment [6,7]. The source and type of a sheen can have a strong effect on these potential impacts and needs to be characterized.

Several imaging-based methods have been developed to detect, classify, and monitor oil sheens in nearshore, ocean, river, marsh, and mudflat ecosystems. Sensors have been used in oil sheen detection and mapping, including radar, laser, UV, visible, infrared, and thermal infrared sensors [8]. Satellite Synthetic Aperture Radar (SAR) images have also been used to detect and map oil sheens over large spatial areas [9–13]. Different machine learning methodologies have been developed to identify oil sheens from these sensor data, including classification [13], object detection [14], and segmentation [15,16]. Convolution neural networks (CNNs) [17] are the most commonly used machine learning algorithm applied for analyzing visible light imagery and have been used for oil sheen monitoring [16,18]. For example, visible light images of oil sheen taken by Unmanned Aerial Vehicles (UAV), along with convolutional neural networks (CNNs), have been used to detect oil spills on the water surface [19,20].

One source of oil sheen is the release of hydrocarbons trapped in subaquatic sediments. This can occur when hydrocarbon-impacted sediments are disturbed with sufficient force to release trapped oil or through processes, such as ebullition (the trapped oil is transported to the surface as an intermediate wetting fluid with gas bubbles [21]). Sheening potential of a sediment is operationally defined as the potential for a sheen to form when the sediment is disturbed, e.g., by a storm or by human disturbance. Currently, there are no automated methods to determine the sheening potential of hydrocarbon-impacted subaquatic sediments. To quantify sheening potential, the sediment must be disturbed in a controlled manner and the water surface in the disturbed area must be monitored for oil sheen formation. To avoid widespread release of oil sheen, this disturbed oil, if released, needs to rise to the surface in a controlled environment, e.g., a small-diameter tube, and then imaged inside the tube at the surface. The appearance of the oil sheen in a small area may be different from that formed over large areas. The large sheen can be easily viewed and detected by satellites and drones with sensors (e.g., radar, laser, UV, visible, infrared, and thermal infrared [8–13]), as the sheen covers a large area and stays for a longer time. Small-area sheen formation may be short lived and can only be observed at a close distance (within several meters). The appearance and color of a small-scale sheen can also be more diverse, as it can be easily influenced by several factors, including oil properties, sheen thickness, light conditions, and the angle of observation [22]. Thus, the current models and image datasets for remote oil monitoring are not appropriate for monitoring localized sheening potential, which is important to understand due to potential community aesthetic concerns. Automating the small-area oil sheen detection process will, therefore, require a robust method to identify sheen formation in various appearances in real time. Further, building a machine learning model for oil sheen detection requires a large dataset of visible image data taken right at the water surface.

The objective of this study is to develop an image-based Oil Sheen Prediction Neural Network (OS-Net) and video-based real-time oil sheen prediction algorithm (VOS-Net) to automatically monitor oil sheen appearance in an oil sheen screening system (OS-SS). We first collected oil sheen image data in a lab-based test system at the distance and angle required for potential future field embodiments. The lab-based test system recorded videos of oil sheen formation from mobilizing embedded oil in sediments with injected air bubbles or injected water. We extracted images from these videos to develop an image-based OS-Net, a CNN based model to predict the existence of oil sheen on water surface images. To develop a powerful (deeper) and robust OS-Net, we employed skip connection in Residual Blocks [23] to address the vanishing gradient problem in deep neural networks. We evaluated different transfer learning strategies to further improve the performance of OS-Net. Finally, we extended the OS-Net to monitor real-time video of oil sheen formation (VOS-Net) to automatically detect the oil sheen in real time in the OS-SS.

The oil sheen prediction algorithm used in this paper is described and experimental results as well as the model performance are discussed. Finally, the advantages and disadvantages of the current model, possible ways to improve the robustness of the algorithm, and other potential applications of this algorithm are described. Our key contributions are: (1) development of OS-Net, a deep convolutional neural network using residual blocks as basic elements and transfer learning to achieve a high test accuracy in oil sheen prediction; (2) exploration of the effectiveness of different strategies of transfer learning, which can provide guidance to other similar limited data tasks; (3) development of the video-based VOS-Net that combines domain knowledge with the machine learning algorithm to improve performance and achieve real-time oil sheen detection from video of transient sheen formation on the water surface; and (4) creation of a new dataset with thousands of images and videos of close-proximity visible light images of oil sheen, which provides data for others to develop close-proximity oil sheen detection models.

2. Related Works

CNNs are a specific type of neural network designed to automatically learn the spatial hierarchies of features from images through convolution kernels in different layers [24]. CNNs have been applied to image classification [25], face recognition [23], object detection [26], segmentation [27], and oil sheen detection [10,16]. Here, we use a CNN as the basis algorithm to classify close-proximity oil sheen images.

A deep neural network is usually more powerful in feature extraction and in learning non-linear decision boundaries. A CNN with more layers can solve more complex problems or improve the model performance [25]. However, a deeper neural network also makes the gradient hard to backpropagate, resulting in the vanishing gradient problem. The deeper the network, the harder earlier layers are to optimize, leading to degraded performance and lower accuracy [23]. Residual Neural Network [25] (ResNet), was introduced in 2015 to solve this problem. The residual block is the core idea of Resnet. By adding skip connections with no trainable parameters across blocks, the residual block creates a "highway" for gradient backpropagation in deep neural networks, which solves the vanishing gradient problem and efficiently optimizes the earlier layers. Here, to efficiently use deeper neural networks with powerful learning ability, we borrow the idea of the residual block to develop the OS-Net.

The transfer learning approach allows for shifting learned knowledge (extract basic feature, aggregate basic features, and form higher-level feature) from one task with sufficient data to improve the performance in another task (target task) with fewer data and improves performance in the target task [28,29]. This approach uses layers of the pre-trained model as the starting point to train the target model. There are two popular strategies when using the pre-trained model approach of transfer learning. The first strategy can use a pre-trained model as a feature extractor to apply some of the pre-trained model's weighted layers to extract features and not update the weights of these layers during training with new data for the new task. The second strategy is fine tuning a pre-trained model. The second strategy is a more involved technique, where previous layers are retrained [30] in addition to replacing the final layer. Transfer learning strategies have been widely applied in image-related tasks [29,30]. Here, we explore transfer learning strategies to determine the most appropriate approach for our purpose. Specifically, we compare three models (OS-Net Without Transfer Learning, OS-Net With Transfer Learning Feature Extraction Strategy, and OS-Net With Transfer Learning Fine-tuning Strategy) to obtain our best OS-Net.

ImageNet [31] is a large (millions of images) and diverse (1000 classes) dataset that can be used to pre-train models for transfer learning. Models pre-trained on ImageNet have been widely applied, including VGG [32], AlexNet [33], ResNet [34], and Inception [35]. These pre-trained models can extract generic visual features (such as color pattern, edges, elementary shapes) efficiently, achieve high accuracy for various visual task, and are easy to access [35,36]. Moreover, these networks have repeatedly been applied to different tasks, from which they were originally trained, to improve the target model's performance [36,37]. Here, we pre-train our OS-Net on a base dataset (ImageNet) and transfer the learned general visual feature extraction ability and adjust the model parameters to fit for the oil sheen prediction task.

3. Methods

The OS-Net proposed here is a deep convolutional neural network with residual blocks as basic elements that provides high accuracy and robust oil sheen detection. The dataset we used is created from images of oil sheen taken at close proximity to the water surface. We improve model performance by using transfer learning with three potential implementations. We also explore the best learning strategy. The main steps consist of data acquisition; data preparation; data preprocessing; OS-Net design, training, and optimizing; model evaluation; and the development of VOS-Net (deployed in OS-SS).

3.1. Data

3.1.1. Data Acquisition

There are currently no publicly available datasets of oil sheen images taken from close proximity to the water surface. Therefore, we created a visible oil sheen dataset using lab simulation videos from the OS-SS prototype. A large quantity and diversity of data was needed to train a robust model [38]. Thus, we first developed our library of close-proximity oil sheen images by creating oil sheen from oily sediments by air sparging or water injection. We recorded the videos using a visible light camera to provide color, shape, and texture of the oil sheen.

Sheen development videos were taken under anticipated field conditions (Figure 1). For the videos, oil was embedded 15 cm deep in a water-sediment column made up of a synthetic sediment mixture. A direct push probe with air or water injection was used to disturb the sediments and create a sheen from oil deposited into sediment in a 4-inch PVC tube. Three variations of the sediment mixture were considered, medium and fine sand mix, fine sand and fine (silt and clay) mix, and only fine (silt and clay) mix. Five crude oils of varying viscosities (2.21 cSt, 5.12 cSt, 66.3 cSt, 76.0 cSt, 469 cSt at 40 °C) were used to cover a range of oil types and ages expected in the field [22]. Different volumes of oil were placed in the sediments to produce sheens of higher or lower thickness as this impacts sheen color and form. The videos were taken using a digital single-lens reflex camera (Nikon DX D7200 with AF-S NKKOR 18–140 mm 1:3.5–5.6 G ED XR Lens) with a cool compact fluorescent lamp (CFL) lightbulb as the light source [22]. The average distance from the lens to the water surface is around 30 cm [22].

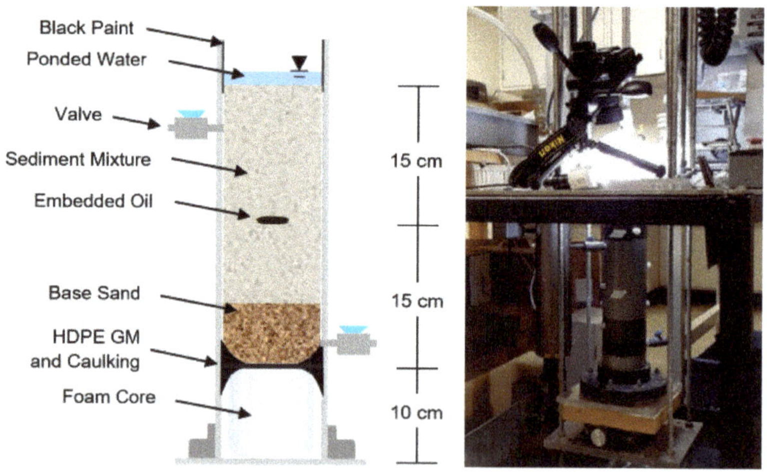

Figure 1. Schematic of water-sediment column experimental design including the lighting and image collection in the laboratory OS-SS setup.

3.1.2. Image Dataset Preparation

Based on the oil sheen videos, one frame of each second of the videos was scraped to create an efficient and nonredundant image dataset. In the data cleaning process, the blurry pictures and pictures where the water surface is obstructed, e.g., by experimenter's hands, were deleted. The photos were labeled manually as "with Sheen" or "no Sheen", based on expert opinion about the presence of the sheen (Figure 2). The total number of pictures extracted from the videos was 3398, which includes 1877 images labeled as "with Sheen" and 1521 images labeled as "no Sheen". After randomizing, 90 percent of the data were used for training and 10 percent for testing. To obtain reliable performance of the method, training and testing datasets were independent.

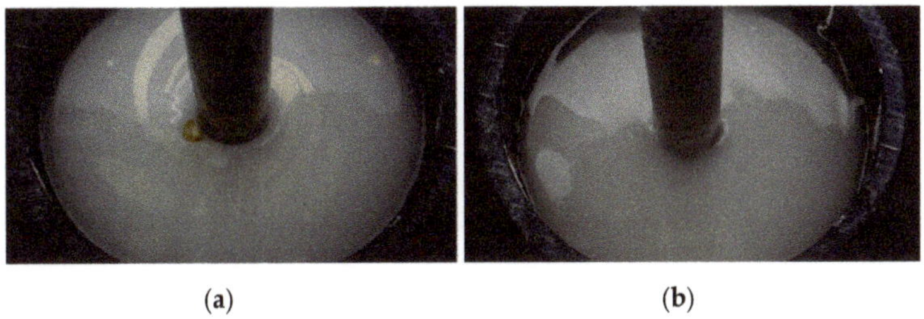

Figure 2. Example of surface water in the OS-SS setup showing a (**a**) "with sheen" image and a (**b**) "no sheen" image used to create the surface sheen image library for the CNN model development.

3.1.3. Data Preprocessing

To develop a model with the best performance, the images were preprocessed before use in the model. The pre-processing methods included redefining the size of the images, data augmentation, normalization, and other operations. A CNN searches for thousands of patterns from the data and the patterns a model can find are related to the image size [39]. As our pre-trained model's input image size is 224 × 224 × 3 (RGB image), the size of oil sheen images would be defined as 224 × 224 × 3 for training and testing to ensure the best transfer of the learned knowledge for our task. We normalized the tensor image using mean and standard deviation to obtain data within a range and reduce the skewness which helps the model to learn faster and more efficiently.

3.2. Oil Sheen Prediction Neural Network (OS-Net)

To build a deep neural network with better performance on oil sheen detection, we use residual block as basic element to overcome the vanishing gradient problem. To further improve the accuracy with limited oil sheen dataset, we use transfer learning strategies and explore the performance of different learning strategy on oil sheen detection.

The OS-Net includes 18 convolutional layers with 8 residual blocks (Figure 3), which is a deep while easy-to-train neural network with a more powerful learning capacity.

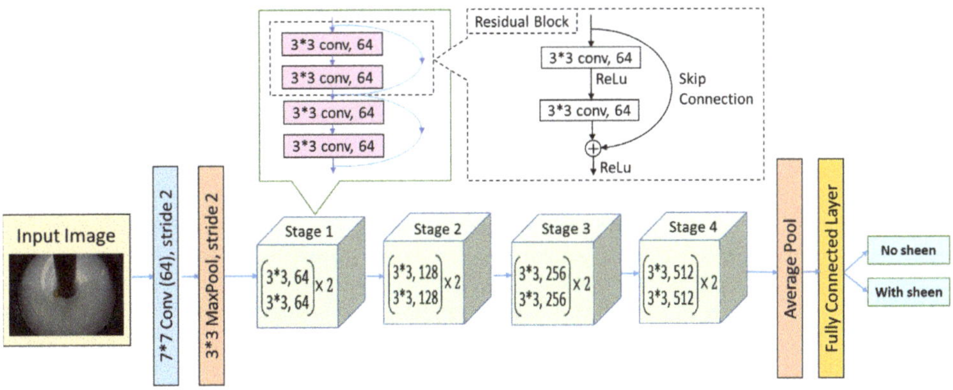

Figure 3. OS-Net architecture.

OS-Net has a traditional classification network architecture (see Figure 3). OS-Net has the first convolutional layer with 64 kernels (size = 7 × 7, stride = 2), followed by a maximum pooling layer (size = 3 × 3, stride = 2) to reduce the input oil sheen image size from 224 × 224 to a feature matrix of 56 × 56. The model backbone consists of 4 stages. Each stage consists of 2 residual blocks and each residual block has two 3 × 3 convolutional layers with skip connection. The penultimate layer is an average pooling layer to down sample the detection of features in feature maps. The last fully connected layer is also the output layer, which provides the classification prediction. We use Softmax to obtain the binary prediction result ('No sheen' or 'With sheen').

To accelerate the training and achieve higher accuracy, we use transfer learning strategy, which borrowed the learned general visual feature extraction knowledge from a larger dataset. The model starts with the pre-trained parameters using the ImageNet dataset and then adjusts these parameters to suit the oil sheen prediction task instead of starting the learning process from scratch with random parameter initialization. We explore the effectiveness of different strategies of transfer learning to gain the best OS-Net. These three strategies all used the same initial OS-Net architecture (Figure 3), but each model was trained differently and compared to the case without transfer learning. (Case A) OS-Net Without Transfer Learning. The model was trained from scratch on close-proximity oil sheen image dataset with random initialized parameters. (Case B) OS-Net With Transfer Learning Feature Extraction Strategy. We used the pre-trained model as a feature extractor to transfer the similar feature extraction ability learned from ImageNet to our model. Specifically, we pre-trained our OS-Net on ImageNet, modified the fully connected layers as shown in Figure 3, froze all the networks except the final layer, and then trained the model on our dataset. (Case C) OS-Net With Transfer Learning Fine-tuning Strategy. We used the pre-trained model as the starting point of training, then fine tuned all parameters for the target task. Specifically, we pre-trained our OS-Net on ImageNet, modified the fully connected layers, unfroze all layers, then trained the model on our oil sheen dataset.

In the training process, cross entropy was used as a loss function to update model weights w. The cross-entropy function is given in Equation (1) [40],

$$L = -\frac{1}{M}\sum_{i=1}^{M} p(x)\log(q(x)) \tag{1}$$

where M represents the number of classes, true probability p is the true label (no oil or with oil), and q is the predicted value of the current model.

The model was trained using the stochastic gradient descent (SGD) optimization algorithm [40,41]. The SGD updates the network parameters (weights and biases) to minimize the loss function. This algorithm is defined as:

$$w_{t+1} = w_t - \alpha\frac{\partial L_t}{\partial w_t} \tag{2}$$

where w donates any trainable variable (W or B), t is the current time step (algorithm iteration), and α is the learning rate.

3.3. OS-Net Performance Evaluation

Four performance evaluation metrics were employed to assess the OS-Net performance, including accuracy (Equation (3)), precision (Equation (4)), recall (Equation (5)), and F1 score (Equation (6)). Accuracy is the ratio of correctly predicted observations to the total observations. Precision is the ratio of correctly predicted positive observations to the total predicted positive observations. Recall is the ratio of correctly predicted positive observations to all observations in actual class. F1 Score is the weighted average of precision and recall [42] and is a reliable and robust indicator for accurate model measurements. The range of the F-1 score is from 0 to 1, with 0 being the worst possible and 1 being the best.

$$\text{Accuracy} = \frac{\text{TP} + \text{TN}}{\text{TP} + \text{FP} + \text{FN} + \text{TN}} \tag{3}$$

$$\text{Precision} = \frac{\text{TP}}{\text{TP} + \text{FP}} \tag{4}$$

$$\text{Recall} = \frac{\text{TP}}{\text{TP} + \text{FN}} \tag{5}$$

$$\text{F1 score } = 2 \times \frac{\text{Precision} \times \text{Recall}}{\text{Precision} + \text{Recall}} \tag{6}$$

where TP refers to true positives, TN refers to true negatives, FP refers to positives, and FN refers to false negatives.

3.4. Real-Time Video Oil Sheen Prediction (VOS-Net)

Based on the OS-Net, we developed a real-time video-based oil sheen prediction system (VOS-Net). The input of the VOS-Net is sheen formation video and the output is the real-time prediction result. Given a video input, the VOS-Net classifies the images every few frames based on image extraction frequency parameter.

The latency of the VOS-Net may delay the prediction results if the model predicts frame by frame. To achieve real-time prediction, an appropriate image extraction frequency will have to be set to offset the delay. When the interval between image extractions is greater than the model inference time, predictions would not be delayed.

While recall is 100% and false negatives are not likely, the precision is not 100% so some false-positive results could occur and a method was developed to address this potential.

We combined oil sheen domain knowledge with our machine learning algorithm to improve the prediction accuracy. When oil leaks from the sediment in OS-SS, the oil will spread out rapidly on the water surface to form sheen. The appearance of oil sheen can sometimes be fleeting, disappearing after a few seconds, or it can persist for a longer period [22]. Thus, if the sheen only appears for one frame (the video includes 60 frames per second), the prediction may be a false positive. We set a filter with kernel size k, where k represents the threshold of positive prediction. We suppressed the positive prediction until k successive positive predictions occurred. The kernel size k can be adjusted to control the sensitivity of VOS-Net to avoid false positives and increase predictive accuracy.

Figure 4 shows how the VOS-Net filter works. The VOS-Net predicts the oil sheen every 60 frames (~1 s) and records the result. Figure 4a shows the ground truth (b) and (c) show the VOS-Net prediction result with the different filters applied. The red line represents the model's prediction of "with sheen", while the green line depicts a model result of "no sheen". The ground truth is when no oil appears in the following frames. Without applying a filter, (b), VOS-Net could have a false positive (red lines). After applying a filter, (c), the VOS-Net can reduce the false-positive incidence and achieve higher accuracy.

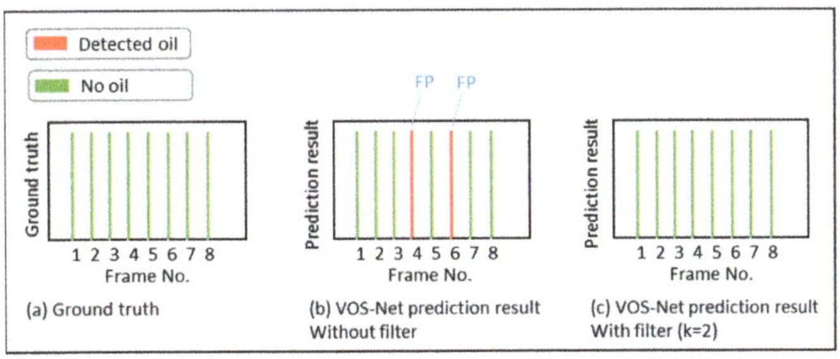

Figure 4. VOS-Net filter performance demonstration. Screening out single false-positive events improves the model accuracy to 99%.

4. Results

4.1. OS-Net Performance

Table 1 shows the accuracy, precision, recall, and F-1 score for three models: (A) OS-Net Without Transfer Learning; (B) OS-Net With Transfer Learning Feature Extraction Strategy; (C) OS-Net With Transfer Learning Fine-tuning Strategy. In model development, we use grid search for hyperparameter tuning for each of the three models mentioned above. Then, we obtain the best hyperparameter settings and the corresponding best model performance (Table 1) for each of the models. The best initial learning rate of the best OS-Net without transfer learning is 0.002, which is larger than the best initial learning rate of the OS-Net with ImageNet pretrained model (0.001). We also use a learning rate schedule to adjust the learning rate during training. The learning rate will decrease rapidly in the first few epochs, then gradually reduce as the iteration continues to make the model more stable. The settings of grid search values for hyperparameters tuning in each model are listed in Table S3. The best hyperparameters for our best OS-Net model are listed here: the momentum factor is 0.9, the initial learning rate is 0.001, epochs are 30, and the batch size is 16.

Table 1. Model evaluation result.

Model	Accuracy (Test)	F-1 Score	Precision	Recall
OS-Net Without Transfer Learning	0.94	0.94	0.93	0.95
OS-Net With Transfer Learning Feature Extraction Strategy	0.94	0.93	0.92	0.94
OS-Net With Transfer Learning Fine-tuning Strategy	0.99	0.99	0.98	1.00

From the results presented in Table 1, the OS-Net With Transfer Learning and the Fine-tuning Strategy approach performed best. Therefore, we selected OS-Net With Transfer Learning and the Fine-tuning Strategy as our final OS-Net.

The OS-Net is robust for the oil sheen classification task in our laboratory-generated dataset, accurately predicting the oil sheen appearance on the water surface in our OS-SS prototype. The accuracy and F-1 score of the OS-Net are up to 99%. The model can efficiently extract the low-level features and high-level features from images. The recall rate is up to 100%, which means that there are no false negatives.

The reason why OS-Net Without Transfer Learning has relatively low accuracy, compared to the final OS-Net, may be a result of the relatively small size of our oil sheen image dataset. When we trained the model from scratch on our oil sheen image dataset, the knowledge that the model can learn is limited, leading to overfitting and a lack of generalizability. The model can only fit well with similar data but not data the model has never seen before. Here, we also tried a "warm-up" training approach to reduce variance in the early stage of training and achieve better performance for the OS-Net without transfer learning model, but the final accuracy was not improved. This result indicates the role of transfer learning is more than just "warming up" the weights better. In contrast, the OS-Net With Transfer Learning and the Fine-tuning Strategy is pre-trained on the ImageNet dataset, which has millions of images, providing the model with a better generalization ability. The model was then fine tuned on the specific oil sheen dataset to adjust its parameters to improve the specific oil sheen prediction task. This suggests that transfer learning is helpful to transfer the extracted basic features from ImageNet to our new task. Even though the high-level features have large visual differences between the two tasks (ImageNet has no oil sheen images), the basic feature (low-level patterns, e.g., edges, color patterns, elementary shapes) extraction ability is shared across tasks.

It is worth noting that the performance of the OS-Net With Transfer Learning Feature Extraction Strategy did not improve significantly compared to the OS-Net Without Transfer Learning. This may be due to the large gap between the target (close-proximity oil sheen) images and the pre-trained model's ImageNet dataset. The types of advanced features (middle or high level) that need to be extracted from the two datasets are not the same, so using the pre-trained model as the feature extractor may not be ideal.

Figure 5 shows sample images with the ground truth label and the OS-Net prediction results. Images with oil sheen or without oil sheen can be correctly classified by the OS-Net. Since the water surface is moving during the video recording and the shape and color of the sheen will be constantly changing during the video, the variance increases. Our OS-Net performs well under these conditions, indicating the robustness and generalizability of our method.

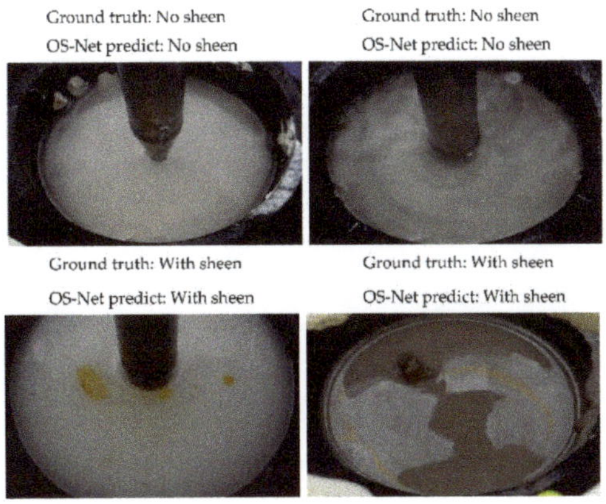

Figure 5. Example oil sheen image ground truth and OS-Net prediction results.

4.2. VOS-Net

Finally, a video-based real-time oil sheen detection algorithm (VOS-Net) was developed, providing users with a visual image with oil sheen detection result. Figure 6 demonstrates the VOS-Net functions. Given a video input, the VOS-Net will show the image extracted on the top window with time, detected frame number, and prediction result of the current frame. Above the video image, a record of the prediction result is shown.

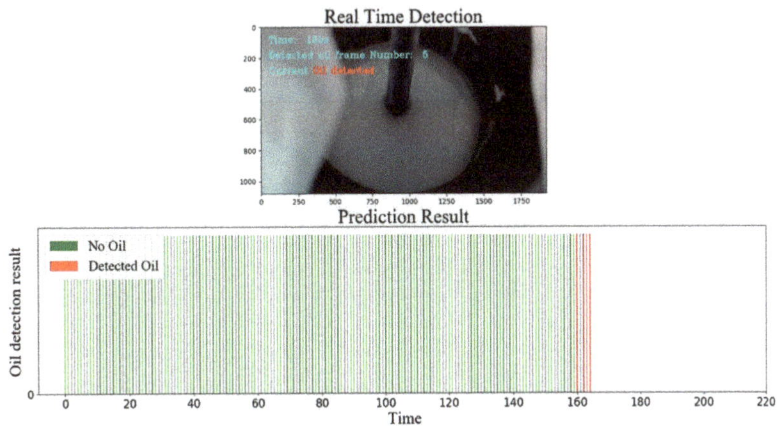

Figure 6. VOS-Net prediction window and result; a demo of VOS-Net result is available at https://youtu.be/kCZnMAYBByk (accessed on 15 July 2022).

The image extraction frequency can be adjusted manually, based on the performance of the users' computer, to offset the latency effect of the VOS-Net. In the test, the model inference time to predict one image is 0.03–0.04 s (2.4 frames on average). The shortest time that oil sheen appears on the water surface in OS-SS is 3 s, based on our lab experiment. In this case, setting the image extraction frequency anywhere from 3 frames to 180 frames, the VOS-Net can achieve real-time prediction on each image. To achieve higher accuracy and avoid response latency, we suggest selecting the image extraction frequency parameter in a range from 30 to 60 (0.5 s to 1 s).

When we applied a filter (k) larger than two, the filter could efficiently suppress false-positive predictions to increase the accuracy of VOS-Net.

5. Discussion and Conclusions

This paper introduces the concept of a subaquatic sediment oil sheen screening system (OS-SS) and develops a novel oil sheen detection system to autonomously monitor the sediment's sheening potential after disturbance with compressed air or water. The OS-SS also provides reliable real-time oil sheen video. Based on the videos from OS-SS, we successfully created a new close-proximity oil sheen image dataset, which includes thousands of oil sheen images (five types of oil sheen formation details under different environmental conditions in the lab). This dataset can become a resource for a broad range of oil-sheen-modeling-related research.

We developed an Oil Sheen Prediction Neural Network (OS-Net), a Convolutional Neural Network (CNN), combined with transfer learning, to achieve real-time image-based oil sheen prediction from video and applied the model in a simulation of the OS-SS. The accuracy of OS-Net was up to 99%. To mitigate false positives from the video stream, we employed a sensitivity filter in the VOS-Net. Thus, our VOS-Net can output accurate feedback to monitor the formation of an oil sheen present given a video stream, even if that formation is short lived in the order of a few seconds to a few tens of seconds. The kernel size, a hyperparameter-insensitive filter, can be adjusted to control the sensitivity of VOS-Net to avoid false positives. Moreover, we determined the image extraction frequency parameters in VOS-Net needed to avoid model prediction latency and provide real-time feedback. To further reduce the model reference time and extend the range of image extraction frequency, we tried to use depth-wise convolution to replace the regular convolution in the OS-Net to reduce the number of parameters and computations. As a result, compared to the original OS-Net, the accuracy of the OS-Net model with depth-wise convolution was reduced by 1.66% and the average latency was reduced by 2.65%. The result suggests that depth-wise convolution can reduce the inference time. How to use depth-wise convolution to reduce model latency while also ensuring accuracy will be a good question to explore in the future. Model quantization, mapping values from a larger set to a smaller one to reduce the floating point, is another approach worth trying to reduce the memory and complexity of computations in the future.

The OS-SS with VOS-Net could be easily applied in the field, potentially being deployed on an autonomous watercraft to map the spatial distribution of sediment sheening potential in rivers, creeks, fords, and shallow waters [43]. The VOS-Net can automatically show the real-time result and record the detection result along with the GPS location to determine the precise location of the sediments with high sheening potential. The OS-SS with VOS-Net could be an efficient tool for mapping problematic regions of subaquatic sediments for further evaluation or remediation. The autonomous nature of the approach also makes it ideal for deployment in difficult-to-reach terrains. Moreover, as our OS-Net is based on visible light images, the requirements and costs for image collection equipment are low. The algorithms could be further explored to be used with a cellphone camera to detect the oil sheen on the water or in other similar scenarios, such as oil drilling or storm runoff in the future.

Furthermore, our results indicated that combining residual blocks with transfer learning helped our OS-Net to overcome the challenge of limited datasets to obtain a deeper and more robust neural network and achieve high accuracy. The appropriate transfer learning strategy was related to the pre-training and task datasets. When there is a significant gap between the two datasets, a fine-tuning transfer learning strategy improved the accuracy of the model. The improved performance of OS-Net is a good example. Data augmentation can also potentially enhance the training dataset for the small dataset. In our case, using random horizontal flip alone improved the model's accuracy by 0.65% (SI Table S2).

Although the model has high accuracy, there are still some limitations. The data distribution likely does not represent the full range of actual data distribution that may be encountered when probing different natural subaquatic environments because of the range of environmental parameters that may be encountered (e.g., different types of oil, different degrees of weathering, different natural organic matter content, and natural biofilm formation). Thus, the domain gap between the lab data during training and the video from actual field samples may influence the accuracy. Future work can overcome these challenges by enlarging the oil sheen dataset collected with different natural conditions, including a range of videos and images taken for field sheen events using the OS-SS.

Supplementary Materials: The following supporting information can be downloaded at: https://www.mdpi.com/article/10.3390/app12178865/s1, Table S1: Input data summary table. Table S2: OS-Net performance under different data augmentation methods. Table S3: OS-Net hyperparameters tuning grid search values. Figure S1: VOS-Net prediction results. The X axis shows the prediction index with time (1 prediction/s) and the Y axis shows the VOS-Net prediction result. The black line represents the region where there was oil sheen present. Ground truth labels are shown at the top of each region.

Author Contributions: J.D., P.B., N.S. and T.P.H.; Data curation, K.S.; Formal analysis, J.D., J.S. and Y.G.; Funding acquisition, P.B., N.S., T.P.H. and G.V.L.; Investigation, K.S. and G.V.L.; Methodology, J.D., K.S. and G.V.L.; Supervision, J.S.; Validation, J.D. and Y.G.; Writing—review & editing, K.S., J.S., Y.G., P.B., N.S., T.P.H. and G.V.L. All authors have read and agreed to the published version of the manuscript.

Funding: This research was funded by Chevron.

Data Availability Statement: The data presented in this study are available on request from the corresponding author. The data are not publicly available due to agreements in place with the study sponsor.

Conflicts of Interest: The authors declare no conflict of interest.

References

1. Neff, J.M. Composition and fate of petroleum and spill-treating agents in the marine environment. In *Synthesis of Effects of Oil on Marine Mammals*; Battelle Memorial Institute: Ventura, CA, USA, 1988; pp. 1–33. [CrossRef]
2. Romero, I.C.; Schwing, P.T.; Brooks, G.R.; Larson, R.A.; Hastings, D.W.; Ellis, G.; Goddard, E.A.; Hollander, D.J. Hydrocarbons in deep-sea sediments following the 2010 deepwater horizon blowout in the Northeast Gulf of Mexico. *PLoS ONE* **2015**, *10*, e0128371. [CrossRef] [PubMed]
3. National Research Council. *Oil in the Sea III: Inputs, Fates, and Effects*; National Academies Press: Washington, CA, USA, 2003. [CrossRef]
4. Picou, J.S.; Gill, D.A.; Dyer, C.L.; Curry, E.W. Disruption and stress in an alaskan fishing community: Initial and continuing impacts of the exxon valdez oil spill. *Ind. Crisis Q.* **2016**, *6*, 235–257. [CrossRef]
5. Adams, A. PAGe 2. *Summary of Information Concerning the Ecological and Economic Impacts of the BP Deepwater Horizon Oil Spill Disaster New England Coral Canyons and Seamounts Area*. NRDC Issue Paper. 19 June 2015. Available online: https://www.nrdc.org/resources/summary-information-concerning-ecological-and-economic-impacts-bp-deepwater-horizon-oil (accessed on 30 July 2021).
6. Effects of Oil Spills: What Impact Does it Have on Wildlife and Humans? Available online: https://www.offshore-technology.com/features/effects-oil-spills/ (accessed on 30 July 2021).
7. Nance, E.; King, D.; Wright, B.; Bullard, R.D. Ambient air concentrations exceeded health-based standards for fine particulate matter and benzene during the deepwater horizon oil spill. *J. Air Waste Manag. Assoc.* **2016**, *66*, 224–236. [CrossRef] [PubMed]
8. Al-Shammari, A.; Levin, E.; Shults, R. Oil spills detection by means of uas and low-cost airborne thermal sensors. *ISPRS Ann. Photogramm. Remote Sens. Spat. Inf. Sci.* **2018**, *4*, 293–301. [CrossRef]
9. Del Frate, F.; Petrocchi, A.; Lichtenegger, J.; Calabresi, G. Neural networks for oil spill detection using ERS-SAR data. *IEEE Trans. Geosci. Remote Sens.* **2000**, *38*, 2282–2287. [CrossRef]
10. Zeng, K.; Wang, Y. A Deep convolutional neural network for oil spill detection from spaceborne SAR images. *Remote Sens.* **2020**, *12*, 1015. [CrossRef]
11. Fiscella, B.; Giancaspro, A.; Nirchio, F.; Pavese, P.; Trivero, P. Oil spill detection using marine SAR images. *Int. J. Remote Sens.* **2010**, *21*, 3561–3566. [CrossRef]
12. Vespe, M.; Greidanus, H. SAR image quality assessment and indicators for vessel and oil spill detection. *IEEE Trans. Geosci. Remote Sens.* **2012**, *50 Pt 2*, 4726–4734. [CrossRef]
13. Topouzelis, K.N. Oil spill detection by SAR images: Dark formation detection, feature extraction and classification algorithms. *Sensors* **2008**, *8*, 6642–6659. [CrossRef]
14. Kubat, M.; Holte, R.C.; Matwin, S. Machine learning for the detection of oil spills in satellite radar images. *Mach. Learn.* **1998**, *30*, 195–215. [CrossRef]

15. Singha, S.; Bellerby, T.J.; Trieschmann, O. Satellite oil spill detection using artificial neural networks. *IEEE J. Sel. Top. Appl. Earth Obs. Remote Sens.* **2013**, *6*, 2355–2363. [CrossRef]
16. De Kerf, T.; Gladines, J.; Sels, S.; Vanlanduit, S. Oil spill detection using machine learning and infrared images. *Remote Sens.* **2020**, *12*, 4090. [CrossRef]
17. Al-Ruzouq, R.; Gibril, M.B.A.; Shanableh, A.; Kais, A.; Hamed, O.; Al-Mansoori, S.; Khalil, M.A. Sensors, features, and machine learning for oil spill detection and monitoring: A review. *Remote Sens.* **2020**, *12*, 3338. [CrossRef]
18. Bukin, O.; Proschenko, D.; Korovetskiy, D.; Chekhlenok, A.; Yurchik, V.; Bukin, I. Development of the artificial intelligence and optical sensing methods for oil pollution monitoring of the sea by drones. *Appl. Sci.* **2021**, *11*, 3642. [CrossRef]
19. Jiao, Z.; Jia, G.; Cai, Y. A new approach to oil spill detection that combines deep learning with unmanned aerial vehicles. *Comput. Ind. Eng.* **2019**, *135*, 1300–1311. [CrossRef]
20. Alharam, A.; Almansoori, E.; Elmadeny, W.; Alnoiami, H. Real time AI-based pipeline inspection using drone for oil and gas industries in Bahrain. In Proceedings of the 2020 International Conference on Innovation and Intelligence for Informatics, Computing and Technologies (3ICT), Sakheer, Bahrain, 20–21 December 2020. [CrossRef]
21. Sale, T.; Hopkins, H.; Andrew, K. Managing Risk at LNAPL Sites. *Am. Pet. Inst. Soil Groundw. Res. Bull.* **2018**, *18*, 51–53.
22. Sitler, K.; Scalia, J.; Sale, T. *Identification and Validation of Screening Methods for Assessment of the Sheening Potential of Embedded Oil in Sediments*; Colorado State University: Fort Collins, CO, USA, 2020.
23. He, K.; Zhang, X.; Ren, S.; Sun, J. Deep residual learning for image recognition. In Proceedings of the IEEE Conference on Computer Vision and Pattern Recognition (CVPR), Las Vegas, NV, USA, 27–30 June 2016; 2016; pp. 770–778.
24. Alzubaidi, L.; Zhang, J.; Humaidi, A.J.; Al-Dujaili, A.; Duan, Y.; Al-Shamma, O.; Santamaría, J.; Fadhel, M.A.; Al-Amidie, M.; Farhan, L. Review of deep learning: Concepts, CNN architectures, challenges, applications, future directions. *J. Big Data* **2021**, *8*, 1–74. [CrossRef]
25. Gu, J.; Wang, Z.; Kuen, J.; Ma, L.; Shahroudy, A.; Shuai, B.; Liu, T.; Wang, X.; Wang, L.; Wang, G.; et al. Recent Advances in Convolutional Neural Networks. *Pattern Recognit.* **2018**, *77*, 354–377. [CrossRef]
26. Zhiqiang, W.; Jun, L. A review of object detection based on convolutional neural network. In Proceedings of the 2017 36th Chinese Control Conference (CCC), Dalian, China, 26–28 July 2017; pp. 11104–11109. [CrossRef]
27. Rehman, S.; Ajmal, H.; Farooq, U.; Ain, Q.U.; Riaz, F.; Hassan, A. Convolutional neural network based image segmentation: A review. *Pattern Recognit. Track. XXIX* **2018**, *10649*, 191–203. [CrossRef]
28. Gao, Y.; Mosalam, K.M. Deep transfer learning for image-based structural damage recognition. *Comput. Civ. Infrastruct. Eng.* **2018**, *33*, 748–768. [CrossRef]
29. Hussain, M.; Bird, J.J.; Faria, D.R. A Study on CNN transfer learning for image classification. *Adv. Intell. Syst. Comput.* **2018**, *840*, 191–202. [CrossRef]
30. Zheng, J.; Yang, G.; Huang, Y.; Liu, L.; Hong, G.; Qiu, Z.; Liu, S. Research of water body turbidity classification model for aquiculture based on transfer learning. *J. Phys. Conf. Ser.* **2021**, *1757*, 012004. [CrossRef]
31. Deng, J.; Dong, W.; Socher, R.; Li, L.-J.; Kai, L.; Li, F.-F. ImageNet: A large-scale hierarchical image database. *IEEE Comput. Soc.* **2010**, *2009*, 248–255. [CrossRef]
32. Simonyan, K.; Zisserman, A. Very deep convolutional networks for large-scale image recognition. *arXiv* **2014**, arXiv:1409.1556.
33. Krizhevsky, A.; Sutskever, I.; Hinton, G.E. ImageNet Classification with Deep Convolutional Neural Networks. *Commun. ACM* **2017**, *60*, 84–90. [CrossRef]
34. Szegedy, C.; Liu, W.; Jia, Y.; Sermanet, P.; Reed, S.; Anguelov, D.; Erhan, D.; Vanhoucke, V.; Rabinovich, A. Going deeper with convolutions. In Proceedings of the IEEE Conference on Computer Vision and Pattern Recognition (CVPR), Boston, MA, USA, 7–12 June 2015; pp. 1–9. [CrossRef]
35. Zhuang, F.; Qi, Z.; Duan, K.; Xi, D.; Zhu, Y.; Zhu, H.; Xiong, H.; He, Q. A comprehensive survey on transfer learning. *Proc. IEEE* **2019**, *109*, 43–76. [CrossRef]
36. Yosinski, J.; Clune, J.; Bengio, Y.; Lipson, H. How transferable are features in deep neural networks? *Adv. Neural Inf. Process. Syst.* **2014**, *4*, 3320–3328.
37. Best, N.; Ott, J.; Linstead, E.J. Exploring the efficacy of transfer learning in mining image-based software artifacts. *J. Big Data* **2020**, *7*, 1–10. [CrossRef]
38. Gong, Z.; Zhong, P.; Hu, W. Diversity in machine learning. *IEEE Access* **2019**, *7*, 64323–64350. [CrossRef]
39. Li, H.; Ellis, J.G.; Zhang, L.; Chang, S.-F. PatternNet: Visual pattern mining with deep neural network. In Proceedings of the 2018 ACM on International Conference on Multimedia Retrieval, Yokohama, Japan, 11–14 June 2018. [CrossRef]
40. Rojas, R. The backpropagation algorithm. *Neural Netw.* **1996**, *40*, 149–182. [CrossRef]
41. Cui, X.; Zhang, W.; Tüske, Z.; Picheny, M. Evolutionary stochastic gradient descent for optimization of deep neural networks. *Adv. Neural Inf. Process. Syst.* **2018**, *2018*, 6048–6058.
42. Yacouby, R.; Axman, D. Probabilistic extension of precision, recall, and F1 score for more thorough evaluation of classification models. In Proceedings of the First Workshop on Evaluation and Comparison of NLP Systems, Punta Cana, Dominican Republic, 20 November 2020; pp. 79–91. [CrossRef]
43. Valada, A.; Velagapudi, P.; Kannan, B.; Tomaszewski, C.; Kantor, G.; Scerri, P. Development of a low cost multi-robot autonomous marine surface platform. *Springer Tracts Adv. Robot.* **2014**, *92*, 643–658. [CrossRef]

applied sciences

MDPI

Article

Tool Wear Monitoring Using Improved Dragonfly Optimization Algorithm and Deep Belief Network

Leo Gertrude David [1], Raj Kumar Patra [2], Przemysław Falkowski-Gilski [3,*], Parameshachari Bidare Divakarachari [4,*] and Lourdusamy Jegan Antony Marcilin [5]

1 Department of Visual Communication, Kumaraguru College of Liberal Arts and Science, Coimbatore 641035, India
2 Department of Computer Science and Engineering, CMR Technical Campus, Hyderabad 501401, India
3 Faculty of Electronics, Telecommunications and Informatics, Gdansk University of Technology, Narutowicza 11/12, 80-233 Gdansk, Poland
4 Department of Electronics and Communication Engineering Nitte Meenakshi Institute of Technology, Bangalore 560064, India
5 Department of ECE, Sathyabama Institute of Science and Technology, Chennai 600119, India
* Correspondence: przemyslaw.falkowski@eti.pg.edu.pl (P.F.-G.); paramesh@nmit.ac.in (P.B.D.)

Abstract: In recent decades, tool wear monitoring has played a crucial role in the improvement of industrial production quality and efficiency. In the machining process, it is important to predict both tool cost and life, and to reduce the equipment downtime. The conventional methods need enormous quantities of human resources and expert skills to achieve precise tool wear information. To automatically identify the tool wear types, deep learning models are extensively used in the existing studies. In this manuscript, a new model is proposed for the effective classification of both serviceable and worn cutting edges. Initially, a dataset is chosen for experimental analysis that includes 254 images of edge profile cutting heads; then, circular Hough transform, canny edge detector, and standard Hough transform are used to segment 577 cutting edge images, where 276 images are disposable and 301 are functional. Furthermore, feature extraction is carried out on the segmented images utilizing Local Binary Pattern (LBPs) and Speeded up Robust Features (SURF), Harris Corner Detection (HCD), Histogram of Oriented Gradients (HOG), and Grey-Level Co-occurrence Matrix (GLCM) feature descriptors for extracting the texture feature vectors. Next, the dimension of the extracted features is reduced by an Improved Dragonfly Optimization Algorithm (IDOA) that lowers the computational complexity and running time of the Deep Belief Network (DBN), while classifying the serviceable and worn cutting edges. The experimental evaluations showed that the IDOA-DBN model attained 98.83% accuracy on the patch configuration of full edge division, which is superior to the existing deep learning models.

Keywords: canny edge detector; deep belief network; dragonfly optimization algorithm; image processing; local binary pattern; tool wear monitoring

Citation: David, L.G.; Patra, R.K.; Falkowski-Gilski, P.; Divakarachari, P.B.; Antony Marcilin, L.J. Tool Wear Monitoring Using Improved Dragonfly Optimization Algorithm and Deep Belief Network. *Appl. Sci.* **2022**, *12*, 8130. https://doi.org/10.3390/app12168130

Academic Editor: Fabio La Foresta

Received: 26 June 2022
Accepted: 8 August 2022
Published: 14 August 2022

1. Introduction

Tool wear is the result of cutting temperature, cutting force, and mechanical friction in the milling process of Computer Numerical Control (CNC) machines [1,2]. Tool wear reduces the workpiece quality and increases the workpiece surface roughness [3]; serious tool wear causes chatter, fracturing, and chipping, which damages both the machine tools and the workpiece, and can also lead to serious processing accidents [4]. Therefore, appropriate monitoring and classification mechanisms could help to decrease the loss caused by tool wear, and to obtain better surface quality [5]. By modifying the structure of solid materials, the machining process is the most significant industrial method for producing semi-finished and final outputs. Cutting parameters are typically utilized to eliminate chips in the substance [6]. However, even though advanced technologies in cutting tool compositions and milling machine control technologies have resulted in extended mechanical properties, the instruments eventually wear down with time [7]. Tool wear invariably has a direct impact on the surface polish and predicted probability of the completed workpiece, resulting in failures [8]. Tool removal and unplanned shutdowns because of worn tools could even result in machine breakdown and product inconsistencies, and eventually in financial loss. As a result, it is critical to avoid unscheduled downtime throughout operation, which has a significant effect on organizational performance [9]. Tool condition monitoring methodologies have been adopted for nearly two decades to minimize this, and they have proved themselves as a basic requirement of modern machine tools [10].

Coated tools are more widely utilized in numerous machining operations than uncoated tools because they have a higher tool performance and increased processability [11]. The remaining thickness of coated tools throughout manufacturing operations might be beneficial for the technician to monitor when using worn tools [12]. To determine the depth of a coating layer, aggressive tools such as microhardness monitoring and metallography would be used, although they are costly, time-consuming, and difficult to utilize [13]. The tool wear monitoring is categorized into direct approaches and indirect approaches based on the hardness of the machining process [14]. The direct approaches based on the measurement of flank wear consists of electrical resistance, radioactivity, and vision inspection [15]. The direct approaches are not conducive to the practical applications, due to the limitation of accessing similar levels of illumination, cutting fluid, and the presence of chips, and high requirements must be reached to measure the environment [16]. Therefore, indirect approaches are effective in monitoring tool wear based on image analysis. The computer vision approach directly measures tool wear, which helps in achieving high levels of reliability and precision [17,18]. Nonetheless, the bulk of tool condition measurement techniques reported in currently published studies have utilized complex, non-production-ready sensing technologies, including force dynamometers [19]. Due to the diversity of signals that must be acquired from the operating equipment, even with sophisticated data-gathering hardware, collecting high-quality datasets to train the machine is challenging and time-consuming [20]. Cutting tools are typically used to decrease cutting temperature and friction during machining processes to increase tool life and surface quality. Water-oil emulsions, with emulsifiers and additives to stop corrosion and bacterial development, make up the majority of cutting fluids. Poisonous compounds are present in the emulsions, machine oils, and the heavy metals that are combined with the fluids during the machining process. In order to lessen the impact on the environment, cutting fluid application must be reduced and hazardous components must be eliminated. In this manuscript, a new Improved Dragonfly Optimization Algorithm – Deep Belief Network (IDOA-DBN) model is proposed for effective tool wear monitoring; the major contributions of this study are listed below:

- After image collection, cutting edge detection is accomplished using the circular Hough transform technique, canny edge detector, and standard Hough transform.
- Subsequently, feature extraction is performed utilizing Speeded up Robust Features (SURF) and Local Binary Pattern (LBP) feature descriptors to extract the texture feature vectors; then, IDOA is proposed to reduce the dimensions of the extracted texture feature vectors, which helps in improving the system complexity and running time of the classifier.
- Lastly, the selected discriminative feature vectors are fed into the DBN to classify serviceable and worn cutting edges. The IDOA-DBN model's effectiveness is evaluated by means of F-score, recall, precision, and accuracy.

This manuscript is structured as follows: recent papers on the topic of tool wear monitoring are reviewed in Section 2. Theoretical descriptions and results of the experimental analysis of the IDOA-DBN model are provided in Sections 3 and 4, respectively. The conclusion of the manuscript is presented in Section 5.

2. Related Works

H. Oo [21] combined Multiple Linear Regression (MLR) and Random Forest Classifier (RFC) techniques to evaluate grinding capacity and detect the wear conditions of robotic belt grinding. Here, five distinct belts exhibiting varied tool wear conditions were used in the proposed simulation process, and 300 observations of grinding belt surface wear were used as training and testing parameters for the belt condition classifier. The fact that the data were fixed was considered one of the limitations. G. Li [22] combined time–frequency, frequency, and time domain feature extraction techniques to extract feature vectors from the vibration and force signals of the CNC machines. Next, a Gradient Boosting Decision Tree (GBDT) approach was applied for selecting the optimal feature vectors, and then classification was carried out using a hybrid method. Here, Prognostics and Health Management (PHM) challenge 2010 dataset was utilized to confirm the projected system. In this case, not every trait was automatically linked to tool wear. In addition, M.T. García-Ordás [23] segmented the cutting edges from the images of edge profile cutting heads, and then a local binary pattern descriptor was used to extract feature values from the segmented wear patch regions. Next, the Support Vector Machine (SVM) classifier was used to classify the cutting edges as serviceable or worn, and it was discovered that the overlapping between these patches began at the bottom of the main edge. G.D. Simon and R. Deivanathan [24] conducted a descriptive statistical evaluation to extract feature vectors from drilling-induced vibration signals, after which, the K-star classification method was used for classifying the tool conditions: bad tool at high speed, bad tool at low speed, good tool at high speed, and good tool at low speed. However, because of its wear resistance properties, the tool proved challenging to cut.

In reference [25], M. Marani utilized a Long Short-Term Memory (LSTM) network for tool flank wear prediction that helped in obtaining better machined surface with low manufacturing cost. Using a validation set, the Root Mean Square Error (RMSE) for the LSTM networks with hidden layers was determined. The findings revealed that the most efficient LSTM contains two layers and eight hidden layers, as well as a lower RMSE. Nonetheless, the prediction behavior of the two-layer LSTM network surpassed the three-layer model. In reference [26], X. Liu utilized raw signals as network inputs for monitoring the tool wear of a high-speed CNC machine. After acquiring the raw sensor signals, feature extraction was accomplished using a parallel residual network for extracting multiscale local feature vectors. Furthermore, a stacked bidirectional LSTM network was utilized to obtain the time series feature vectors related to the tool wear properties. The suggested LSTM had maximum convergence speed, but, at the same time, it exhibited high training loss. In reference [27], Z. Huang developed a Deep Convolutional Neural Network (DCNN) model for tool wear monitoring based on multidomain feature fusion. However, when used as a predictor for machine vibration, the DCNN algorithm would have been unable to detect tool wear in actual environments. X.C. Cao [28] combined a CNN model and Derived Wavelet Frames (DWF) for tool wear state prediction utilizing machine spindle vibration signals. The reconstituted subsignals were then layered into a 2-D signal vector to replicate the design of a 2-D CNN, while simultaneously maintaining additional information. Unfortunately, this raised the network's complex nature, making it even harder to optimize and increase the likelihood of overfitting.

X. Liao [29] combined an SVM classifier and Genetic Algorithm (GA) to predict tool wear state. Initially, wavelet packet decomposition techniques, frequency domain, and time domain statistics were used to extract cutting force signal. Finally, the SVM classifier with Grey Wolf Optimizer (GWO) was developed to obtain the state recognition results. On the training dataset, the performance accuracy was indeed the lowest. To increase tool wear prediction accuracy, X. Wu [30] used Bidirectional LSTM (BiLSTM) and Singular Value Decomposition (SVD) methods. The measuring input was received from the cutting force output. The raw cutting force data were then recreated using the Hankle matrix, and then, the signal features were extracted using the SVD of the regenerated matrix. This technique properly detects the tool's wear phase; however, it is unable to estimate the tool's wear rating. In [31], S. Laddada integrated an improved extreme learning machine and complex continuous wavelet transform to investigate the condition of cutting tools, and to predict their remaining lifetime. The suggested technique entails combining the Complex Continuous Wavelet Transform (CCWT) and the Improved Extreme Learning Machine (IELM). Moreover, since it was premised on the use of past data, its reliability is low.

By reviewing the abovementioned existing studies, it is clear that deep-learning-based classification techniques perform substantial data reductions, which may cause information loss. Additionally, with such techniques it is difficult to develop feature descriptors for precise tool wear estimation, due to the variations in cutting force. Therefore, a novel feature optimization method based DBN is proposed in this manuscript, which is free from three major concerns: overfitting, training, and testing efficiency.

3. Methodology

Deep Belief Networks (DBN) are used to establish tool wear prediction models due to their superior learning speeds and fast rates of convergence to the optimal outcomes, regardless of whether the sample data are small or large. This improves the modeling accuracy and efficiency of the tool wear monitoring system. DBN parameters are typically manually adjusted, which results in low predictive accuracy and efficiency. The Improved Dragonfly Optimization Algorithm (IDOA) is then designed to optimize the computational complexity and running time of the DBN. Therefore, the combination of IDOA and DBN provides better results when compared to other approaches, which lack efficiency and accuracy, especially when the sample becomes very large, and exhibit prolonged computing times. Thus, this research proposes that the combination of IDOA and DBN is suitable for tool wear prediction. In terms of tool wear classification, the proposed IDOA-DBN model includes five stages: image collection (dataset); cutting edge detection (circular Hough transform, canny edge detection, and standard Hough transform); feature extraction (SURF and LBP feature descriptors); feature optimization (IDOA); classification (DBN). The workflow of the IDOA-DBN model is specified in Figure 1.

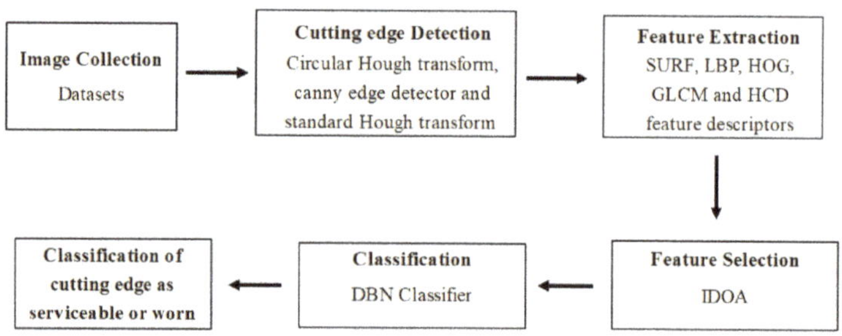

Figure 1. Workflow of the IDOA-DBN model.

3.1. Image Collection

In this manuscript, a dataset with 254 cutting edge images of an edge-profile-milling machine is utilized for investigating the effectiveness of the proposed model. In this study, the tool head is cylindrical in shape and comprises 30 inserts, which are arranged in six groups of five inserts. In this dataset, a Dalsa Genie m1280 1/3 camera with Azure-2514 mm lens is used for capturing the images, with a pixel resolution of 2592 $\times$ 1944 and focal length of 25 mm. Three Light-Emitting Diode (LED) bar lights are used to achieve independent illumination conditions in the environment [23]. The sample-acquired images are presented in Figure 2.

Figure 2. Acquired sample images from cutting edge dataset.

3.2. Cutting Edge Detection

After image acquisition, the circular Hough transform technique is used for detecting the screws located in the inserts, and the canny edge detector is employed for detecting the edges of the inserts [32,33]. The definition of the image is expanded to include ellipsoids, circles, and expressions with powers of three and higher. Irises are localized using the circular Hough transform; we also suggest computing the initial derivatives of the brightness of the image and thresholding the resulting values to generate the points of the parametric form. Furthermore, the standard Hough transform technique is utilized for detecting the vertical lines, and finally, the cutting edge of the insert is extracted from the acquired images [34]. Among 577 cutting edge images, the 276 images are labeled as worn edges, and 301 images are labeled as serviceable edges. The cutting edges of six cutting tools are presented in Figure 3.

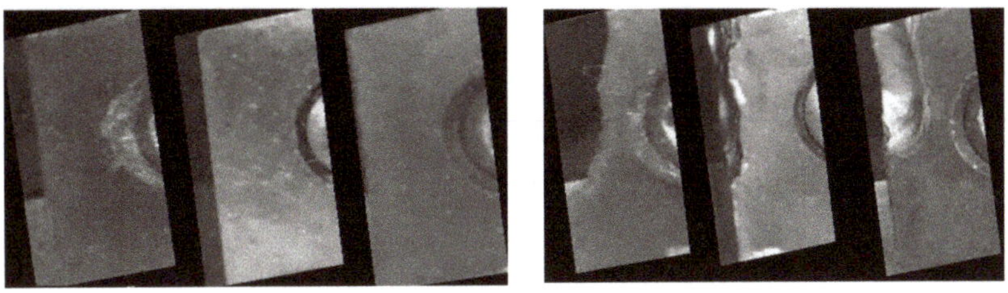

(**a**) (**b**)

Figure 3. Detected portion of cutting edges: (**a**) serviceable edges and (**b**) edges with high wear.

3.3. Feature Extraction

After collecting the images, a patch-based technique is applied for validating the insert wear level, which completely relies on dividing the cutting edge images into wear patches with dissimilar orientations, shapes, and sizes. Then, the wear patches are further categorized into serviceable wear or disposable wear. The different patch configurations for the cutting edges are full edge division, homogeneous grid division, small edge division, half edge division, and two-band division. After dividing the cutting edge images, feature extraction is carried out utilizing SURF and LBP feature descriptors.

3.3.1. Speeded Up Robust Features (SURF)

The SURF feature descriptor is a robust and fast algorithm for comparison, local, and similarity invariant representation of the images [35]. The SURF feature descriptor completely relies on its faster computation of operators utilizing box filters. While the Hessian matrix determinant is maximum, the SURF feature descriptor detects the blob-like structure. Let us consider the point $x = (x, y)$ in the acquired image I; the Hessian matrix $H(x, \sigma)$ with scale σ is determined at x using Equation (1):

$$H(x,\sigma) = \begin{bmatrix} L_{xx}(x,\sigma) & L_{xy}(x,\sigma) \\ L_{xy}(x,\sigma) & L_{yy}(x,\sigma) \end{bmatrix} \quad (1)$$

where $\frac{\partial^2}{\partial x^2}$ indicates Gaussian second-order derivation, $L_{xx}(x,\sigma), L_{yy}(x,\sigma), \text{and} L_{xy}(x,\sigma)$ states convolution of $\frac{\partial^2}{\partial x^2}$ at point. In the SURF feature descriptor, the scale spaces are categorized into octaves to recognize the interest points at different scales, where each octave includes a series of intervals. The convolutional window scale with the parameters octave (o) and interval (i), is specified in Equation (2), and the relationship between scale σ and window size is mathematically stated in Equation (3):

$$L = 3 \times (i \times 2^o + 1) \quad (2)$$

$$L = \sigma \times 9/1.2 \quad (3)$$

Furthermore, the SURF key points are calculated utilizing Equation (4):

$$DoH(x, L) = max(\sum_{k_i=i-1}^{i+1} \sum_{k_x=x-2^o}^{x+2^o} \sum_{k_y=y-2^o}^{y+2^o} DoH(k_x, k_y, o, k_i)) \geq \lambda \quad (4)$$

where DoH indicates determinant of the Hessian matrix and λ denotes positive threshold value. If the Hessian matrix trace value is higher than zero, a bright blob with scale $L = 3 \times (i \times 2^o + 1)$ is detected at (x, y). Around 1288 feature vectors are extracted from the acquired images using the SURF feature descriptor.

3.3.2. Local Binary Pattern (LBP)

LBP is a texture feature descriptor that extracts pixel-wise information from the acquired images. The LBP feature descriptor considers the number of neighbors (p) in each pixel (c) within the radius (r). If the grey-level value of p is higher than or equal to c, the value of one is assigned, or zero otherwise. By summing up the values to the power of two, the LBP of every pixel is calculated using Equation (5) [36]:

$$LBP_{p.r} = \sum_{p=0}^{p-1} s(g_p - g_c)2^p, s(x) = \{\begin{matrix} 1 & if\ x \geq 0 \\ 0 & if\ x < 0 \end{matrix} \quad (5)$$

where p indicates the number of neighbors, g_p represents the value of the pth neighbor, and g_c denotes the center pixel value; the LBP feature descriptor extracts 3470 texture feature vectors. Using the feature-level fusion method, the SURF and LBP feature vectors are integrated; thus, the model becomes multidimensional, increasing its complexity.

3.3.3. Histogram of Oriented Gradients (HOG)

HOG is a shape identifier that determines the layout and structure of a local character [37]. This HOG describes how the intensity gradient spreads across the region, which is used to determine the character. The HOG is often based on the gradient direction accumulating via the pixels of a small spatial region, such as a single cell. The HOG descriptor in each cell collects the gradient direction's local 1-D histogram. This process is performed by constructing the local histogram over a large spatial region, typically a block of cells, and using the data to normalized all of the block's cells. The detection window is placed over the overlapping grid in this case. Equation (6) represents the pixel's horizontal and vertical gradient:

$$\begin{cases} G_p = I'(p+1,q) - I'(p-1,q) \\ G_q = I'(p,q+1) - I'(p,q-1) \end{cases} \quad (6)$$

The magnitude and orientation of HOG gradients are expressed in Equations (7) and (8), respectively:

$$G(p,q) = \sqrt{G_p^2 + G_q^2} \tag{7}$$

$$\theta(p,q) = \text{argtan}\left(\frac{G_q}{G_p}\right) \tag{8}$$

The weights of the gradient magnitudes that are in the direction of interest are combined to generate the cell's histogram vector. Furthermore, the HOG's output vector is made up of normalized cells from many detection window elements and frames.

3.3.4. Grey-Level Co-Occurrence Matrix (GLCM)

The GLCM is a powerful feature descriptor that evaluates the spatial link between two pixels to assess the textural qualities of an image. Because the spacing and orientation between pixels are altered, the simultaneous presence of pixel pairs can be determined. There are 14 types of features presented in the GLCM. Here, some of the textural features are derived [38]. Successful description is based on the textural information gained from the GLCM.

The instantaneous occurrence probability of two pixels in the GLCM is stated as $P(i,j,\delta,\theta)$. It contains the pixel with, respectively, grey-level i and j as $f(x,y)$ and $(x+\Delta x, y+\Delta y)$; θ is defined as declination, and δ is stated as distance. The statistical formulation is stated in Equation (9):

$$P(i,j,\delta,\theta) = \left\{ (x,y), \begin{array}{c} |[(x,y),(x+\Delta x,y+\Delta y)]|f(x,y)=i, \\ f(x+\Delta x,y+\Delta y)=j; x=0,1,\ldots,N_x-1; \\ y=0,1,\ldots,N_y-1 \end{array} \right\} \tag{9}$$

where $i,j = 0,1,\ldots L-1$; image pixel coordinates are represented as $x$and$y$; the grey-level image is stated as L; numbers of columns and rows are stated as $N_x$andN_y. As a result, the appropriate feature selection is performed with IDOA, which decreases the model's operating performance and computational efficiency.

3.3.5. Harris Corner Detection (HCD)

The Harris corner point's features are utilized to separate the background and foreground. Rotating the window in every orientation along a corner point should produce a significant shift in luminance. The point intensity of Harris's point in the fingerprint foreground areas is much greater than in the background regions. The Harris method, originally proposed by Harris et al., is a variation or extension of the Moravec corner detection algorithm that extracts corner points using grey disparity between images [39]. Assuming that a greyscale image I and window $w(x,y)$ are considered to be removed (through movements of u in the x direction, and v in y direction), the intensity variation is calculated as Equation (10):

$$E(u,v) = \sum_{x,y} w(x,y)[I(x+u,y+v) - I(x-y)]^2 \tag{10}$$

where $I(x,y)$and$I(x+u,y+v)$ are intensities at position (x,y) and at moving window $x+u,y+v$, correspondingly. Because windows with corners have been explored with a wide range of intensity, we decided to focus on them. As a result, the component from Equation (10) is rewritten as Equation (11):

$$\sum_{x,y} [I(x+u,y+v) - I(x-y)]^2 \tag{11}$$

Equation (12) is attained by Taylor extension:

$$E(u,v) \approx \sum_{x,y} [I(x,y) + uI_x + vI_y - I(x,y)]^2 \quad (12)$$

Escalating Equation (12) and eliminating the appropriate products leads to Equation (13):

$$E(u,v) \approx \sum_{x,y} u^2 I_x^2 + 2uvI_xI_y + v^2 I_y^2 \quad (13)$$

By evaluating Equation (13) in matrix form, we attain Equation (14):

$$E(u,v) \approx [u\ v] \left(\sum_{x,y} w(x,y) \begin{matrix} I_x^2 & I_xI_y \\ I_xI_y & I_y^2 \end{matrix} \right) \begin{bmatrix} u \\ v \end{bmatrix} \quad (14)$$

Alternatively, consider $M = \sum_{x,y} w(x,y) \begin{matrix} I_x^2 & I_xI_y \\ I_xI_y & I_y^2 \end{matrix}$.
In this instance, Equation (14) is revised and presented as Equation (15):

$$E(u,v) \approx [u\ v] M \begin{bmatrix} u \\ v \end{bmatrix} \quad (15)$$

The eigenvalues of matrix M are stated as λ_1 and λ_2, which creates a rotationally invariant description. At this time, three points need to be considered:

- As soon as an eigenvalue (λ_1 or λ_2) is high and significantly superior, an edge occurs.
- Pixel point is stated to be in a flat area when $\lambda_1 \approx \lambda_2$ and its values are low.
- Pixel point is stated to be in a corner when $\lambda_1 \approx \lambda_2$ and its values are high.
- A score can be determined for every window to evaluate whether it is likely to represent a corner, as shown in Equation (16):

$$R = \det(M) - k(trace(M))^2 \quad (16)$$

where $\det(M) = \lambda_1, \lambda_2$, $trace(M) = \lambda_1 + \lambda_2$, and constant k is fixed in the range of 0.04 to 0.06. A window is obtained through score R that is greater than a positive threshold rate, i.e., considered to consume a corner.

3.4. Feature Optimization

After feature extraction, the IDOA is proposed for discriminative feature selection, where IDOA is a metaheuristic optimization algorithm that mimics the dynamic and static behaviors of dragonflies. The feature selection process is considered a problem of global combinatorial optimization, which seeks to reduce the quantity of features and redundant, noisy, and redundant data, while producing a uniform level of classification accuracy. As a result, this research introduces the Improved Dragonfly Optimization Technique (IDOA), an optimization algorithm that performs better. The discrete search space consists of all feasible arrangements of attributes chosen from the dataset. It might be possible to list every potential subset of characteristics, given the limited number of features. The improved dragonfly employs more group knowledge to influence its own behavior, ensuring that the group is diverse, and creating a balance between the stages of exploration and exploitation to increase the algorithm's search efficiency. This feature selection technique is typically quicker and more efficient, and minimizes overfitting, which eliminates redundant and noisy data to find a subset of relevant features using the strength of the IDOA to improve classification outcomes. The two major phases in the IDOA are exploitation and exploration, which are modeled statically or dynamically, either by avoiding the enemy or searching for food [40]. Usually, the swarms have three behaviors: cohesion, alignment, and separation. Furthermore, two additional behaviors are added to these three behaviors in the IDOA: avoiding the enemy and moving towards food. The purpose of including these two behaviors in the IDOA is to increase the survival time of the swarm. In this algorithm, two vectors are considered: the position and step for updating the position of dragonflies in a search space. The step vector is also considered as the speed that determines the direction of dragonflies. After the step vector calculation, the position vector is updated.

In the IDOA, the coefficients (cohesion, alignment, separation, food factor, inertia coefficient, enemy factor, and iteration number) are enabled to perform exploitative and exploratory behaviors. The cohesion coefficient is high and the alignment coefficient is low in the exploitation process; conversely, the cohesion coefficient is low and the alignment coefficient is high in the exploration process. In the conventional DOA, the Levy flight mechanism is used to enhance the probabilistic behavior, randomness, and the discovery of artificial dragonflies. Hence, the Levy flight mechanism improves the DOA efficacy to a certain extent. However, the step size control is contrary to the nature of the Levy flight mechanism. The agents have to move outside the search space, if a long step is considered. To overcome these issues, the Brownian motion (P_g), is considered in the IDOA for enhancing the probabilistic behavior, randomness, and the discovery of dragonflies. The Brownian motion (P_g) is mathematically determined in Equations (17) and (18):

$$P_g = \frac{1}{s\sqrt{2\pi}} exp\left(-\frac{(dimension - agents)^2}{2s^2} \right) \tag{17}$$

$$s = \sqrt{\frac{m_t}{m_s}}, \text{ and } m_s = 100 \times m_t \tag{18}$$

where $m_t = 0.01$ indicates the motion time of an agent, and m_s specifies the number of sudden motions. The parameter settings of the IDOA are as follows: the number of search agents is five, the search domain is [0–1], the dimension is equal to the extracted feature vectors, and the number of iteration is 20. The proposed IDOA selects 3476 feature vectors, which are used as input values in the DBN for classification.

3.5. Classification

After the selection of discriminative feature vectors, tool wear batch classification is carried out using DBN. The DBN is one of the effective deep learning models that consists of a number of Restricted Boltzmann Machines (RBMs) for data classification [41]. The learned activation unit of the first RBM is the input for succeeding RBMs in the stacks. In addition, the DBN is an undirected graphical technique, where the visible variables are linked to the hidden units using undirected weights. However, the DBNs are constrained; there is no connection within the visible and hidden variables. The probability distribution, p_d, of visible variables (m), hidden units (n) and energy function ($E(m,n;\theta)$) is mathematically depicted in Equation (19):

$$- logp_d(m,n)\alpha E(m,n;\theta) = -\sum_{i=1}^{|V|}\sum_{j=1}^{|Q|} w_{ij}m_i n_j - \sum_{i=1}^{|V|} b_i m_i - \sum_{j=1}^{|Q|} a_j n_j \tag{19}$$

where $\theta = (w,b,a)$ indicates parameter set, b_i and a_j denote bias, w_{ij} represents the symmetric weight between the visible variables (m), and α represents learning rate. In the DBN model, the numbers of hidden and visible layers are considered as $|Q|$ and $|V|$. The conditional probability distribution of visible variables (m) and hidden units (n) is defined in the Equations (20) and (21):

$$p_d(n_j|m;\theta) = sigm \sum_{i=1}^{|V|} w_{ij}m_i + a_j \tag{20}$$

$$p_d(m_i|n;\theta) = sigm \sum_{j=1}^{|Q|} w_{ij}n_i + b_j \tag{21}$$

where $sigm(M) = \left(\frac{1}{1+e^{-m}}\right)$ represents sigmoid activation function, and the parameter θ represents learned utilizing contrastive divergence. In the DBN classifier, the parameter θ is obtained utilizing RBM, and it is defined by $p_d(n|\theta)$ and $p_d(m|n,\theta)$. The probability of creating a visible variable is stated in Equation (22):

$$p_d(m) = \sum_n p_d(n|\theta) p_d(m|n,\theta) \tag{22}$$

The term $p_d(m|n,\theta)$ is maintained after determining θ from an RBM; then, $p_d(n|\theta)$ is exchanged using consecutive RBMs that treat the previous RBM hidden layer as a visible value. The parameter settings of the DBN are as follows: the transfer function is sigmoid function, the dropout rate is 0.1, the batch size is 0.5, the learning rate is 0.01, maximum iteration is 100, and initial and final momentum are 0.5 and 0.9. Figure 4 shows the architecture of the IDOA-DBN process.

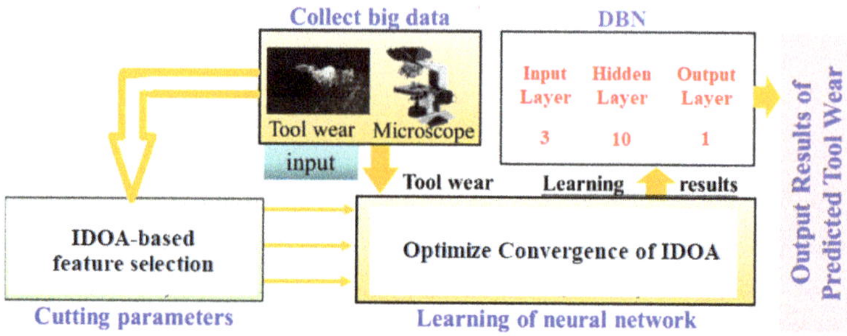

Figure 4. Architecture of the IDOA-DBN process.

4. Experimental Results

In the present research, the cutting edges were divided into a number of subregions by the proposed method (wear patches, or WP). Each WP was defined using textural descriptors, and were categorized as worn or serviceable using a Deep Belief Network. Lastly, the number of WPs designated as worn determines whether a cutting edge is serviceable or worn. This study separated each training set image into patches that were later identified as being in a worn or serviceable zone. The objective of this research was to develop a classification model that evaluates individual patches and, based on its predictions, renders judgement on the degree of tool wear. Therefore, manual division prevents potentially flawed patch extractions that can result in the creation of suboptimal classifiers. After manually extracting the patches, the images were left with 896 patches, 466 of which were serviceable and 430 of which were worn. The proposed IDOA-DBN model's effectiveness was validated using the MATLAB 2020a software tool on a system configuration with 16 GB random access memory, Linux operating system, and a 4 TB hard disk. The IDOA-DBN model's efficacy was analyzed by implementing many classification techniques and feature optimization algorithms, and tested using performance metrics such as F-score, recall, precision, and accuracy. In this manuscript, the confusion matrix was used to calculate the performance metrics. Here, the serviceable class was a negative class and the worn class was a positive class. The confusion matrix is clearly depicted in Table 1.

Table 1. Confusion matrix.

Classes	Prediction Class	
	Serviceable	Worn
Serviceable	*TN*	*FP*
Worn	*FN*	*TP*

Precision, recall, accuracy, and F-score were the metrics used in this research to evaluate the methodology. The confusion matrix depicted in Figure 4 was created by allocating the worn class as the positive class, and the serviceable class as the negative class. The performance metric, the F-score, was determined as the harmonic mean of recall and precision, and it was estimated using Equation (23). Similarly, the recall was defined as the fraction of worn inserts, and it was computed utilizing Equation (24). In the tool wear batch classification, the performance metric, recall, played a vital role, since the cost of misclassifying a serviceable cutting edge was lower than the cost of misclassifying a worn cutting edge:

$$F - score = \frac{2TP}{FP + 2TP + FN} \times 100 \quad (23)$$

$$Recall = \frac{TP}{TP + FN} \times 100 \quad (24)$$

Similarly, the performance metric, precision, was defined as the fraction of inserts classified as worn that were actually worn. The accuracy was determined by the successful prediction for the total number of samples, where *TP*, *TN*, *FP*, and *FN* were defined as true positive, true negative, false positive, and false negative. Figure 5 shows the confusion matrix. The mathematical expressions of precision and accuracy are defined in Equations (25) and (26):

$$Precision = \frac{TP}{TP + FP} \times 100 \quad (25)$$

$$Accuracy = \frac{TP + TN}{TP + TN + FP + FN} \times 100 \quad (26)$$

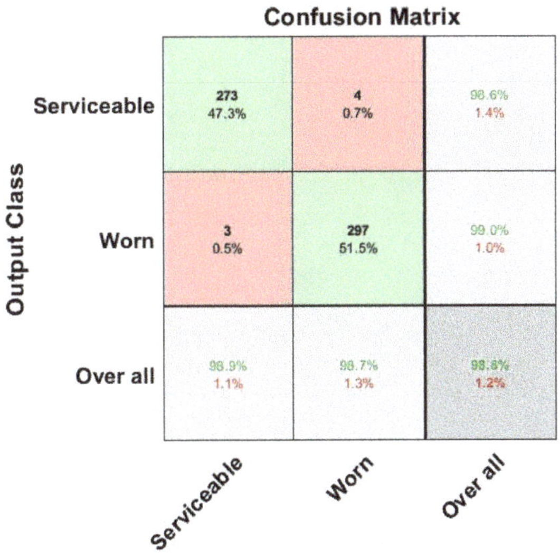

Figure 5. Confusion matrix.

4.1. Quantitative Evaluation

The proposed IDOA-DBN model's effectiveness was validated utilizing dissimilar feature optimization techniques, such as Grasshopper Optimization Algorithm (GOA), genetic algorithm, Particle Swarm Optimization (PSO) algorithm, DOA, and IDOA, by means of precision, accuracy, F-score, and recall. In this study, the performance analysis was conducted for different patch configurations, such as Full Edge Division (FED), Homogeneous Grid Division (HGD), Small Edge Division (SED), Half Edge Division (HED), and

Two-Band Division (TBD), respectively. By viewing Table 2, the IDOA with DBN classifier obtained efficient performance in tool wear classification compared to other existing optimizers, such as GOA, genetic algorithm, PSO, and DOA.

Table 2. Experimental results of various feature optimization techniques with the IDOA.

Patch Configuration	Optimizer	F-Score (%)	Recall (%)	Precision (%)	Accuracy (%)
FED	GOA	80.90	78.90	80.87	79.08
	Genetic	82.34	86.30	85.60	83.49
	PSO	87.90	88.56	88.70	88.60
	DOA	90.80	92.30	93.94	94.59
	IDOA	95.30	94.53	96.70	98.83
HGD	GOA	75.39	67.69	70.70	80.88
	Genetic	77.78	72.30	74.50	81.20
	PSO	85.49	75.60	78.90	85.60
	DOA	86.06	83.20	83.50	88.88
	IDOA	86.20	87.90	88.88	90.87
SED	GOA	77.50	70.90	68.90	87.30
	Genetic	75.40	78.78	80.82	83.40
	PSO	85.06	83.40	84.50	86.67
	DOA	89.80	90.76	89.88	91.20
	IDOA	93.20	92.03	94.50	96.78
HED	GOA	78.49	76.40	80	84.50
	Genetic	77	78.90	83.42	85.55
	PSO	87.74	88.36	82.20	89.28
	DOA	89.76	87.30	89.87	92.20
	IDOA	90.45	93.20	94.56	95.60
TBD	GOA	77.65	78.43	70.60	78.09
	Genetic	78.70	82.49	74.05	80.93
	PSO	84.50	85.60	85.60	86.70
	DOA	90.10	89.40	92.30	91.29
	IDOA	92.03	94.57	95.55	94.56

The IDOA with DBN classifier achieved an F-score of 95.30%, recall of 94.53%, precision of 96.70%, and accuracy of 98.83% for FED patch configuration. Regarding HGD patch configuration, the IDOA with DBN classifier attained an F-score of 86.20%, recall of 87.90%, precision of 88.88%, and accuracy of 90.87%. For the SED, HED, and TBD patch configurations, the proposed IDOA with DBN achieved 93.20%, 90.45%, and 92.03% F-scores; 92.03%, 93.20%, and 94.57% recall values; 94.50%, 94.56%, and 95.55% precision values; and 96.78%, 95.60%, and 94.56% classification accuracy.

AlexNet and ResNet models are capable of preserving the useful information in the network without overtrain or loss the features, while some of the recent CNN architecture pooling layers tend to lose useful features by overtraining the features. Furthermore, AlexNet and ResNet are suitable to handle large-scale data analysis. Therefore, in this study, these two models were preferred for the comparative analysis. In Table 3, it can be seen that the performance evaluation was conducted using various classification techniques, such as AlexNet, Autoencoder, ResNet-14, ResNet-18, and DBN, on five different patch configurations, namely FED, HGD, SED, HED, and TBD, by means of F-score, recall, precision, and accuracy. By viewing Table 3, the combination of DBN classifier with IDOA obtained maximum performance in tool wear classification as compared to other classifiers, such as AlexNet, Autoencoder, ResNet-14, and ResNet-18. Compared to other classification techniques, the DBN has two major advantages: reductions in both the over-smoothing problem and the training data fragmentation problem.

Table 3. Experimental results of various classification techniques with DBN.

Patch Configuration	Classifier	F-Score (%)	Recall (%)	Precision (%)	Accuracy (%)
FED	AlexNet	83.95	80.96	85.85	84.58
	Autoencoder	86.77	87.60	87.67	87.40
	ResNet-14	88.90	89.96	89.70	88.68
	ResNet-18	93.86	90.80	94.96	95.58
	DBN	95.30	94.53	96.70	98.83
HGD	AlexNet	78.50	77.60	76.70	83.90
	Autoencoder	79.78	78.90	77.57	84.20
	ResNet-14	84.40	79.93	79.98	85.50
	ResNet-18	85.48	84.78	86.70	87.87
	DBN	86.20	87.90	88.88	90.87
SED	AlexNet	76.70	74.90	76.98	88.30
	Autoencoder	78.90	76.78	82.80	89.40
	ResNet-14	86.06	85.40	88.50	89.69
	ResNet-18	88.80	88.70	93.84	92.22
	DBN	93.20	92.03	94.50	96.78
HED	AlexNet	79.40	77.78	80.90	85.50
	Autoencoder	81.20	78.98	85.49	86.75
	ResNet-14	88.78	84.86	86.26	89.29
	ResNet-18	88.76	89.30	90.88	93.20
	DBN	90.45	93.20	94.56	95.60
TBD	AlexNet	80.65	79.40	79.67	80.49
	Autoencoder	82.70	84.40	84.75	80.99
	ResNet-14	86.60	86.78	87.68	88.90
	ResNet-18	91.70	88.89	92.34	94.11
	DBN	92.03	94.57	95.55	94.56

4.2. Comparative Evaluation

The comparative investigation between the proposed IDOA-DBN model and the existing image texture analysis model [23] is presented in Table 4. M.T. García-Ordás, et al. [23] used circular Hough transform technique, canny edge detector, and standard Hough transform technique for segmenting the cutting edges of acquired images. Furthermore, the texture feature extraction was accomplished using conventional, completed, and adaptive LBP descriptors. The extracted feature vectors were fed into the SVM classifier for classifying the cutting edges as serviceable or worn. The experimental evaluation showed that the proposed IDOA-DBN model attained efficient performance in tool wear batch classification related to the existing image texture analysis model [23] on five different patch configurations. The feature optimization utilizing IDOA is an integral part of this manuscript that significantly reduces the computational complexity and running time of the DBN by selecting the discriminative feature vectors. The IDOA-DBN model took 43 s to process the whole dataset, which is better than other deep learning models, and the complexity of the proposed model is linear when selecting the discriminative feature vectors. Table 4 shows the comparison results of various patch configurations. Figure 6 shows the graphical presentation of the various patch configurations.

The researchers tuned the network model's design and hyperparameters for the signal matrix data, leading to a high recognition accuracy. Furthermore, the suggested tool wear state technique's optimized neural network with double neurons is incompatible with all of these datasets. Based on the design structure provided in the literature, calculations were performed to discover a suitable system model, and the obtained accuracy results are presented in Table 5. The suggested IDOA-DBN approach delivers superior identification accuracy (98.83%) and a more compact network structure than existing DWF-CNN [28] techniques, which achieved 98.7%. Figure 7 shows the graphical illustration of accuracy with existing DWF-CNN [28] methods.

Table 4. Results of various patch configurations.

Patch Configurations	Model	F-Score (%)	Precision (%)	Accuracy (%)
FED	Texture analysis [23]	85.50	93.40	86.40
	IDOA-DBN	95.30	96.70	98.83
HGD	Texture analysis [23]	81.50	80	81.20
	IDOA-DBN	86.20	88.88	90.87
SED	Texture analysis [23]	90.30	89.70	90.30
	IDOA-DBN	93.20	94.50	96.78
HED	Texture analysis [23]	87.20	91	87
	IDOA-DBN	90.45	94.56	95.60
TBD	Texture analysis [23]	90.90	93.40	90.90
	IDOA-DBN	92.03	95.55	94.56

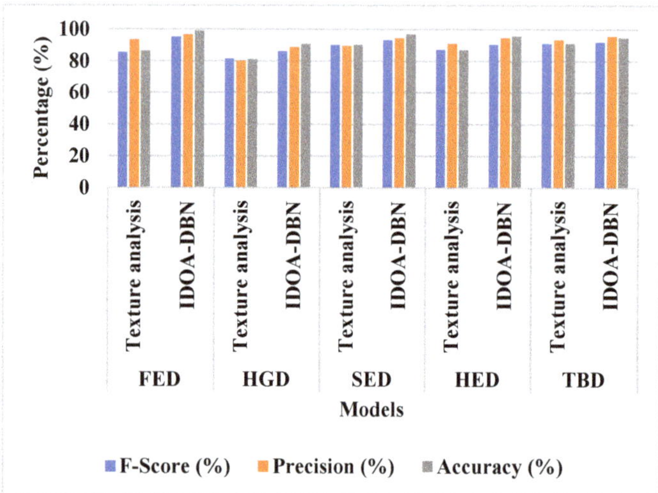

Figure 6. Graphical presentation of the various patch configurations with existing method.

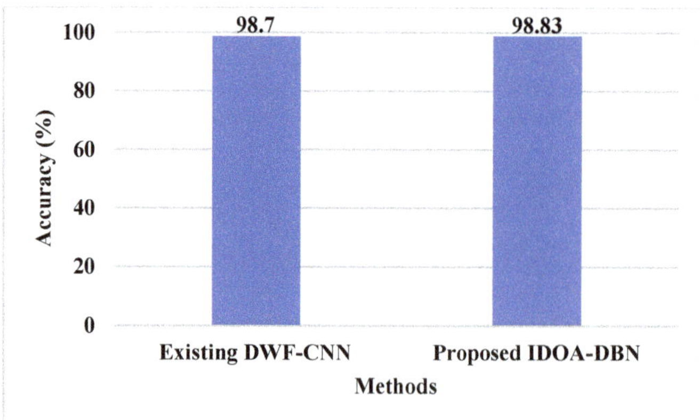

Figure 7. Graphical illustration of accuracy with existing method.

Table 5. Comparative analysis of accuracy with existing method.

Method	Accuracy (%)
Existing DWF-CNN [28]	98.7
Proposed IDOA-DBN	98.83

The prediction impacts of BiLSTM methods based on SVD features and time domain features were evaluated to demonstrate the effectiveness of IDOA-DBN features in tool wear detection. SVD-BiLSTM [30] was utilized for experimental studies to further validate the functionality of the IDOA-DBN tool wear prediction system. Because all learning methods have the same architecture, only the type of network layers needed to be changed; the performance parameters were maintained, and all design variables were SVD features. Mean Absolute Error (MAE), Mean Absolute Percentage Error (MAPE), and Mean Square Error (MSE) values were computed to generate time domain features of the same scale before the SVD features. Table 6 lists the outcomes of the prediction results in terms of the training and testing sets. The proposed IDOA-DBN model was more suitable in the error assessment of the training set, and computation was easier. Therefore, IDOA-DBN produced better training outcomes in the training set in the same time frame. Despite this, existing SVD-BiLSTM [30] lacks cell state and has inadequate long-term memory capacity, which is evident in the test set error values produced by SVD-BiLSTM [30].

Table 6. Comparative analysis of prediction results in training and testing set.

Method	Training Set			Testing Set		
	MAE	MAPE	MSE	MAE	MAPE	MSE
SVD-BiLSTM [30]	2.4948	2.4112	3.6782	4.3044	1.8896	5.0873
Proposed IDOA-DBN	2.4625	2.3729	3.5988	4.2553	1.8309	4.1032

In this study, IDOA-DBN was proposed and considered to be the appropriate design in all data and test sets, because the IDOA-DBN can memorize all data features. Consequently, in both the training set and the test set, IDOA-DBN is superior to SVD-BiLSTM [30]. The model's input data were constructed using a time step of five, which means that the cutting edge signals for each of the previous five sampling periods were matched with the tool wear value for the 5th sampling period. The first 265 sets of samples were used as the model training sample out of 315 total measured data, which were divided into 311 sets of sample data. The remaining 46 groupings of samples were employed as a model test sample. Figure 8 shows the graphical illustration of the prediction results.

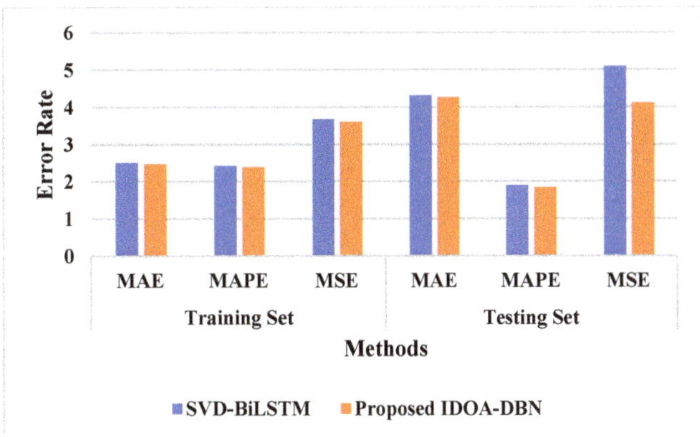

Figure 8. Graphical illustration of prediction results with existing method.

5. Conclusions

In this manuscript, an IDOA-DBN model is implemented as an effective tool wear monitoring technique. Since Deep Belief Network (DBN) performs better than other models in terms of learning speed, and quickly converges to the best results, regardless of the size of the sample data, it was utilized to create tool wear prediction models. This increased the tool wear monitoring system's modeling efficiency and accuracy. Due to the manual adjustment of DBN parameters, prediction accuracy and efficiency were generally low. In order to reduce the computational complexity and operating time of the DBN at this point, the Improved Dragonfly Optimization Algorithm (IDOA) was proposed. The inclusion of the IDOA in the proposed model significantly diminished the computational complexity and running time of the classifier. As indicated in the resulting phase, the proposed IDOA-DBN model obtained superior performance values for tool wear monitoring compared to the existing deep learning models, such as AlexNet, Autoencoder, ResNet-14, and ResNet-18, on different patch configurations. The IDOA-DBN model attained a maximum classification accuracy of 98.83% on the patch configuration of full edge division. As a future extension, a hybrid feature selection algorithm should be included in the proposed model to further enhance the tool wear monitoring technique.

Author Contributions: The paper investigation, resources, data curation, writing—original draft preparation, writing—review and editing, and visualization were performed by L.G.D. The paper conceptualization, software, were conducted by R.K.P. and L.J.A.M. The validation and formal analysis, methodology, supervision, project administration, and funding acquisition of the version to be published were conducted by P.F.-G. and P.B.D. All authors have read and agreed to the published version of the manuscript.

Funding: This research received no external funding.

Institutional Review Board Statement: Not applicable.

Informed Consent Statement: Not applicable.

Data Availability Statement: Not applicable.

Conflicts of Interest: The authors declare no conflict of interest.

References

1. Hesser, D.F.; Markert, B. Tool wear monitoring of a retrofitted CNC milling machine using artificial neural networks. *Manuf. Lett.* **2019**, *19*, 1–4. [CrossRef]
2. Guo, J.; Li, A.; Zhang, R. Tool condition monitoring in milling process using multifractal detrended fluctuation analysis and support vector machine. *Int. J. Adv. Manuf. Technol.* **2020**, *110*, 1445–1456. [CrossRef]
3. Bergs, T.; Holst, C.; Gupta, P.; Augspurger, T. Digital image processing with deep learning for automated cutting tool wear detection. *Procedia Manuf.* **2020**, *48*, 947–958. [CrossRef]
4. Wu, X.; Liu, Y.; Zhou, X.X.; Mou, A. Automatic identification of tool wear based on convolutional neural network in face milling process. *Sensors* **2019**, *19*, 3817. [CrossRef] [PubMed]
5. Martínez-Arellano, G.; Terrazas, G.; Ratchev, S. Tool wear classification using time series imaging and deep learning. *Int. J. Adv. Manuf. Technol.* **2019**, *104*, 3647–3662. [CrossRef]
6. Peng, R.; Pang, H.; Jiang, H.; Hu, Y. Study of tool wear monitoring using machine vision. *Autom. Control Comput. Sci.* **2020**, *54*, 259–270. [CrossRef]
7. Qiao, H.; Wang, T.; Wang, P. A tool wear monitoring and prediction system based on multiscale deep learning models and fog computing. *Int. J. Adv. Manuf. Technol.* **2020**, *108*, 2367–2384. [CrossRef]
8. Stavropoulos, P.; Papacharalampopoulos, A.; Souflas, T. Indirect online tool wear monitoring and model-based identification of process-related signal. *Adv. Mech. Eng.* **2020**, *12*, 1687814020919209. [CrossRef]
9. Liu, T.; Zhu, K.; Wang, G. Micro-milling tool wear monitoring under variable cutting parameters and runout using fast cutting force coefficient identification method. *Int. J. Adv. Manuf. Technol.* **2020**, *111*, 3175–3188. [CrossRef]
10. Chen, N.; Hao, B.; Guo, Y.; Li, L.; Aqib Khan, M.; He, N. Research on tool wear monitoring in drilling process based on APSO-LS-SVM approach. *Int. J. Adv. Manuf. Technol.* **2020**, *108*, 2091–2101. [CrossRef]
11. Liu, T.; Zhu, K. A switching hidden semi-Markov model for degradation process and its application to time-varying tool wear monitoring. *IEEE Trans. Ind. Inform.* **2020**, *17*, 2621–2631. [CrossRef]

12. Marani, M.; Zeinali, M.; Kouam, J.; Songmene, V.; Mechefske, C.K. Prediction of cutting tool wear during a turning process using artificial intelligence techniques. *Int. J. Adv. Manuf. Technol.* **2020**, *111*, 505–515. [CrossRef]
13. Jamshidi, M.; Rimpault, X.; Balazinski, M.; Chatelain, J.F. Fractal analysis implementation for tool wear monitoring based on cutting force signals during CFRP/titanium stack machining. *Int. J. Adv. Manuf. Technol.* **2020**, *106*, 3859–3868. [CrossRef]
14. Zhang, X.; Han, C.; Luo, M.; Zhang, D. Tool wear monitoring for complex part milling based on deep learning. *Appl. Sci.* **2020**, *10*, 6916. [CrossRef]
15. García-Ordás, M.T.; Alegre-Gutiérrez, E.; González-Castro, V.; Alaiz-Rodríguez, R. Combining shape and contour features to improve tool wear monitoring in milling processes. *Int. J. Prod. Res.* **2018**, *56*, 3901–3913. [CrossRef]
16. Dou, J.; Xu, C.; Jiao, S.; Li, B.; Zhang, J.; Xu, X. An unsupervised online monitoring method for tool wear using a sparse auto-encoder. *Int. J. Adv. Manuf. Technol.* **2020**, *106*, 2493–2507. [CrossRef]
17. Antić, A.; Popović, B.; Krstanović, L.; Obradović, R.; Milošević, M. Novel texture-based descriptors for tool wear condition monitoring. *Mech. Syst. Signal Process.* **2018**, *98*, 1–15. [CrossRef]
18. Zhang, X.; Pan, T.; Ma, A.; Zhao, W. High efficiency orientated milling parameter optimization with tool wear monitoring in roughing operation. *Mech. Syst. Signal Process.* **2022**, *165*, 108394. [CrossRef]
19. Ong, P.; Lee, W.K.; Lau, R.J.H. Tool condition monitoring in CNC end milling using wavelet neural network based on machine vision. *Int. J. Adv. Manuf. Technol.* **2019**, *104*, 1369–1379. [CrossRef]
20. Kuntoğlu, M.; Aslan, A.; Yurievich Pimenov, D.; Ali Usca, Ü.; Salur, E.; Kumar Gupta, M.; Mikolajczyk, T.; Giasin, K.; Kapłonek, W.; Sharma, S. A review of indirect tool condition monitoring systems and decision-making methods in turning: Critical analysis and trends. *Sensors* **2020**, *21*, 108. [CrossRef]
21. Oo, H.; Wang, W.; Liu, Z. Tool wear monitoring system in belt grinding based on image-processing techniques. *Int. J. Adv. Manuf. Technol.* **2020**, *111*, 2215–2229. [CrossRef]
22. Li, G.; Wang, Y.; He, J.; Hao, Q.; Yang, H.; Wei, J. Tool wear state recognition based on gradient boosting decision tree and hybrid classification RBM. *Int. J. Adv. Manuf. Technol.* **2020**, *110*, 511–522. [CrossRef]
23. García-Ordás, M.T.; Alegre-Gutiérrez, E.; Alaiz-Rodríguez, R.; González-Castro, V. Tool wear monitoring using an online, automatic and low cost system based on local texture. *Mech. Syst. Signal Process.* **2018**, *112*, 98–112. [CrossRef]
24. Simon, G.D.; Deivanathan, R. Early detection of drilling tool wear by vibration data acquisition and classification. *Manuf. Lett.* **2019**, *21*, 60–65. [CrossRef]
25. Marani, M.; Zeinali, M.; Songmene, V.; Mechefske, C.K. Tool wear prediction in high-speed turning of a steel alloy using long short-term memory modelling. *Measurement* **2021**, *177*, 109329. [CrossRef]
26. Liu, X.; Liu, S.; Li, X.; Zhang, B.; Yue, C.; Liang, S.Y. Intelligent tool wear monitoring based on parallel residual and stacked bidirectional long short-term memory network. *J. Manuf. Syst.* **2021**, *60*, 608–619. [CrossRef]
27. Huang, Z.; Zhu, J.; Lei, J.; Li, X.; Tian, F. Tool wear predicting based on multi-domain feature fusion by deep convolutional neural network in milling operations. *J. Intell. Manuf.* **2020**, *31*, 953–966. [CrossRef]
28. Cao, X.C.; Chen, B.Q.; Yao, B.; He, W.P. Combining translation-invariant wavelet frames and convolutional neural network for intelligent tool wear state identification. *Comput. Ind.* **2019**, *106*, 71–84. [CrossRef]
29. Liao, X.; Zhou, G.; Zhang, Z.; Lu, J.; Ma, J. Tool wear state recognition based on GWO-SVM with feature selection of genetic algorithm. *Int. J. Adv. Manuf. Technol.* **2019**, *104*, 1051–1063. [CrossRef]
30. Wu, X.; Li, J.; Jin, Y.; Zheng, S. Modeling and analysis of tool wear prediction based on SVD and BiLSTM. *Int. J. Adv. Manuf. Technol.* **2020**, *106*, 4391–4399. [CrossRef]
31. Laddada, S.; Si-Chaib, M.O.; Benkedjouh, T.; Drai, R. Tool wear condition monitoring based on wavelet transform and improved extreme learning machine. *Proc. Inst. Mech. Eng. Part C J. Mech. Eng. Sci.* **2020**, *234*, 1057–1068. [CrossRef]
32. Okokpujie, K.; Noma-Osaghae, E.; John, S.; Ajulibe, A. An improved iris segmentation technique using circular Hough transform. In *IT Convergence and Security*; Springer: Singapore, 2018; pp. 203–211.
33. Yang, Y.; Zhao, X.; Huang, M.; Wang, X.; Zhu, Q. Multispectral image-based germination detection of potato by using supervised multiple threshold segmentation model and Canny edge detector. *Comput. Electron. Agric.* **2021**, *182*, 106041. [CrossRef]
34. Jothi, A.; Jayaram, S.; Dubey, A.K. Intra-ocular lens defect detection using generalized Hough transform. In Proceedings of the 6th International Conference on Reliability, Infocom Technologies and Optimization, Trends and Future Directions, ICRITO, Noida, India, 20–22 September 2017; pp. 177–181.
35. Yuk, E.H.; Park, S.H.; Park, C.S.; Baek, J.G. Feature-learning-based printed circuit board inspection via speeded-up robust features and random forest. *Appl. Sci.* **2018**, *8*, 932. [CrossRef]
36. Islam, M.A.; Yousuf, M.S.I.; Billah, M.M. Automatic plant detection using HOG and LBP features with SVM. *Int. J. Comput. (IJC)* **2019**, *33*, 26–38.
37. Zhou, W.; Gao, S.; Zhang, L.; Lou, X. Histogram of oriented gradients feature extraction from raw Bayer pattern images. *IEEE Trans. Circuits Syst. II Express Briefs* **2020**, *67*, 946–950. [CrossRef]
38. Wang, J.S.; Ren, X.D. GLCM based extraction of flame image texture features and KPCA-GLVQ recognition method for rotary kiln combustion working conditions. *Int. J. Autom. Comput.* **2014**, *11*, 72–77. [CrossRef]
39. Bakheet, S.; Al-Hamadi, A.; Youssef, R. A fingerprint-based verification framework using Harris and SURF feature detection algorithms. *Appl. Sci.* **2022**, *12*, 2028. [CrossRef]

40. Mirjalili, S. Dragonfly algorithm: A new meta-heuristic optimization technique for solving single-objective, discrete, and multi-objective problems. *Neural Comput. Appl.* **2016**, *27*, 1053–1073. [CrossRef]
41. Yang, Y.; Zheng, K.; Wu, C.; Niu, X.; Yang, Y. Building an effective intrusion detection system using the modified density peak clustering algorithm and deep belief networks. *Appl. Sci.* **2019**, *9*, 238. [CrossRef]

applied sciences

MDPI

Article

HCI-Based Wireless System for Measuring the Concentration of Mining Machinery and Equipment Operators

Jerzy Jagoda [1,*], Mariusz Woszczyński [1], Bartosz Polnik [1] and Przemysław Falkowski-Gilski [2,*]

1 KOMAG Institute of Mining Technology, Pszczynska St. 37, 44-101 Gliwice, Poland; mwoszczynski@komag.eu (M.W.); bpolnik@komag.eu (B.P.)
2 Faculty of Electronics, Telecommunications and Informatics, Gdansk University of Technology, Narutowicza St. 11/12, 80-233 Gdansk, Poland
* Correspondence: jjagoda@komag.eu (J.J.); przemyslaw.falkowski@eti.pg.edu.pl (P.F.-G.)

Abstract: Maintaining stable and reliable working conditions is a matter of vital importance for various companies, especially those involving heavy machinery. Due to human exhaustion, as well as unpredicted hazards and dangerous situations, the personnel has to take actions and wisely plan each move. This paper presents a human–computer interaction (HCI)-based system that uses a concentration level measurement function to increase the safety of machine and equipment operators. The system has been developed in response to the results of user experience (UX) analyses of the state of occupational safety, which indicate that the most common cause of accidents is the so-called insufficient concentration while performing work. The paper presents the reasons for addressing this issue and a description of the proposed electroencephalography (EEG)-based solution in the form of a concentration measurement system concept. We discuss in-field measurements of such a prototype solution, together with an analysis of obtained results. The method of implementing a wireless communication interface is also provided, along with a visualization application.

Keywords: EEG (electroencephalography); HCI (human–computer interaction); mining; occupational safety; swarm intelligence; UX (user experience); wireless networks

Citation: Jagoda, J.; Woszczyński, M.; Polnik, B.; Falkowski-Gilski, P. HCI-Based Wireless System for Measuring the Concentration of Mining Machinery and Equipment Operators. *Appl. Sci.* **2023**, *13*, 5396. https://doi.org/10.3390/app13095396

Academic Editor: Alessandro Lo Schiavo

Received: 10 March 2023
Revised: 20 April 2023
Accepted: 24 April 2023
Published: 26 April 2023

1. Introduction

In human–computer interaction (HCI), electroencephalogram (EEG) signals can be added as an additional input to machines. An integration of real-time EEG-based human emotion recognition algorithms in human–computer interfaces can affect the user's experience (UX), making it, i.e., more complete or engaging, less or more stressful, depending on the target of the applications. Currently, the most accurate EEG-based emotion recognition algorithms are very subject-dependent. Therefore, a training session is required each time before running the application [1].

The Polish Higher Mining Inspectorate's analysis of the state of safety, presented in the report from 2015, entitled "Assessment of the state of occupational safety, mine rescue and general safety in connection with mining and geological activities", shows that one of the most frequent causes of accidents is the so-called insufficient concentration during work [2].

An analysis by specialists from the Cracow University of Technology and the AGH University of Science and Technology on the operator–machine system states that "On the basis of the research carried out, it can be concluded that the parameters related to the psychophysical condition of the operator (experience, fatigue, state of health and motivation of the operator) have the greatest influence on the performance of the operator-machine earthmoving system (...)" [3]. The need to review the operator's level of concentration also exists in other industries.

In order to improve the level of safety in the operation of machinery and equipment [4,5], a method has been developed to minimize the impact of errors made by machine and equipment operators as a result of insufficient concentration on the work being performed.

Any way of eliminating human error can significantly improve the safety of industrial operations. The results of this work can lead to an actual increase in the safety of work of operators of heavy machinery and equipment.

After analyzing domestic and foreign solutions, it was decided to implement a system based on the measurement of the electroencephalographic signal [6,7]. During the conceptual work, a hardware model of the system was built and tests were carried out on a selected group of car drivers and people performing simple manual activities. The selected group of people participating in the study was intended to best represent the profile of an operator of mining machinery and equipment. Conducting similar tests in real conditions would require specific permits and appropriate certification of devices. This paper presents the results of the model, field tests and their analyses. In the future, the system may become a component of the integrated visualization, diagnostic and control system, realized as part of the work carried out at KOMAG Institute of Mining Technology (ITG KOMAG) [8–11].

This paper is organized as follows: Section 1 provides an introduction to the investigated topic. Section 2 presents a review of related works and the advancements and possibilities of utilizing HCI- and EEG-based solutions in various occupational safety scenarios. Section 3 refers to the materials and methods section, including the concept of a dedicated concentration measurement system, as well as utilized hardware and software layer during our studies. Section 4 presents results of a measurement campaign, involving different levels of stress and required skills. Section 5 includes a discussion, presenting the concept of a dedicated system for the mining industry or any occupation linked with heavy machinery and hazardous working conditions. Section 6 concludes this manuscript and provides an insight into future study directions.

2. Related Works

As we know, everyone experiences stress in life. It should be noted that moderate stress can be beneficial to humans, whereas excessive stress proves to be harmful to their health. In order to monitor stress, different methods can be utilized. In this work [12], an algorithm for stress level recognition from EEG is proposed. The experiment was carried out on nine subjects, in which a Stroop color-word test was used as a stressor to induce four levels of stress. Authors have proposed and analyzed a number of different feature combinations and classifiers. By combining fractal dimension and statistical features, with the aid of Support Vector Machine (SVM) as the classifier, 4 levels of stress could be recognized with an average accuracy of approx. 67%, whereas 3 levels of stress could be detected with an accuracy of approx. 75%, and 2 levels of stress could be recognized with an accuracy of approx. 85%. The levels of stress were visualized in real-time. As pointed out, such a system could be applied for stress monitoring of, i.e., air traffic controllers, operators, etc.

A similar study [13] assessed working memory load during computer use with neural network pattern recognition applied to EEG spectral features. It involved a group of eight participants performing high-, moderate-, and low-load working memory tasks. Frontal theta EEG activity increased and alpha activity decreased with increasing load. As observed, such changes probably reflected task difficulty-related increase in mental effort and the proportion of cortical resources allocated to task performance. Test data segments from high and low load levels were discriminated with approx. 95% accuracy, whereas more than 80% of test data segments associated with a moderate load could be discriminated from high- or low-load data segments. As pointed out, results support the possibility of using EEG-based methods for monitoring cognitive load during human-computer interaction.

According to [14], in case of tasks requiring sustained attention, human alertness varies on a minute time scale. This can imply serious consequences in crucial occupations, ranging from, i.e., air traffic control to monitoring of nuclear power plants. Changes in the electroencephalographic power spectrum accompany these fluctuations in the level of alertness, as assessed by measuring simultaneous changes in EEG and performance on an auditory monitoring task. By combining power spectrum estimation, principal component analysis and artificial neural networks (ANN), the authors show that continuous, accurate, noninvasive, and near real-time estimation of an operator's global level of alertness is indeed feasible. In this experiment, data were recorded from two central scalp sites. This could enable a practical system for noninvasive monitoring of the cognitive state of human operators in attention-critical settings.

In [15], EEG recordings were obtained from a group of 16 individuals interacting with a personal-computer-based flight simulation. Their tasks have been labeled as low, moderate, and high difficulty. With higher task difficulty, frontal theta activity increased, whereas alpha activity decreased. Index values were computed for every 4 s of task data. Across participants, mean task load index values increased systematically with growing task difficulty and differed significantly between various task versions. Results of this study may be utilized in case of multivariate EEG-based methods to monitor task loading during naturalistic computer-based work.

Another paper [16] describes a brain–computer interface (BCI) system, which uses a set of adaptive linear preprocessing and classification algorithms for single-trial detection of error-related negativity (ERN). We use the detected ERN as an estimate of a subject's perceived error during an alternative forced choice visual discrimination task. The detected ERN was used to correct subject errors. Results have shown average improvement in subject performance by 21% when errors were automatically corrected via the BCI.

3. Materials and Methods

3.1. Concept of a Concentration Measurement System

The concept of the system, in which the operator's electromagnetic brain signal is read and analyzed by a measurement module, is shown in Figure 1. After processing the data, a decision is made to forward the signal to the machine's control panel in order to generate a possible warning signal if the operator's minimum concentration threshold is exceeded [6].

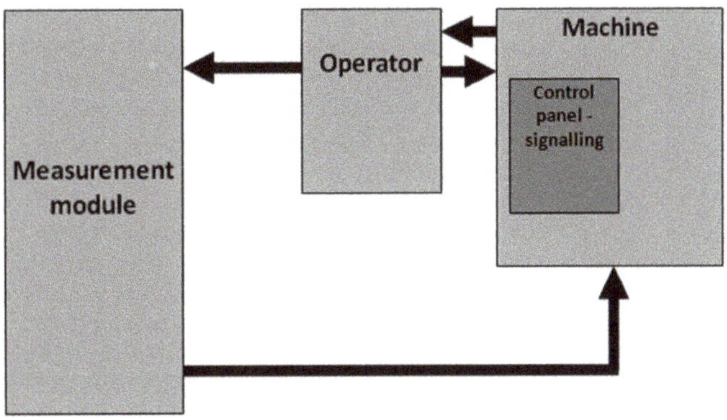

Figure 1. Block diagram of the operator concentration measurement system model.

It was decided to use EEG signal because the electroencephalographic analysis makes it possible to determine the bioelectrical activity of the cerebral cortex [17], which, depending on the frequency, provides the following information:

- Alpha waves (frequency range 8–13 Hz, amplitude 20–100 μV)—these waves occur in the case of adults with complete relaxation and eyes closed. The waveform is sinusoidal in nature, with the largest amplitudes received above the visual cortex. The alpha rhythm is blocked during attention (especially visual attention) and mental effort.
- Beta waves (frequency range 14–35 Hz, amplitude up to 20 μV)—these waves occur during states of activity, information processing, attention or arousal. They are received by the frontal-central region.

- Theta waves (frequency range 4–7 Hz, up to 30 μV)—they occur mainly in case of children. When it comes to adults, they appear in a state of deep meditation, sleep, trance, hypnosis, dreaming, intense emotion, etc.
- Delta waves (frequency range 0.5–4 Hz, amplitude 75–200 μV)—very characteristic for deep sleep. They are collected from the entire surface of the skull.
- Gamma waves (frequency range 35–100 Hz)—these waves that occur in stressful situations during stage fright, anxiety and so-called emergency situations.

This work focuses on measuring the EEG signal using the biofeedback method, in which the measured signal is given back to the patient in the form of, i.e., a visualization or computer game. The feedback contains information about which brain waves are predominant at a given moment. These waves are used, for example, to control a virtual element when brain activity increases in the desired frequency band. The corresponding correlation of alpha, beta and theta wave activity allows the identification of the subject's level of concentration.

3.2. Hardware and Software Layer

In order to verify the assumptions of the system model concept, tests were carried out to check, in particular, the correctness of the choice of the measurement of the EEG signal emitted by the operator's brain in order to identify his/her level of concentration. In our study, we utilized the MindWave Mobile device 1, coming from NeuroSky (San Jose, CA, USA) [18]. We intended to verify whether the single-electrode EEG signal measurement is sufficient to monitor the concentration level of the operator of machinery and equipment.

The MindWave Mobile is equipped with a single, dry frontal electrode named Fp1 (according to the international 10–20 standard defining electrode placement points during EEG signal analysis) and an ear-mounted reference clip.

NeuroSky's product is powered by a single AAA battery, which provides between 6 and 8 h of continuous operation. The data transmission rate realized by the device is equal to 250 kbit/s. The device operates in the frequency range of 2.420–2.471 GHz. The maximum signal strength is equal to 6 dBm. The sampling frequency for measurements carried out using the NeuroSky MindWave Mobile is set to 512 Hz [6].

In addition to the raw EEG signal, the device transmits the level of focus (attention) and relaxation (meditation). Both ratios are dimensionless. The manufacturers of the equipment do not provide details of their determination, only information on their correlation with alpha, beta and theta waves are given [19].

For the purpose of this study, a custom-build software was developed to record the data, i.e., time, focus and meditation level, and save the data for further analysis. The software is compatible with any kind of device running the Android operating system. It contains summation filters that process the data from the data logger and correlate it according to the time base with the events that were gathered by the video recorder.

Figure 2 shows the block diagram of the test bed for the operator's concentration measurement system. The testbed was equipped with an EEG measurement tool, namely the MindWave Mobile device, from which data were recorded in a data logger, which is a mobile application running on a mobile Android-powered terminal. The workstation also consisted of a video recorder, which allowed the measurement results to be correlated with specific recorded events, namely visual information. It should be noted that all devices were synchronized in time.

Tests were carried out for two variants of machine operators: car drivers and people performing simple manual tasks. Drivers of passenger cars consisted of o groups: people who have a driving license for more than 10 years, and people who have a driving license for less than 1 year. People performing precise manual activities were assemblers of SMD electronic components with 10 years of experience.

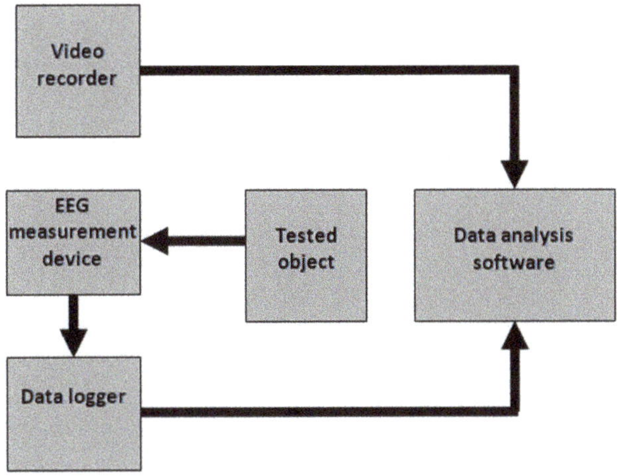

Figure 2. Block diagram of the operator's concentration measurement system model.

During our study, the following system functions were verified:

- Whether the system is able to identify the response to specific stimuli while driving (short-distance driving).
- Whether the system can identify measurement differences (driving two drivers on the same route with different driving skills).
- Whether it is possible to identify distractions over short and long periods of time (long-distance driving and simple manual tasks).

During the field tests, the following parameters were recorded and monitored: time, level of concentration and level of meditation (relaxation).

4. Results

The measurement data from the data logger were subjected to a summation filter in the data analysis software and correlated with the corresponding events obtained by the video recorder. An example of the oscillogram obtained during tests is shown in Figure 3. It represents the level of concentration of a driver during two short trips. Four events were listed that could have affected his/her level of concentration.

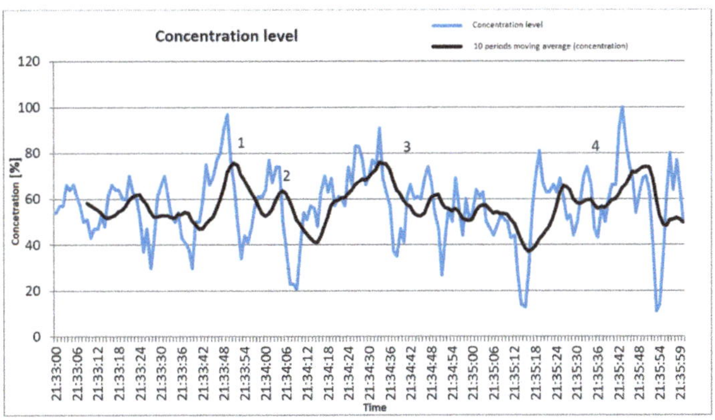

Figure 3. Level of driver concentration.

Events 1, 2 and 3 occurred in the evening when there was little traffic on the road. Locations were identified where the driver reacted to passing vehicles (events 1, 2 and 3), as shown in Figure 4. Increased levels of concentration were observed when approaching a junction, recorded as event number 4.

Figure 4. Example of recorded event—passing another vehicle.

The recording of concentration levels was repeated on route 2, which is characterized by increased traffic (motorway). Elevated levels were recorded on 9 occasions. One of these was an access to a section where there were difficulties due to ongoing road works, as shown in Figure 5.

Figure 5. Example of a recorded event—unexpected road works.

The study found that the system detects the test driver's reaction to certain driving situations in the form of an increased level of concentration.

Tests were also carried out on a route of about 700 km long. Drivers drove in mixed areas, including built-up areas, expressways and motorways. Figure 6 shows the time intervals during which the drivers drove on motorways and expressways.

In case of interval 1, a decrease in the average concentration level of the drivers up to the stop was observed. For intervals 2 and 3, a decrease in the average concentration level was also observed (after a stop, the vehicle was driven by the same person). It was found that the average concentration level for compartment 1 increased in a short period of time when there was a change of driver (person).

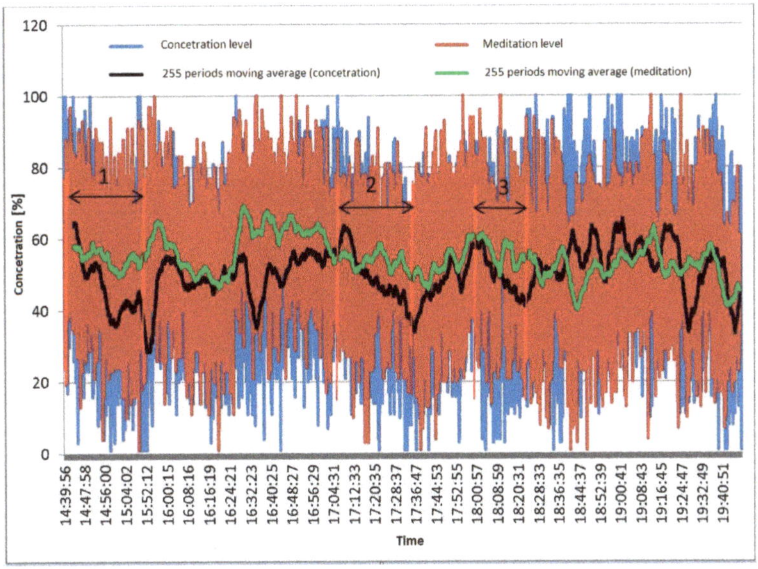

Figure 6. Average concentration level of drivers on a motorway or expressway.

The average concentration level recorded on the motorway (Section 1) was equal to 45.8% (driving time of approx. 1 h 12 min), while the average concentration level calculated at the same time on the mixed section (urban expressways) was equal to 50.3%, as shown in Figure 7.

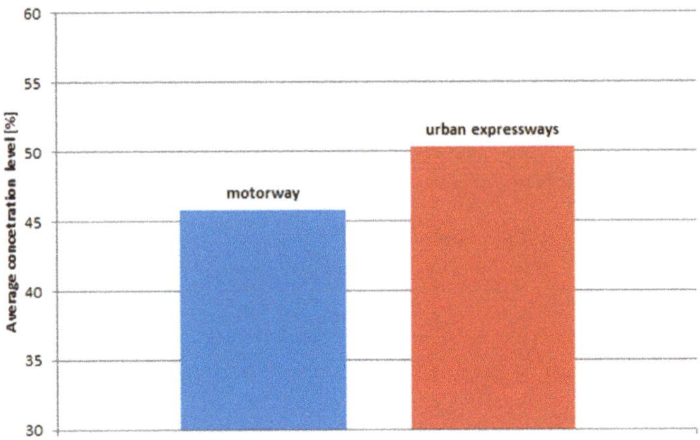

Figure 7. Average concentration level of drivers on a motorway or expressway—first data part.

The average concentration level calculated on motorway (Sections 2 and 3) was equal to 50% and 49% respectively. The corresponding average value of the concentration level recorded in the urban area was equal to 58%, as shown in Figure 8.

The level of concentration of drivers with varying levels of driving skills was also verified, as shown in Figures 9–11. As observed, the average level of concentration while driving for inexperienced drivers was equal to 53.6%.

Figure 8. Average concentration level of drivers on a motorway or expressway—second data part.

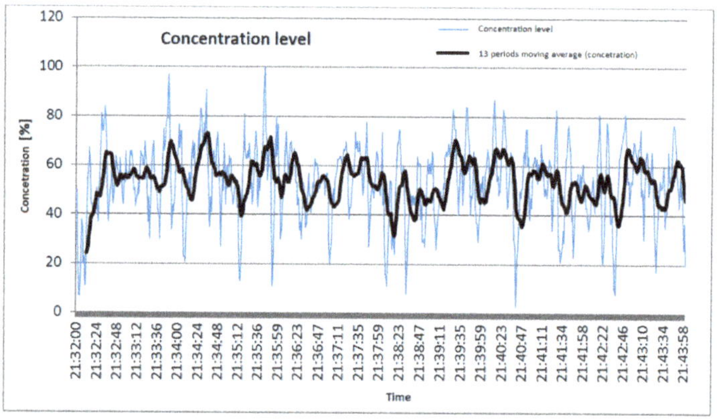

Figure 9. Concentration of a less experienced drivers.

The average level of concentration while driving in case of an experienced driver, as shown in Figures 11 and 12, was equal to 39.5%.

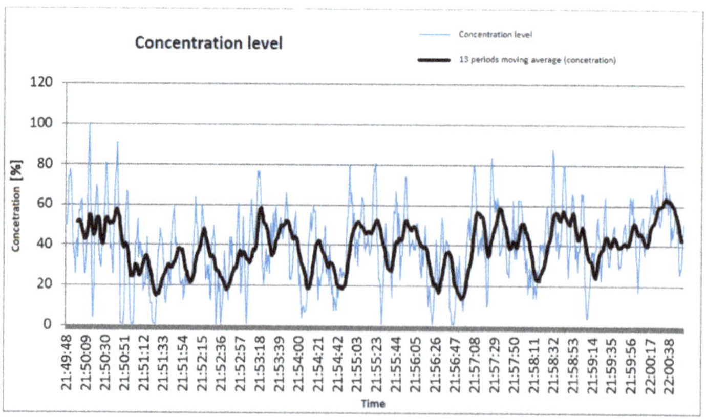

Figure 10. Concentration level of an experienced driver.

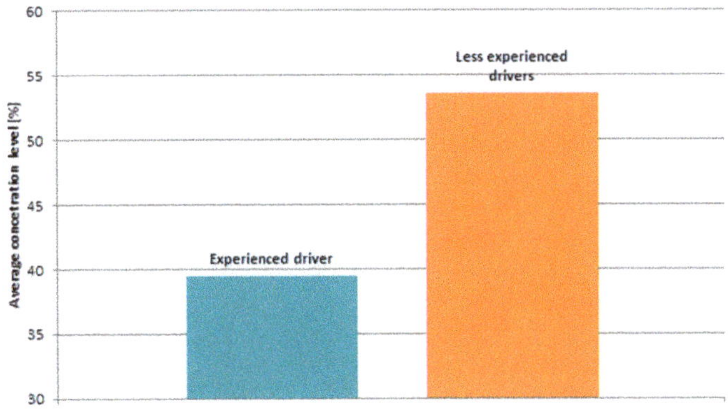

Figure 11. Concentration level of an experienced driver and a less experienced drivers.

It was found that the system was able to detect differences in driving in urban areas (more driving events occurring) and motorway areas (fewer driving events occurring) after averaging the focus level values. The system also detected a difference in focus between experienced and inexperienced drivers. During long, monotonous driving sessions, there was a large drop in the driver's average concentration levels.

A study was also carried out on the concentration of people performing the activity of moving small objects from one container to another using tweezers. As shown in Figure 12, the task lasted 3 min, where the concentration level of those performing simple activities rose to a peak and then fell. The situation was repeated, but the maxima reached in the second minute were smaller than the previous ones.

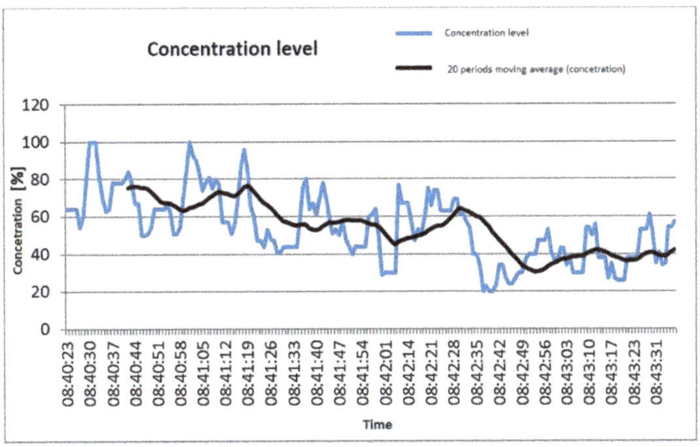

Figure 12. Concentration level of people performing simple manual tasks.

The level of concentration was highest when there were few items left in the first container and decreased when the subject began to move items from a full container to an empty one.

5. Discussion

One of the main limitations of EEG monitoring is the recording electrode; its principle of operation and application to the scalp has changed very little since EEG's first introduction as a clinical tool. With the evolution of amplifiers and data acquisition systems, modern-day EEG systems now have the capability to record for multiple hours or even days with little intervention [20].

The research carried out has confirmed that it is possible to detect the level of concentration of machine and vehicle operators during their work. As the working characteristics of a mining machine operator are very different from those of a car driver, it will be necessary to carry out specific studies in order to propose a dedicated system concept designed for this group of people.

It is assumed that, for reasons of user comfort, the measuring elements should not interfere with the work. It was therefore decided to place them inside the operator's helmet, as shown in Figure 13.

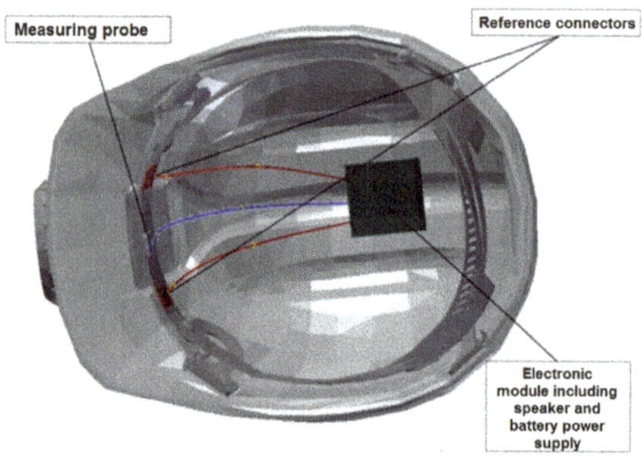

Figure 13. Prototype of a device for measuring EEG signals for machine operators.

The electronic module shall be mounted in the upper part of the miner's helmet, and the measuring probe will be guided to two points on the operator's forehead, which are the reference points for measuring the EEG signals. A suitable housing shall be used as the EEG signal inputs need to be protected against electromagnetic pulses (EMP) in accordance with the IEC 61000-4-2 recommendation [21].

The system model was built on the basis of a freely available market solution, MindWave Mobile, which analyses the acquired EEG signal and performs its evaluation by generating a focus level signal. In the proposed 3D prototype model of the system, it will be necessary to develop a hardware solution and algorithms that will allow appropriate analysis of the acquired EEG signal. It will also be necessary to verify the proposed placement of the measurement electrodes.

The readable field strength of the electromagnetic field emitted by the operator's brain ranges from a few to tens of microvolts (μV). It was therefore necessary to use a signal amplification circuit. After amplification, the signal was filtered in order to extract specific frequencies. It is planned to use a band-pass filter to extract the frequencies of alpha waves (8–13 Hz) and theta waves (4–8 Hz) from the amplified EEG signal [22–24]. These waves carry information about the level of focus and meditation. In our prototype solution, we intend to use a single-electrode solution for EEG signal measurement. The electrode will be placed on the forehead of the operator. There is a possibility of interfering with the measurement in a single-electrode system. However, this matter was verified during the carried out tests. In addition, the location indicated for the measurement of alpha waves was the occipital region, while theta waves were found in the temporal region. However, these waves were also present in the frontal region, as confirmed by this study. However, if the results of the prototype tests under real-time operating conditions will indicate too high level of interference, additional measuring electrodes will be used.

The operator's focus level data will be delivered in a wireless manner to the main server in an underground mine environment and then transmitted on the surface to the supervisory visualization system, shown in Figures 14 and 15.

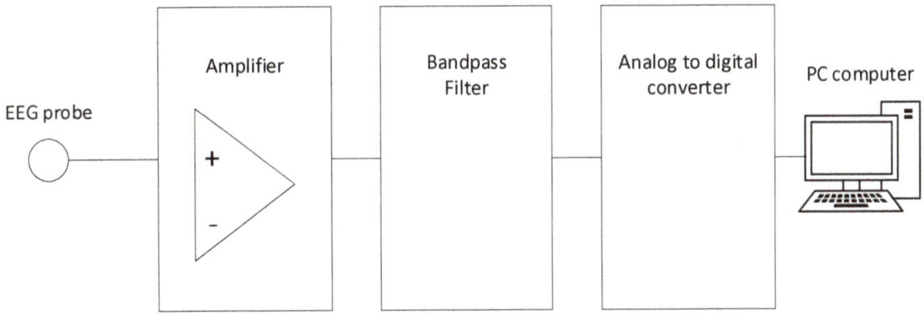

Figure 14. Proposed block diagram of the EEG signal measurement system of the prototype solution.

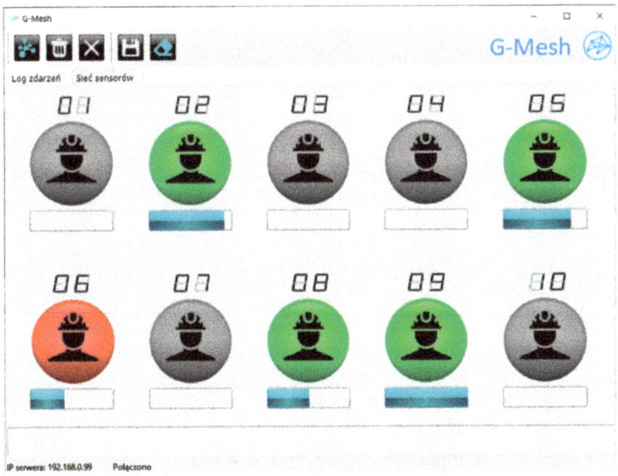

Figure 15. Interface of the G-Mesh monitoring application (gray—inactive device; green—high level of concentration; red—low level of concentration).

The device mounted inside a helmet will be one of the nodes in a wireless network with a mesh topology, where the routing process will be performed by algorithms based on swarm intelligence (SI). These algorithms are characterized by the fact that, regardless of the number of variables in the solution space, they adapt to the constraints and help to solve the optimization problem. The use of these algorithms makes it possible to maintain the transmission parameters of the data transmitted in sensor networks at an appropriate level of quality. Factors that need to be properly addressed include throughput, low latency, and high reliability. Sensor networks based on swarm intelligence algorithms are self-organizing and multi-redundant. Additional information may be found in [25–30].

To sum up, a 3D model of the device prototype was developed and a graphical interface of the software for visualizing measurement data was also introduced, along with guidelines for building a sensor network.

6. Conclusions

This paper presents a system utilizing a focus level measurement function to improve machine operator's safety. In this study, a dedicated solution was proposed, evaluated and discussed. Additionally, field tests were conducted to verify the design assumptions. The experiments were conducted on two groups of people with skills similar to a mining machine operator: a car driver and a person performing manual work requiring high level of concentration. The conclusions of the research were as follows:

- The recorded concentration levels were correlated with particular road events (increased concentration level value)—the driver (operator) reacts with increased concentration levels to particular road events. The measurement of the EEG signal could be used to identify the operating process of a machine or equipment.
- During a long monotonous drive, a decrease in concentration level was observed—EEG signal measurement of the machine operator was able to identify a decrease in work performance. It was therefore possible to generate an arousal signal when a reduced level of concentration on the activity being performed was detected.
- During manual tasks, the recorded concentration level decreased with every minute of work performed—manual tasks requiring a high level of concentration through repetition were also capable of causing a decrease in concentration level on the work performed. A system that uses the measurement of the machine operator's EEG signal could increase occupational safety by generating alarm (arousal) signals when a decrease in concentration level is detected.

Our research confirmed that the proposed concept of a concentration measurement system could be used to build a prototype device to control the concentration level of heavy machinery operators, especially in the oil and mining industry.

Guidelines for the construction of a prototype system were presented, along with a 3D model showing the proposed installation of a single-electrode device in a mining helmet. The concept of a single-electrode measurement channel was also introduced. The amplitudes and frequencies of the EEG signal, which will be used in the prototype to identify the concentration level and meditation of the operator, were discussed. The concept of a reliable wireless data transmission link to the master unit was also discussed.

Further work will be undertaken in the near future to adapt the system to the requirements of working in underground mines, which may involve building a proprietary hardware solution for EEG signal analysis and conducting tests on operators of mining machinery and equipment. The results of such experiments will allow the establishment of minimum focus values (thresholds) that must not be exceeded in order to maintain stable and safe working conditions. If these thresholds would be exceeded, appropriate voice or graphic information could be generated to increase the operator's concentration level. Our research could also determine the impact of the operator's concentration level on the level of occupational safety. Additional inspiration for future research directions can be found in [31–34].

Author Contributions: Conceptualization, J.J., B.P. and M.W.; methodology, J.J., B.P. and M.W.; software, J.J. and M.W.; validation, J.J. and M.W.; formal analysis, J.J., B.P. and M.W.; investigation, J.J., B.P. and M.W.; resources, J.J., B.P. and M.W.; data curation, J.J. and M.W.; writing—original draft preparation, J.J., M.W., B.P. and P.F.-G.; writing—review and editing, J.J., M.W., B.P. and P.F.-G.; visualization, J.J. and M.W.; supervision, J.J., B.P. and M.W.; project administration, J.J., M.W. and P.F.-G.; funding acquisition, P.F.-G. All authors have read and agreed to the published version of the manuscript.

Funding: This research received no external funding. The APC was funded by Gdansk University of Technology.

Institutional Review Board Statement: The study was conducted in accordance with the Declaration of Helsinki and approved by the Institutional Review Board (or Ethics Committee) of KOMAG Institute of Mining Technology.

Informed Consent Statement: Informed consent was obtained from all subjects involved in the study. Written informed consent has been obtained from the patient(s) to publish this paper.

Data Availability Statement: Not applicable.

Conflicts of Interest: The authors declare no conflict of interest.

References

1. Lan, Z.; Sourina, O.; Wang, L.; Liu, Y. Real-time EEG-based emotion monitoring using stable features. *Vis. Comput.* **2016**, *32*, 347–358. [CrossRef]
2. Polish Higher Mining Office. Available online: www.wug.gov.pl/download/6178.pdf (accessed on 8 March 2023).
3. Radziszewska-Zielina, E.; Sobotka, A.; Plebankiewicz, E.; Zima, K. Preliminary identification and evaluation of parameters affecting the performance of the operator-machine earthmoving system. *Build. Arch.* **2013**, *12*, 53–60.
4. Grieger, A.; Sawicki, T.; Sędłak, P.; Janicki, W.; Lewaszkiewicz, Ł. Safety situation in the workplace as function of a machine operator's level of training. *Logistika* **2015**, *5*, 909–916.
5. Jukiewicz, M.; Merkisz, J.; Orszulak, B. Using mindwave to biomeasurment attention factor in research field related with passenger vehicle simulator. *Logistika* **2014**, *3*, 2650–2656.
6. Jagoda, J. System pomiaru koncentracji operatora maszyn i urządzeń górniczych. *Masz. Górn.* **2017**, *2*, 71–80.
7. Wedde, F. BeeAdHoc: An energy efficient routing algorithm for mobile ad hoc networks inspired by bee beahavior. In Proceedings of the Genetic and Evolutionary Computation Conference, Washington, DC, USA, 25–29 June 2005.
8. Jasiulek, D.; Świder, J. Mechatronic systems in mining roadheaders—Examples of solutions. *Pom. Automat. Robot.* **2013**, *1*, 121–127.
9. Jasiulek, D.; Stankiewicz, K.; Woszczyński, M. Intelligent self-powered sensors in the state-of-the-art control systems of mining machines. *Arch. Min. Sci.* **2016**, *61*, 907–915. [CrossRef]
10. Jendrysik, S.; Jasiulek, D.; Stankiewicz, K. System sterowania ścianowym przenośnikiem zgrzebłowym. *Masz. Górn.* **2015**, *1*, 29–32.
11. Kostka, M.; Krzak, Ł.; Gawliński, A.; Jasiulek, D.; Latos, M.; Rogala-Rojek, J.; Stankiewicz, K.; Bartoszek, S.; Jendrysik, S.; Jura, J. Systemy monitoringu, diagnostyki i sterowania maszyn górniczych. *Masz. Gór.* **2015**, *3*, 88–96.
12. Hou, X.; Liu, Y.; Sourina, O.; Tan, Y.R.E.; Wang, L.; Mueller-Wittig, W. EEG based stress monitoring. In Proceedings of the 2015 IEEE International Conference on Systems, Man, and Cybernetics, Hong Kong, China, 9–12 October 2015.
13. Gevins, A.; Smith, M.E.; Leong, H.; McEvoy, L.; Whitfield, S.; Du, R.; Rush, G. Monitoring working memory load during computer-based tasks with EEG pattern recognition methods. *Hum. Factors* **1998**, *40*, 79–91. [CrossRef]
14. Jung, T.P.; Makeig, S.; Stensmo, M.; Sejnowski, T.J. Estimating alertness from the EEG power spectrum. *IEEE Trans. Biomed. Eng.* **1997**, *44*, 60–69. [CrossRef] [PubMed]
15. Smith, M.E.; Gevins, A.; Brown, H.; Karnik, A.; Du, R. Monitoring task loading with multivariate EEG measures during complex forms of human-computer interaction. *Hum. Factors* **2001**, *43*, 366–380. [CrossRef] [PubMed]
16. Parra, L.C.; Spence, C.D.; Gerson, A.D.; Sajda, P. Response error correction-a demonstration of improved human-machine performance using real-time EEG monitoring. *IEEE Trans. Neural Syst. Rehabil. Eng.* **2003**, *11*, 173–177. [CrossRef]
17. Broniec, A.; Chodak, J. Control of a simple electrical device by means of an EEG signal. *Automation* **2009**, *13*, 1059–1067.
18. NeuroSky. Available online: https://neurosky.com/biosensors/eeg-sensor/biosensors/ (accessed on 8 March 2023).
19. Paszkiel, S. EEG signal acquisition using Neurosky Mindwave Mobile for control processes implemented from the android system level. *Pozn. Univ. Technol. Acad. J.* **2015**, *84*, 237–244.
20. Ives, J.R. New chronic EEG electrode for critical/intensive care unit monitoring. *J. Clin. Neurophysiol.* **2005**, *22*, 119–123. [CrossRef]
21. International Electrotechnical Commission. *IEC 61000-4-2 Recommendation*; Electromagnetic Compatibility (EMC)—Part 4-2: Testing and Measurement Techniques—Electrostatic Discharge Immunity Test; International Electrotechnical Commission: Geneva, Switzerland, 2008.
22. Rana, K.D.; Vaina, L.M. Functional roles of 10 Hz alpha-band power modulating engagement and disengagement of cortical networks in a complex visual motion task. *PLoS ONE* **2014**, *9*, e107715. [CrossRef]
23. Brown, T.; Johnson, R.; Milavetz, G. Identifying periods of drowsy driving using EEG. *Ann. Adv. Automot. Med.* **2013**, *57*, 99–108.
24. Malar, E.; Gauthaam, M.; Kalaikamal, M.; Muthukrishnan, S. The EEG based driver safety system. *IACSIT Int. J. Eng. Technol.* **2012**, *4*, 340–343. [CrossRef]
25. Jagoda, J.; Hetmańczyk, M.; Stankiewicz, K. Dispersed, self-organizing sensory networks supporting the technological processes. *Min. Mach.* **2021**, *2*, 13–23.
26. Jagoda, J.; Stankiewicz, K. Intelligent routing algorithms in the complex sensors network for control energy storage. In Proceedings of the 20th International Conference on Advanced Batteries, Accumulators and Fuel Cells, Brno, Czech Republic, 25–28 August 2019.
27. Stankiewicz, K. A method for the self-organization of a sensor network in belt conveyor exploitation. *Maint. Probl.* **2016**, *3*, 145–154.
28. Smolarek, A.; Malinowski, T. Protokoły trasowania w sieciach ad hoc. *Zesz. Nauk. Wyższ. Szkoł. Informat.* **2012**, *8*, 47–60.
29. Di Caro, G.; Ducatelle, F.; Gamberdella, L.M. Swarm intelligence for routing in mobile ad hoc networks. In Proceedings of the 2015 IEEE Swarm Intelligence Symposium, Pasadena, CA, USA, 8–10 June 2005.
30. Lech, M.; Berry, B.M.; Topcu, C.; Kremen, V.; Nejedly, P.; Lega, B.; Gross, R.E.; Sperling, M.R.; Jobst, B.C.; Sheth, S.A.; et al. Direct electrical stimulation of the human brain has inverse effects on the theta and gamma neural activities. *IEEE Trans. Biomed. Eng.* **2021**, *68*, 3701–3712. [CrossRef] [PubMed]
31. Marks, V.S.; Saboo, K.V.; Topçu, Ç.; Lech, M.; Thayib, T.P.; Nejedly, P.; Kremen, V.; Worrell, G.A.; Kucewicz, M.T. Independent dynamics of low, intermediate, and high frequency spectral intracranial EEG activities during human memory formation. *NeuroImage* **2021**, *245*, 118637. [CrossRef] [PubMed]

32. Cimbalnik, J.; Dolezal, J.; Topcu, C.; Lech, M.; Marks, V.S.; Joseph, B.; Dobias, M.; Van Gompel, J.; Worrell, G.; Kucewicz, M.T. Intracranial electrophysiological recordings from the human brain during memory tasks with pupillometry. *Sci. Data* **2022**, *9*, 6. [CrossRef] [PubMed]
33. Stańczak, L.; Kaniak, W. Occupational health and safety management in hard coal mines in the aspect of dust hazard. *Min. Mach.* **2021**, *2*, 53–62.
34. Stankiewicz, K. Mechatronic systems developed at the KOMAG. *Min. Mach.* **2020**, *2*, 59–68.

applied sciences

MDPI

Article

A Case Study on Implementing Agile Techniques and Practices: Rationale, Benefits, Barriers and Business Implications for Hardware Development

Paweł Weichbroth

Department of Software Engineering, Faculty of Electronics, Telecommunications and Informatics, Gdansk University of Technology, 80-233 Gdańsk, Poland; pawel.weichbroth@pg.edu.pl; Tel.: +48-58-347-29-89

Abstract: Agile methodologies, along with the corresponding tools and practices, are claimed to facilitate teams in managing their work more effectively and conducting their work more efficiently while fostering the highest quality product within the constraints of the budget. Therefore, the rate of awareness and adoption of Agile frameworks both within and outside the software industry has increased significantly. Yet, the latest studies show that the adoption of Agile techniques and practices are not one-size-fits-all, and highlight the challenges, risks, and limitations regarding numerous domains. In this regard, the state-of-the-art literature provides comprehensive reading. However, in the case of hardware manufacturing, it seems to be sparse and fragmented. To fill this gap, the goal of this study is to analyze and present an in-depth account of the implementation of mix agile-oriented tools and practices. To tackle this goal, a single industry case study was undertaken, based on the primary data obtained through the interview protocol and the secondary data extracted from the project's documentation. The findings concern three areas. First, the rationale behind the implementation of agile for hardware development is explained. Second, the implemented agile techniques and practices are identified, as well as the supporting tools through which their adoption was successfully undertaken. Third, the areas positively impacted by their application are highlighted with the corresponding evaluation measures deployed; moreover, the barriers to adopting Agile practices encountered, and the benefits gained from particular techniques, are further discussed. The presented findings might be of great importance for both researchers and practitioners who are searching for empirical evidence regarding Agile-oriented implementations. Finally, in terms of both benefits and barriers, business implications for hardware development are formulated. Alongside this, numerous open issues and questions present interesting research avenues that concern, in particular, the effectiveness of collaboration and areas of communication through the lens of agile techniques and practices.

Keywords: agile; implementation; hardware development; case study; user experience

Citation: Weichbroth, P. A Case Study on Implementing Agile Techniques and Practices: Rationale, Benefits, Barriers and Business Implications for Hardware Development. *Appl. Sci.* **2022**, *12*, 8457. https://doi.org/10.3390/app12178457

Academic Editors: Przemysław Falkowski-Gilski, Tadeus Uhl, Zbigniew Lubniewski and Zhongjie Wang

Received: 13 June 2022
Accepted: 21 August 2022
Published: 24 August 2022

1. Introduction

The frustrations around failures of software development projects in the 1990s led to the search for more effective and cost-efficient ways of organizing and managing tasks and resources [1]. In early 2001, a group of software leaders met and discussed the possibilities of how to improve human productivity and performance, as well as how to bring value to the customers by delivering the expected software products [2]. At that meeting, the participants neither introduced nor used "agile" during discussions, instead the words "light" and "lightweight" were more common, reflecting their attitudes to the issue of bringing new software to market faster [3]. Eventually, two key opportunities were identified, namely shortening the time of delivery of the first version of the working software, and getting feedback from the users to confirm their requirements [4]. This approach laid significant foundations for the Agile methodology as we know it today focused on speed to market, rapid feedback and continuous improvement [5].

On the other hand, the hardware businesses have also come across many obstacles and hardships in their way of updating existing and developing new products. The reasons for their failures can be divided into three categories: technical, financial, and sales/marketing [6]. It should be noted here that the second and third categories do not fall into the scope of our study, and will therefore not be further tackled. The first group of technical issues occurs during the development and early manufacturing stages often concern [6]:

- underestimating the product development;
- underestimating the complexity of scaling to mass manufacturing;
- poor quality testing;
- product overcomplexity;
- overpromising to customers.

To deal with these issues successfully, numerous methodologies, frameworks, and single methods have already been adopted by applying a different set of assumptions and settings, due to the unique needs of both the product and manufacturer. However, recent studies show that hardware development is enacted through agile methodologies [7], or just arbitrarily selected agile practices [8]. Similarly, our study falls into the scope of this modern but still insufficiently explored research domain. More specifically, the goal of this paper is to analyze and present an in-depth account of the implementation of agile techniques and practices, based on an industry case study of a project undertaken by the Tele and Radio Research Institute (TRRI) [9] and managed together with Meritus Software Systems [10].

The rest of the paper is structured as follows. In the Section 2, the research background and related work are outlined. In Section 3, the research design is specified. Next, in Section 4, the findings of the study are presented, followed by the conclusions given in Section 5.

2. Research Background and Related Work

2.1. Theoretical Background

In general, Agile is the umbrella term for a family of management practices [11]. Nowadays, many other organizations are actively adopting agile project management [12] as an iterative approach to delivering a project throughout its life cycle [13]. The reasons for implementing Agile practices concern the more flexible and adaptive working structures [14], active stakeholders and extensive user participation throughout the project [15], higher performance visibility [16], and transparency [17], to name just a few.

Considering the software industry, the Agile Alliance [18] is one of the most prominent organizations devoted to organizations and individuals that apply the values and principles of Agile. This non-profit, membership organization was founded on the Manifesto for Agile Software Development (ASD) [19]. By nature, ASD refers to a group of software development methods based on iterative and incremental development [20], where requirements and solutions evolve through collaboration between self-organizing cross-functional teams [19]. The Agile Manifesto expresses four key values, defined as follows [21]:

1. Individuals and interactions over processes and tools;
2. Working software over comprehensive documentation;
3. Customer collaboration over contract negotiation;
4. Responding to change over following a plan.

It should be noted and empathized here that, in the case of the fourth value, an agile view on the project means that responding to changes, instead of neglecting or even ignoring them, benefits the project by providing additional value.

Moreover, these four values are broken down and detailed through twelve supporting principles, formulated in the following way [22]:

1. Our highest priority is to satisfy the customer through the early and continuous delivery of valuable software;

2. Welcome changing requirements, even late in development. Agile processes harness change for the customer's competitive advantage;
3. Deliver working software frequently, from a couple of weeks to a couple of months, with a preference for the shorter timescale;
4. Businesspeople and developers must work together daily throughout the project;
5. Build projects around motivated individuals. Provide them with the environment and support they need and trust them to get the job accomplished.
6. The most efficient and effective method of conveying information to and within a development team is face-to-face conversation;
7. Working software is the primary measure of progress;
8. Agile processes promote sustainable development. The sponsors, developers, and users should be able to maintain a constant pace indefinitely;
9. Continuous attention to technical excellence and good design enhances agility;
10. Simplicity, the art of maximizing the amount of incomplete work, is essential.
11. The best architectures, requirements, and designs emerge from self-organizing teams;
12. At regular intervals, the team reflects on how to become more effective, then tunes and adjusts its behavior accordingly.

Based on the above principles, several agile methodologies have been built upon [23], including Scrum [24], Kanban [25], and Extreme Programming (XP) [26]. According to the 15th State of Agile Report [27], Agile adoption within software development teams increased from 37% in 2020 to 86% in 2021, where Scrum was the most popular (66%), with an additional 15% of the two following derivations, namely ScrumBan (9%) and Scrum/XP (6%); the minority concern Kanban (6%), Iterative (4%), XP (1%), Lean Startup (1%), and others (5%), whereas 2% of the respondents did not know which Agile methodology was adopted.

More specifically, over the last two decades, several agile practices have been developed that rely on the interaction of self-organizing teams with cross-functional skillsets. Globally, across a broad range of industries in the software development domain [28], the most commonly used tools by the individuals are [27]:

- Daily standups (87%), a team update meeting held each day [29], where everyone involved is asked to report on what has been completed, and commit the future work to do; moreover, the occurred impediments (blocking factors) are discussed; typically, in Scrum, the daily standup is a 15-min time-boxed event [30];
- Retrospectives (83%), a meeting that is held at the end of an iteration [31], with the aim of self-reflecting on what went well and what could be improved for the next iteration;
- Sprint/iteration planning (83%), where a sprint, only used in relation to Scrum [32], is a predefined, time-boxed period in which a team works to accomplish a set amount of work [33], and an iteration, in the context of an Agile project, is a time-box during which development takes place [34].

Considering agile planning, in particular involving workflows, tasks, and user stories, the three the most frequent practices are [27]:

- Kanban boards (77%), a visual-organization tool that tracks the workflow by depicting work items as cards at various stages of a development process, represented by labeled columns; in the simplest form, the board columns fall into three buckets: at the starts of the board is the "To Do" (or "Ready") column which contains all the cards that are next up, the "Doing" (or "In Progress") column include all the cards that are currently being worked on, and "Done" contains all cards that have been finished [35];
- Task boards (67%), in Scrum, a task board is a visual display of the progress of the Scrum team during a sprint [36]; typically, a board is divided into four columns, namely: starting from the left, "Stories", contains a list of all the user stories in the current sprint backlog, "To Do", contains subtasks of the stories that work has not started on, "In progress" depicts all the tasks on that work has already begun, and "Done", presents all the tasks that have been completed [37];

- Spreadsheets (66%), stands for a computer application used to create and manipulate spreadsheets, electronic documents in which data is arranged in tabular form and can be manipulated and used in calculations, along with formulas that relate the data. Currently, there is a plethora of spreadsheet applications on the software market, including both commercial and non-commercial, available for both desktop computers and mobile devices.

Having said that, one might ask: what are the benefits of being agile? There is no one simple answer to this question. However, the report from the 15th State of Agile survey shows that the implementation of agile has positively impacted areas within the organizations such as: managing changing priorities (70%), visibility (70%), business/IT alignment (66%), delivery speed/time to market (64%), team productivity (60%), and team morale (60%), just to quote a few. The impact was measured by various indicators, spanning from customer/user satisfaction (59%), business value (58%), business objectives achievements (50%), on-time delivery (48%), quality (48%), and productivity (41%), just to name a few.

2.2. Agile-Oriented Frameworks for Hardware Development

SAFe. The Scaled Agile Framework (SAFe), created by Dean Leffingwell, is a set of organizational and workflow patterns for implementing and scaling practices and techniques at the enterprise level in accordance with lean principles [38]. SAFe is a body of knowledge that incorporates frameworks such as Scrum, Extreme Programming, Kanban and Lean at the team level [39]. It advocates four levels: Value stream, Program, Portfolio and Team level [40]. The principal values of SAFe, which are key to SAFe's effectiveness, concern: alignment, built-in quality, transparency, and program execution [41]. In this line of thinking, the Scaled Agile organization provides six universal principles, outlining how SAFe applies to hardware development [42]:

1. Organize Around Value;
2. Assume Variability and Preserve Options;
3. Build Incrementally, Integrate Frequently;
4. Design for Change;
5. Perform Work in Small Batches;
6. Build Continuous Integration for Hardware Development.

While SAFe is built on top of the agile methodology, harnessing its advantages, in particular, is based on ten immutable, underlying Lean-Agile principles; the above six principles have been arbitrarily selected and reformulated in the context of the hardware domain. Nevertheless, there is open space for creativity and innovation, since, as stated by the authors, "they can be used to guide hardware engineers to create and adopt their own best practices" [42].

Scrum for Hardware. Scrum for Hardware (or Scrum for HW) is a framework that combines Scrum with modular architecture and Lean/XP practices [43], customizing the agile practices and techniques for hardware and system design. The Scrum in Hardware Guide defines 8 areas to consider before, during and after implementation, namely:

1. Uncertainty in Hardware Projects;
2. Stubs and Mock-Ups;
3. Modularize;
4. Continuously Integrate;
5. Interface Design;
6. Test and Data-Driven Development;
7. Team;
8. Working Product at the End of Each Sprint.

In general, the guide can be used by any organization, small or large, regardless of its field of activity, structure, or organization.

Hybrid Stage-Gate. The Stage-Gate (SG) process is a conceptual and operational road map for moving a new-product project from idea to launch and beyond [44]. In the simplest format, SG "consists of two major components:

(1) a series of stages, where the project team undertakes the work, obtains the needed information, and does the subsequent data integration and analysis, followed by;
(2) gates, where go/kill decisions are made to continue to invest in the project" [44].

An overview of the general Stage-Gate system is depicted in Figure 1.

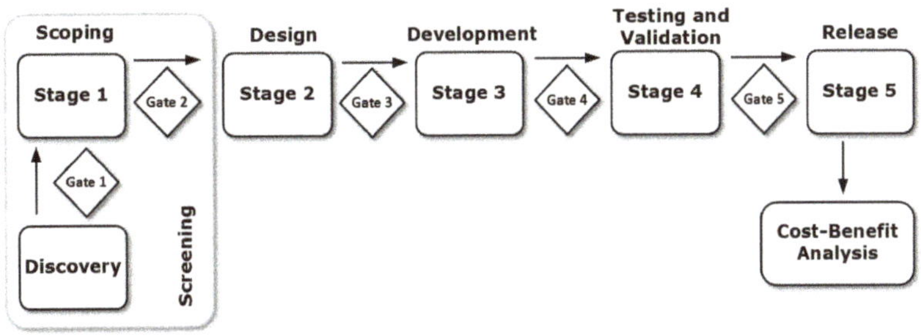

Figure 1. Overview of the generic Stage-Gate system. Source: [44].

Stage-Gate is a macro model, designed to support selecting the projects to be followed by the model, and once selected, to map out the key stages, best practices, management techniques, roles, and responsibilities. On the other hand, there is an agile approach that brings micromanagement practices and techniques to the table, which are proven to foster adaptability, encourage creativity, and improve productivity. Cooper and Sommer introduced a hybrid approach, which incorporates Agile methods for physical product development [45], by integrating them with the SG system, and then utilized them for manufactured or physical product development (hardware). In other words, they argue that "Agile is normally applied as the project management method within the stages in Stage-Gate", while "the stages remain, and Agile is applied within some stages" [45]. Moreover, Scrum is a favored methodology since it has been indicated as "the most appropriate for hardware development" for all case studies uncovered so far in industry.

Modified Agile for Hardware Development (MAHD). Based on the Agile principles, Simpson and Hinkle [46] developed the Modified Agile for Hardware Development (MAHD) Framework, devoted to the manufacturing industry, in particular for products and portfolios that include a combination of electronics, mechanical components, as well as software elements [47]. By design, MAHD takes into account the following physical products requirements [48]:

- allow for flexibility to change, but also freeze designs;
- build quality plans and manage supply chains;
- define electrical and mechanical product attributes;
- develop documentation;
- develop flexible prototyping and validation product strategies;
- share resources and external partners.

MAHD promotes four foundation principles to follow, namely:

1. Short development cycles to drive learning and adapting to change;
2. Accountable, autonomous, and focused teams;
3. Validating work at the end of each development cycle;
4. Applying intelligent, rapid prototyping strategies.

Interestingly, while the first and the second principles are focused on the human factor, the remaining two are targeted at the product. Such a dichotomy seems to be reasonable and justified, taking into consideration the work breakdown structure in general.

2.3. Related Work

The rationale behind adopting agile (Scrum) practices to integrate the development of both hardware and software is provided by Lima et al. who claim that "the development of devices that combine hardware and software has created new challenges", since "the new built devices have a short life cycle and frequently require upgrading" [49]. In this study, the authors introduced a development process so-called "Agile and Co-Design", which consists of seven phases and expected outcomes: Conception (product vision), Setting-up (specifications), Design (deliverables), Definition Of Done (checklist), Testability (integration tests), Implementation (hardware and software), Verification (evaluation). The findings show the application of their approach enabled deep interaction between teams representing different areas of knowledge, and fostered knowledge exchange and innovation creation.

In a similar vein of cautionary optimism, Laanti stated that "the benefits of incremental and iterative development of hardware may be as high as with software development" [50]. However, afterward, the obstacles are articulated: "the idea of incremental design was causing problems", since "people felt the hardware should have been created and tested as one entity", and "it is not flexible the same way software is". On the other hand, the application of iterations "can be used to mature the design".

The broad spectrum of issues and interests regarding the potential, transition and applicability of agile development in the hardware sector is given by Atzberger and Kristin [51], and later by Atzberger et al. [52]. The four categories of the findings concern:

- constraints of physicality, the authors claim that "the greatest hindrance is to realize potentially shippable increments in one iteration", and as a consequence "the technical feasibility to produce prototypes" is hardly achievable in certain industrial branches, due to "the inability to break down the product into modules" [51];
- agile mindset, an individual attitude toward understanding, collaborating, learning, adapting and growing in the spirit of trust, responsibility and ownership, is pointed out as a major challenge for not only individuals, but also for the company;
- team distribution, the inability to work co-located resulted in a higher degree of communication needs between physically dispersed project participants (teams);
- scaling, the ability to apply the principles, practices and outcomes in other projects (horizontal scaling), or at other levels of the organization (vertical scaling), in order to change the way people approach their own work and working with others.

Naturally, there are numerous other studies that have reported on the implementation of the agile approach, and its consequences on the project effectiveness and deliverables. Moreover, there are several well-known and admired industry corporations that are claimed to cultivate agile principles and practices, including Spacex [53], Tesla [54], General Motors [55], and Toyota [56], just to name a few. Nevertheless, agile is not a silver bullet that enables everyone to become adaptive, innovative, and resilient; neither is it an excuse to stop thinking, nor a panacea for solving all issues.

3. Research Design

3.1. Case Setting

Tele and Radio Research Institute (TRRI) is located in Poland (Warsaw) and operates in the business industry. TRRI's mission is to conduct "comprehensive and interdisciplinary scientific examinations and development activities on highly advanced technologies and innovations, in particular with regard to ICT systems, electronics, electronic installation and Industry 4.0 solutions" [57]. In 2019, the Institute became a part of The Łukasiewicz Research Network Institute, which, associating 26 research institutes with 8000 employees, located in 12 cities across Poland, is the third largest research network in Europe [58]. For the sake of clarity and simplicity, henceforth, TRRI will be called the "project partner" (PP), or just a "partner".

Meritus Software Systems (Meritus) is a software producer with its headquarters in Poland (Warsaw). Established in 1990, for over 30 years the company has earned the reputation of being a trustworthy and loyal business partner among cooperating companies. Meritus is a team of enthusiastic computer specialists devoted to developing highly efficient and usable software by following the modern patterns and trends in the development of IT systems. The landmark software products are: Pinquark Warehouse Management System, Geo Train AI, merSoft ERP, and skanujfakture.pl [10]. Henceforward, Meritus will be called the "project leader" (PL), or simply the "leader".

3.2. Project Settings

The current study is entirely based on a project, performed in cooperation between the partner and the leader, entitled "An Industry 4.0 Mobile Process Management System Supported by Artificial Intelligence" (hereafter called the "project"). The project's budget was 1.2 million euro and was conducted between January 2019 and December 2021. The aim of the project was to develop and implement a cloud-based process management system, facilitated by six AI modules, available for both web-based and mobile users. The scope of the project falls into the smart factory domain, thus following the modern trend of automation and data exchange in manufacturing technologies, including cyber-physical systems, cloud and cognitive computing, and the Internet of Things.

It should be noted here that by definition, a "smart factory" refers to the workers and machinery related to the execution of their tasks; by design, a smart factory is flexible and re-configurable, low-cost, changeable, and thereby agile [59].

3.3. Project Context and Organization

To set up the context of the study, one of the most relevant solutions is to provide the scope of the project, which serves as a measurable and exogenous variable [60]. In this line of thinking, first the requirements are briefly provided, regarding the hardware deliverables. More specifically, their features and properties, with the embedded software developed specifically for their needs, are discussed with the project schedule illustrating a summary of the physical device development process.

Remote monitoring and control of processes in technological areas, and thus the need to transmit large amounts of data while maintaining confidentiality, requires the use of efficient communication solutions that are characterized by high-performance embedded microcontrollers, while also maintaining a low price and reliable long-term operational time in specific environmental conditions that are often extreme and unfavorable for the operation of electronic systems of large-scale integration. To this end, the goal of the project was the development of a universal, cost-effective communication modem, which should be flexible enough to adapt to the changeable requirements imposed by data acquisition from different technological areas.

The guidelines for the hardware and software platforms were based on maximizing the functionality of the module solution for use in various technological areas with a strong emphasis on aspects of Machine to Machine (M2M) communication [61] and Internet of Things (IoT) devices, especially devoted to industrial applications.

The hardware platform of the communication modem model was made in two iterations. In the first iteration, a model of the modem was made to test the functionality of the selected OSD335x solution and to check the validity of the solution in terms of the connection of the printed circuit board elements. The second, taking into account functional corrections and comments after EMC/LVD EMC tests.

First production stage [main unit]. The communication module model design is divided into two blocks. The main unit block, integrating the SOM and Ethernet communication circuits, and the second on a separate PCB with communication circuits and connectors. This division was due to the effective distribution into functional blocks and the technological requirements for the PCB for the individual functional blocks. With the proposed division, a reduction in PCB manufacturing costs was achieved, as well as the fact that in case of an error resulting from incorrect functional assumptions, it can be corrected manually or the printed circuit can be improved with only one module, which reduces the manufacturing costs.

First production stage [base unit]. The base unit includes connectors and add-on circuits required to communicate with sensors and measurement sensors. In the design of the communication modem model, it was assumed that the basic communication interface to the sensors and measurement sensors will be USB 2.0, with a high-speed, 480 MBit/s communication interface. USB 2.0 is a universal and standardized serial communication interface. One differential port and a possible 5 V power supply are required for data exchange. The USB 2.0 protocol is flexible and allows for the use of standardized data exchange protocols such as classes e.g., ACM, HID, mass storage as well as the definition of a custom class. It allows for a maximum of 256 devices on a single bus via HUB circuits.

Second production stage [main unit]. The second iteration required amendments to components of the modem, including the microprocessor and Ethernet physical layer hardware unit layout, to optimize the size of the device to be more compact. The microprocessor was changed to a version with built-in e-MMC memory. The change allowed for the design to be optimized for component availability. e-MMC memories are among the difficult to access electronic components. The change in the Ethernet physical layer layout was dictated by replacing the chip with a more modern one and adding Media-Independent Interface (MII) support, which will make it possible to use Programmable Real-Time Units (PRUs) to implement the EtherCAT (Ethernet Control Automation Technology) and Profibus (PROcess FIeld BUS) protocols. While the former protocol was chosen as it fulfills both hard and soft real-time computing requirements, since it combines the efficient and relatively high-speed message transmission of Ethernet networks [62], the latter was used because of its high speed and architecture, designed and optimized especially for communication between automation systems and decentralized devices [63].

Second production stage [base unit]. The changes that were made in the second iteration were related to the optimization of the PCB dimensions. Comments after EMC testing were also introduced. The power supply for the module requires that a noise filter be added to the input and common mode ferrites be added to the Ethernet and USB interface lines.

3.4. Research Methods

First and foremost, content analysis was used as the primary research method. By definition, according to Mackey, content analysis is "any technique for making inferences by systematically and objectively identifying special characteristics of messages" [64]. In this case, we followed the guidelines defined by Krippendorff [65], along with their practical assessment elaborated by Stemler [66]. Therefore, the following six questions have been addressed and answered.

1. **Which data are analyzed?** Due to the project complexity, considering the number of planned tasks, time frames and the total number of hardware and software deliverables, the further investigation was strictly oriented toward issues and concerns related to the project management. More specifically, the focus was on the identification, analysis and evaluation of only those processes that regarded planning, organizing, leading and controlling the efforts of the project members, engaged on the side of both the leader and the partner.
2. **How are they defined?** The data throughout the project was collected in a few digital repositories, including the data gathered by the online tools throughout the project duration, as well as the documents library available in the cloud storage.
3. **What is the population from which they are drawn?** The population is limited to one project, in particular two data repositories owned by the leader and the partner.
4. **What is the context relative to which the data are analyzed?** The context is specified by the goal of the study, which precisely indicates the scope and objectives for the researcher to follow. In particular, the data (information) must concern the implementation of agile practices throughout the project. In other words, a prece-

dent or unrepeatable application of any agile practice was not considered for further investigation due to the limited analysis and evaluation.

5. **What are the boundaries of the analysis?** Obviously, when the study concerns a single case, the results are not meant to be generalized. Therefore, it is correct to name generalizability as a limitation of the current, and any other, qualitative research.

To sum up, the content analysis was aimed at identifying, analyzing, and evaluating the project's documentation, including documents, reports, spreadsheets, information from work communication and collaboration tools, to extract the agile practices implemented throughout the project.

The second research method, applied to validate the obtained results from the documentation study, was an in-depth interview. According to Boyce and Neale, the in-depth interviewing technique involves "conducting intensive individual interviews with a small number of respondents to explore their perspectives on a particular idea, program, or situation" [67]. Moreover, regarding the interview format, the standardized open-ended interview was adopted, since such a design allows the interviewees to "contribute as much detailed information as they desire" [68]. Moreover, the interview was supported by a questionnaire (see Appendix A). For the needs of this study, the process for conducting in-depth interviews consisted of five steps:

1. **Plan**. The identified stakeholders involved two representatives, one from the leader, and one from the partner. To reduce the bias resulting from insufficient experience and knowledge, the inclusion criteria were as follows:
    a. at least five years of professional experience;
    b. involvement in the project throughout its duration;
    c. managerial role in the project.

The two respondents were both males with over 20 years of professional experience in the IT sector. In the project, the first respondent (R1) worked in the position of project manager, while the second respondent (R2) held the role of research and development manager. Therefore, they were both positively classified;

2. **Develop instruments**. The interview protocol and the questionnaire were designed around the three constructs of interest for this research, namely: (1) the rationale behind adopting the agile approach for the project, (2) the practices and tools applied, and finally (3) the performance evaluation. Appendix A shows the details regarding both data collection instruments;
3. **Collect data**. The phone interviews took place in the period from April to May 2022. Alongside safety and health issues due to the outbreak of COVID-19, this approach is now becoming widely used [69]. The data collection follow-up concerned an email request for each of the respondents to independently fill in the questionnaire;
4. **Analyze data**. To "make sense" of the collected data, a thematic analysis was conducted on the information gathered via the interview, while the survey data were compared and summarized by using the Google Sheets application;
5. **Disseminate findings**. The target audience does not only involve the project stakeholders, but also both computer science researchers and industry practitioners. To reach such a broad audience, our intention is to publish our work in open access mode, which means that our paper will be free of charge, digital and online, and free of most copyright and licensing restrictions. In addition, the paper preprint will also be posted in an electronic format and made available to the public on at least one preprint service.

4. Findings

The findings are scrutinized with regards to each question, where a strict partitioning is given to distinguish the three different areas of interest.

Part I. Agile Adoption Rationale.

Q1. What were the most important reasons for adopting Agile in the consortium? Having in mind the experience and knowledge with respect to the reasons for the adoption of Agile in software development, there is strong agreement between the respondents regarding three of its outcomes:

- accelerate software delivery;
- improve business and IT alignment;
- reduce project risk.

Moreover, they also agree that Agile adoption enhances:

- the ability to manage changing priorities;
- software quality;
- improve engineering discipline.

However, there is no consensus on whether the agile-oriented organization (team) is able to better respond to volatile market conditions or enhance delivery predictability. Moreover, it seems that the respondents also disagree on the positive impact of agility on the team morale.

Part II. Agile Methodology, Techniques and Practices Adoption.

Q2. Which agile approach was adopted? Both respondents described the applied methodology as "iterative," emphasizing the split of the development process into multiple explicit interactions, each assumed to deliver required features.

Q3: Which of the following Agile techniques and practices were used? Both respondents declared that three agile techniques were used during the project, namely:

- release planning;
- short iterations;
- sprint/iteration planning.

Moreover, they both indicated the lack of using activities for determining which products or projects to work on, in which order, and for how long, termed agile portfolio planning. Second, in the case of estimation techniques, the usage of planning poker was not confirmed either for estimating user stories or tasks. Third, daily standups were not the practice adapted to discuss a project's progress at a high level. Fourthly, the respondents declared that it was found unnecessary to release increments frequently. Last but not least, Kanban, being a transparent and visual system for real-time communication of the workload capacity, was not implemented throughout the project.

Q4. Which Agile planning and delivery tools did you use? From the set of 22 tools possible to select, the respondents agreed on three tools, including:

- task board;
- automated build tool;
- unit test tool.

On the other hand, consensus was also reached on neglecting the majority of the planning and delivering tools, including the product road-mapping, along with the family of management tools related to requirements, product, and portfolio, as well as the customer idea.

Moreover, neither index cards nor timecards were used in the project. While the former are simply the tokens representing user stories, task breakdowns, backlog items or even reminders, and the latter are the method of recording and tracking the amount of an employee's time that is spent on each job. Second, no static analysis was performed along with story mapping to formulate new hardware features. In the case of work management, the Kanban board was not found useful.

Considering the system evaluation from the perspective of the external user, no automation test tools were deployed in the project. Interestingly, other tools that are highly appreciated from the software vendor's perspective were not found suitable in the case of a hardware manufacturer. These are:

- bug tracking;
- continuous integration tool;
- release/ deployment automation tool;
- wireframes.

Ultimately, in the case of project and knowledge management, the agile project management and Wiki tools were only declared to be used by the project leader, whereas the refactoring tool was strictly used by the project partner.

Q5. Which agile planning tools did you use? While there were 21 tools to choose from, unexpectedly, each of the respondents indicated only one different tool: Atlassian Jira, in the case of the project leader, and Microsoft Excel, in the case of the project partner. The list of the other not implemented tools can be found in Appendix A.

Part III. Agile Adoption Evaluation.

Q6. Has the implementation of agile positively impacted each of the following areas within your company? Considering the 13 areas that were positively impacted by the implementation of agile practices and supported by the corresponding tools, there was no agreement between the respondents in any of these. Keeping in mind the question stated above, the first respondent identified three areas:

- managing changing priorities;
- predictability;
- risk reduction;

which were assessed as neutral (intact) by the second respondent.

On the other hand, the second respondent declared five other areas, positively impacted by the agile implementation, namely:

- delivery speed/time to market;
- engineering discipline;
- software quality;
- team morale;
- team productivity.

However, both respondents had a neutral view on three areas, including: business/IT alignment, software maintainability, and visibility.

Q7. How has your organization measured the success of Agile delivery? Both respondents agreed on one thing only, which is that by managing the scope and schedules with an Agile-centric approach, their teams were able to deliver business value, measured by tangible and working product features, added in a fixed period of time.

Moreover, the second respondent indicated that within his organization, another ten measures were successfully implemented: budget vs. actual cost, customer retention customer and user satisfaction, defects into production, defects over time, iteration burndown, planned vs. actual release dates, team morale, test pass/fail over time, and velocity.

Q8. What are the most significant barriers to adopting Agile practices in the consortium? There was one common barrier indicated when adopting agile practices which concerned the lack of business (customer, or product) understanding.

Nevertheless, considering the other twelve barriers, in the case of the following four, namely:

- fragmented tooling and project-related data/measurements;
- general organization resistance to change;
- pervasiveness of traditional development methods;
- unwilling to admit mistakes and learn from delivery failure;

the answers were the opposite, while their total absence concerned the project leader, their presence affected the project partner.

Moreover, the latter also reported another two barriers encountered during the adoption of Agile practices regarding inconsistent processes and practices across teams, and organizational culture at odds with agile values.

Q9. What are the benefits of adopting the release planning (TP10) technique? In total, eleven benefits were submitted under each respondent's independent evaluation, of which only the following three were found to be significant to a similar extent:

- increased efficiency by highlighting the goals or deadline for release;
- making high-quality plans;
- increased visualization potential of release planning problems.

With some doubt from the first respondent, and strong or moderate belief from the second, the other five benefits of using the release planning technique concerned:

- eliminating waste in the Planning Process;
- setting clear expectations about the objectives and product features;
- helping team members stay on track and complete their tasks on time;
- increased flexibility and decreased development lead time;
- supporting the identification and mitigation of potential issues and risks.

Ultimately, both respondents were not fully convinced of whether the release planning enabled the early identification of feature dependencies, and disagreed on the increased motivation of the developers, as well as on the mediation of multiple opinions.

Q10. What are the benefits of adopting the short iterations (TP12) technique? There was mutual agreement that the shorter the iteration, the less process overhead is needed to keep things on course. Partial agreement of using short iterations was reached for the following seven benefits:

- provide more-frequent feedback from customers than long iterations;
- afford the team more opportunities to reflect on and improve their work practices;
- deliver a rapid response to changes in priority without disrupting work in process;
- foster scope management since it is easier to move the small backlog items around than the large ones;
- eliminate internal process overheads by running short iterations since user stories are small units of work and the team can follow the iteration status in a simpler way;
- support product adaptability since it is easy to introduce new features, change existing features, and make any other functional change;
- support the team's ability to work on several features in parallel.

Finally, a different view was presented on the claim that short iterations enabled early problem detection since agile teams recognized process smells and acted on them immediately.

Q11. What are the benefits of adopting sprint/iteration planning (TP14)?

Unquestionably, sprint (iteration) planning was declared the enabler that positively influenced the product quality. Both respondents, representing the project leader and the project partner, partially agreed that using the sprint (iteration) planning technique brought the following benefits:

- helps teams to control projects;
- effective management of the product backlog;
- provides an opportunity for team members to participate in planning and hence better understand the work assignments;
- provides opportunities to build team cohesion and identification;
- helps in understanding the user stories;
- serves to validate something we do not understand, or something misinterpreted,
- enables setting precise estimates for the task;
- fosters knowledge-sharing and skill-improvement.

However, the respondents doubted whether the planning meetings helped to define the goals for each team member.

Q12. What are the benefits of using a task board (PD17)? Full agreement was reached with regards to all three benefits submitted to the evaluation in relation to this question. Thus, it was declared that using a task board definitely:

- ensures efficient diffusion of information relevant to the whole team;
- enables the team to stay focused on the work progress;
- allows the team to represent any relevant information.

On the other hand, it is a pity that the respondents did not provide any benefits other than the predefined ones.

Q13. What are the benefits of using an automated build tool (PD3)? Without any doubt, both respondents pointed out that using an automated build tool improved product quality by supporting bug and error detection. Moreover, they also agreed that using such a tool contributed to:

- saving time and money;
- by keeping a history of builds and releases, it helps in investigating bugs and errors;
- an increase in developers' productivity since build automation ensures fast feedback;
- acceleration of the product delivery by eliminating redundant tasks.

Q14. What are the benefits of using a unit test tool (PD20)? The two highest appraised benefits of using a unit test tool are concerned with fostering a higher quality of individual components, that in consequence, lead to increasing the overall system resiliency; moreover, it is also claimed to increase the testers' productivity. Additionally to these, the remaining advantages are related to:

- early detection of specification deficiencies;
- cost reduction;
- easier source code refactoring.

Both respondents agreed that the sum of the benefits achieved was greater than the sum of the costs encountered. Among the main challenges of running an Agile development project with remote teams, there was one in common: keeping the final goal in mind. In other words, team members sometimes forgot that "progress takes place when they work by this rule". Moreover, a few other issues were individually raised that concerned the differences between team members, regarding their knowledge, skills, and agile mindset, especially in the case of teams from organizations from different business sectors.

Last but not least, in the current case study, the organization and governance of the undertaken project (see Figure 1) were very similar to the Hybrid Stage-Gate approach. However, none of the respondents were familiar with its theoretical foundations, in contrast to the agile methodologies and frameworks. In general, considering the latter, the rationale behind its adoption was concerned with the aim of preventing and minimizing the risk of project failure. More specifically, the adoption of agile practices and techniques provided rewarding opportunities due to their emphasis on instant and close collaboration and communication, resulting in reducing, or even eliminating the information gap.

5. Discussion

This study contributes to the existing literature by reporting and adding an important new investigation regarding agile hardware development. The growing interest among researchers in revisiting existing agile methodologies in the new domains has opened up new frontiers. In light of the evidence derived from our case study, a better understanding of implementing agility in hardware manufacturing is provided, drawing on the reasons for and against its adoption, and indicating the practices, techniques and tools applied. Moreover, the areas positively impacted by the agile implementation are highlighted with the corresponding evaluation measures deployed, along with the barriers and obstacles encountered. These findings may be of great importance for planning and specifying possible future agile implementations.

Regarding the other related studies, Atzberger and Paetzold argue that "adaptations to the field of hardware development are inevitable, therefore, principles and practices are necessary for the companies to circumvent the hardware-related issues (. . .)" [51]. In this vein of argument, Sharifi and Zhang claim that "agility in manufacturing may be achieved through the implementation and integration of appropriate practices which provide the required abilities for a company to respond properly to changes" [70]. This claim has been confirmed by Montero et al., since "it is possible to use principles and practices of agile hardware development to clarify and promote the robustness" of manufacturing processes [8]. Our findings are consistent with those similar reported observations. By documenting the case of agile implementation within the hardware domain, we deliver one more piece of evidence for supporting the agile-oriented mindset.

Nevertheless, the current study suffers from two main limitations associated with the research methods applied. First, since we are aware that there is no way to generalize the results obtained from this study, there is a need to design and perform more research in similar settings. Second, the data analysis was reliant on the project documentation (secondary data) and interviews (primary data) with the selected project stakeholders, which followed the project closure. Therefore, both data sources were qualitative in nature. In the case of the former, relevant reports were carefully selected and evaluated, whereas the latter was deliberately scrutinized to validate the study's results. Nevertheless, there is a risk of bias related to the human factor which could occur during the process of information analysis and synthesis. To sum up, more research is needed to confirm (or disprove) our findings.

6. Conclusions

In this paper, we report on selected outcomes from an agile implementation in a hardware manufacturing project. More specifically, this research is an attempt to answer the questions regarding the feasibility and validity of the implementation of agile-oriented techniques and practices, supported by the deployment of relevant tools.

The business implications, in terms of benefits resulting from the implementation of the agile techniques and practices to hardware development, concern three main areas:

- achieving an effective risk mitigation strategy;
- greater adaptability and innovation capacity;
- effective communication and collaboration.

On the other hand, the business implications in terms of challenges are related to human nature, which is manifested by:

- resistance to change;
- a lack of or inadequate management support;
- poor communication and collaboration.

Moreover, with the agile approach and the change in the way of thinking and working, an organization is able to create an environment that is open to new opportunities arising for its employees, including training, promotion, flexible working, job rotation, and job enrichment. Nevertheless, any manager should ask herself/himself: is my team ready to become agile? Yet, the mindset shift is the hardest to make for leaders. During the transition, the key element is to change their leadership style from telling the team what to do to mentoring the team.

Several open issues and questions present interesting research avenues. In particular, there are still very few studies that have investigated domains not related to the IT industry. These poorly explored landscapes, which provide valuable opportunities for future research, can provide more lessons regarding the effectiveness of collaboration and communication through the lens of agile techniques and practices.

Funding: This research is the result of the project entitled "An Industry 4.0 Mobile Process Management System Supported by Artificial Intelligence", co-financed by the European Union from the European Regional Development Fund under the Regional Operational Program of the Mazovia Voivode ship for 2014–2020, under grant agreement no. RPMA.01.02.00-14-b528/18.

Institutional Review Board Statement: Not applicable.

Informed Consent Statement: Not applicable.

Data Availability Statement: Not applicable.

Acknowledgments: Special thanks are due to the two respondents for providing comprehensive and valuable answers to the questionnaire questions.

Conflicts of Interest: The author declares no conflict of interest. The funders had no role in the design of the study; in the collection, analyses, or interpretation of the data; in the writing of the manuscript, or in the decision to publish the results.

Appendix A. The Questionnaire

The questionnaire began with a short introduction that briefly describes the project. The first three parts are concerned with the following areas of the project undertaken: planning, execution, and evaluation. The last part aimed to collect the respondents' demographic data.

Part I. Agile Adoption Rationale and Selection.

Q1. What were the most important reasons for adopting Agile?

Code	Item	Answer [1]
O1	Accelerate software delivery	
O2	Better manage distributed teams	
O3	Better respond to volatile market conditions	
O4	Enhance ability to manage changing priorities	
O5	Enhance delivery predictability	
O6	Enhance software quality	
O7	Improve business and IT alignment	
O8	Improve engineering discipline	
O9	Improve project visibility	
O10	Improve team morale	
O11	Increase software maintainability	
O12	Increase team productivity	
O13	Reduce project cost	
O14	Reduce project risk	
O15	Other? Please specify:	

[1] The answers for the items (excluding O15) were predefined with the following list of five items: (1) Definitely Not, (2) Probably Not, (3) Possibly, (4) Probably Yes, (5) Definitely Yes.

Part II. Agile Methodology, Techniques and Practices Adoption.

Q2. Which agile approach was adopted?

(a) Scrum.
(b) ScrumBan.
(c) Kanban.
(d) Iterative.
(e) I don't know.
(f) None.

Q3. Which of the following Agile techniques and practices did you use?

Code	Item	Answer [2]
TP1	Agile Portfolio planning	
TP2	Agile/Lean UX	
TP3	Common work area	
TP4	Daily standup	
TP5	Dedicated customer/product owner	
TP6	Frequent releases	
TP7	Kanban	
TP8	Planning Poker/Team Estimation	
TP9	Product roadmapping	
TP10	Release planning	
TP11	Retrospectives	
TP12	Short iterations	
TP13	Single team	

Code	Item	Answer [2]
TP14	Sprint/Iteration planning	
TP15	Sprint/iteration reviews	
TP16	Story mapping	
TP17	Other? Please specify:	

[2] The answers for the items (excluding TP17) were predefined with the following list of five items: (1) Definitely Not, (2) Probably Not, (3) Possibly, (4) Probably Yes, (5) Definitely Yes.

Q4. Which Agile planning and delivery tools did you use?

Code	Item	Answer [3]
PD1	Agile Project Management Tool	
PD2	Automated Acceptance Test Tool	
PD3	Automated Build Tool	
PD4	Bug Tracker	
PD5	Continuous Integration Tool	
PD6	Customer Idea Management Tool	
PD7	Index Cards	
PD8	Kanban Board	
PD9	Product and Portfolio Management (PPM) Tool	
PD10	Product Roadmapping	
PD11	Refactoring Tool	
PD12	Release/Deployment Automation Tool	
PD13	Requirements Management Tool	
PD14	Spreadsheet	
PD15	Static Analysis	
PD16	Story Mapping Tool	
PD17	Task board	
PD18	Timecards	
PD19	Traditional Product Management Tool	
PD20	Unit Test Tool	
PD21	Wiki tool	
PD22	Wireframes	
PD23	Other? Please specify:	

[3] The answers for the items (excluding PD23) were predefined with a list of two items: (1) No, and (2) Yes.

Q5. Which agile planning tools did you use?

Code	Item	Answer [4]
PT1	Atlassian Jira	
PT2	Atlassian Jira Align	
PT3	Axosoft	
PT4	Azure DevOps	
PT5	Azure DevOps Server (Microsoft TFS)	
PT6	Broadcom Rally (CA Agile Central)	
PT7	Bugzilla	
PT8	Digital.ai Agility (formerly VersionOne)	
PT9	Digital.ai TeamForge (CollabNet TeamForge)	
PT10	Google Docs	
PT11	Hansoft	
PT12	HP Agile Manager	
PT13	HP Quality Center/ALM	
PT14	IBM Rational Team Concert	
PT15	In-house / self-made solution	
PT16	LeanKit	

Code	Item	Answer [4]
PT17	Microsoft Excel	
PT18	Microsoft Project	
PT19	Pivotal Tracker	
PT20	Targetprocess	
PT21	Trello	
PT22	Other? Please specify:	

[4] The answers for the items (excluding PT22) were predefined with a list of two items:(1) No, and (2) Yes.

Part III. Agile Adoption Evaluation.

Q6. Has the implementation of agile positively impacted each of the following areas within your company?

Code	Item	Answer [5]
IP1	Business/IT alignment	
IP2	Cost reduction	
IP3	Delivery speed/time to market	
IP4	Engineering discipline	
IP5	Managing changing priorities	
IP6	Managing distributed teams	
IP7	Predictability	
IP8	Risk reduction	
IP9	Software maintainability	
IP10	Software quality	
IP11	Team morale	
IP12	Team productivity	
IP13	Visibility	
IP14	Other? Please specify:	

[5] The answers for the items (excluding IP14) were predefined with the following list of five items: (1) Strongly disagree, (2) Disagree, (3) Neither agree nor disagree, (4) Agree, and (5) Strongly agree.

Q7. How has your organization measured the success of Agile delivery?

Code	Item	Answer [6]
ME1	Budget vs. actual cost	
ME2	Burn-up chart	
ME3	Business value delivered	
ME4	Cumulative flow chart	
ME5	Customer retention	
ME6	Customer/user satisfaction	
ME7	Cycle time	
ME8	Defect resolution	
ME9	Defects into production	
ME10	Defects over time	
ME11	Earned value	
ME12	Estimation accuracy	
ME13	Individual hours per iteration/week	
ME14	Iteration burndown	
ME15	Planned vs. actual release dates	
ME16	Planned vs. actual stories per iteration	
ME17	Product utilization	
ME18	Release burndown	
ME19	Revenue sales impact	
ME20	Scope change in a release	

Code	Item	Answer [6]
ME21	Team morale	
ME22	Test pass/fail over time	
ME23	Velocity	
ME24	WIP, Work in Progress	
ME25	Other? Please specify:	

[6] The answers for the items (excluding ME25) were predefined with a list of two items:(1) No, and (2) Yes.

Q8. What are the most significant barriers to adopting Agile practices?

Code	Item	Answer [7]
BA1	Fragmented tooling and project-related data/measurements	
BA2	General organization resistance to change	
BA3	Inadequate management support and sponsorship	
BA4	Inconsistent processes and practices across teams	
BA5	Insufficient training and education	
BA6	Lack of business/customer/product understanding	
BA7	Lack of skills/experience with agile methods	
BA8	Minimal collaboration and knowledge pairing	
BA9	Not enough leadership participation	
BA10	Organizational culture at odds with agile values	
BA11	Pervasiveness of traditional development methods	
BA12	Regulatory compliance or government issue	
BA13	Unwilling to admit mistakes and learn from delivery failure	
BA14	I don't know	
BA15	Other? Please specify:	

[7] The answers for the items (excluding BA15) were predefined with the following list of five items: (1) Strongly disagree, (2) Disagree, (3) Neither agree nor disagree, (4) Agree, and (5) Strongly agree.

Q9. What are the benefits of adopting the release planning (TP10) technique?

Code	Item	Answer [8]
B1-TP10	Eliminating Waste in the Planning Process	
B2-TP10	Increased Flexibility and Decreased Development Lead Time	
B3-TP10	Increased Motivation of the Developers	
B4-TP10	Mediation of multiple opinions	
B5-TP10	Increased efficiency by highlighting the goals or deadline for release	
B6-TP10	Making high-quality plans	
B7-TP10	Increased visualization potential of release planning problems	
B8-TP10	Setting clear expectations about the objectives and product features	
B9-TP10	Early identification of feature dependencies	
B10-TP10	Helping team members stay on track and complete their tasks on time	
B11-TP10	Supporting the identification and mitigation of potential issues and risks	
B12-TP10	Other? Please specify: ...	

[8] The answers for the items (excluding B12) were predefined with the following list of five items: (1) Definitely Not, (2) Probably Not, (3) Possibly, (4) Probably Yes, (5) Definitely Yes.

Q10. What are the benefits of adopting the short iterations (TP12) technique?

Code	Item	Answer [9]
B1-TP12	Provide more-frequent feedback from customers than long iterations	
B2-TP12	Afford the team more opportunities to reflect on and improve their work practices	
B3-TP12	Deliver a rapid response to changes in priority without disrupting work in process	
B4-TP12	Enable early problem detection since agile teams recognize process smells and act on them immediately	
B5-TP12	Foster scope management since it is easier to move the small backlog items around than the large ones	
B6-TP12	Eliminate internal process overheads by running short iterations since user stories are small units of work and the team can follow the iteration status in a simpler way.	
B7-TP12	The shorter the iteration, the less process overhead is needed to keep things on course.	
B8-TP12	Support product adaptability since it is easy to introduce new features, change existing features, and make any other functional change	
B9-TP12	Support the team's ability to work on several features in parallel	
B10-TP12	Other? Please specify: …	

[9] The answers for the items (excluding B10) were predefined with the following list of five items: (1) Definitely Not, (2) Probably Not, (3) Possibly, (4) Probably Yes, (5) Definitely Yes.

Q11. What are the benefits of adopting sprint/iteration planning (TP14)?

Code	Item	Answer [10]
B1-TP14	Helps teams control projects	
B2-TP14	Effective management of the product backlog	
B3-TP14	Positively influences product quality	
B4-TP14	Provides an opportunity for team members to participate in planning and hence better understand the work assignments	
B5-TP14	Provides opportunities to build team cohesion and identification	
B6-TP14	Helps in understanding the user stories	
B7-TP14	Serves to validate something we do not understand, or something misinterpreted	
B8-TP14	Planning meetings help to define the goals for each team member	
B9-TP14	Enables setting precise estimates for the task	
B10-TP14	Fosters knowledge-sharing and skill-improvement	
B11-TP14	Other? Please specify: …	

[10] The answers for the items (excluding B11) were predefined with the following list of five items: (1) Definitely Not, (2) Probably Not, (3) Possibly, (4) Probably Yes, (5) Definitely Yes.

Q12. What are the benefits of using a task board (PD17)?

Code	Item	Answer [11]
B1-PD17	Ensures efficient diffusion of information relevant to the whole team	
B2-PD17	Enables the team to stay focused on the work progress	
B3-PD17	Allows the team to represent any relevant information	
B4-PD17	Other? Please specify: …	

[11] The answers for the items (excluding B4) were predefined with the following list of five items: (1) Definitely Not, (2) Probably Not, (3) Possibly, (4) Probably Yes, (5) Definitely Yes.

Q13. What are the benefits of using an automated build tool (PD3)?

Code	Item	Answer [12]
B1-PD3	Saving time and money	
B2-PD3	By keeping a history of builds and releases, it helps in investigating bugs and errors	
B3-PD3	Increases developers' productivity since build automation ensures fast feedback	
B4-PD3	Accelerates product delivery by eliminating redundant tasks	
B5-PD3	Improves product quality by supporting bug and error detection	
B6-PD3	Other? Please specify: ...	

[12] The answers for the items (excluding B6) were predefined with the following list of five items: (1) Definitely Not, (2) Probably Not, (3) Possibly, (4) Probably Yes, (5) Definitely Yes.

Q14. What are the benefits of using a unit test tool (PD20)?

Code	Item	Answer [13]
B1-PD20	Early detection of the specification deficiencies	
B2-PD20	Fostering higher quality of individual components, and thus increase overall system resiliency	
B3-PD20	Increases testers' productivity	
B4-PD20	Cost reduction	
B5-PD20	Easier refactoring the source code	
B6-PD20	Other? Please specify: ...	

[13] The answers for the items (excluding B6) were predefined with the following list of five items: (1) Definitely Not, (2) Probably Not, (3) Possibly, (4) Probably Yes, (5) Definitely Yes.

Q15. In your opinion, considering the implementation of agile techniques and practices, is the sum of the benefits achieved greater than the sum of the costs encountered?

(1) Definitely Not, (2) Probably Not, (3) Possibly, (4) Probably Yes, (5) Definitely Yes.

Q16. In your opinion, what are the main challenges of running an Agile development project with remote teams?

Q17. What have been the most valuable lesson(s) learned in easing the adoption of agile practices and techniques?

Part IV. The Respondent's Data.

Q18. What is your age?

Q19. What is your sex?

Q20. How many years of experience in the IT sector do you have?

Q21. What was your role in the project?

Q22. Were you involved in the project throughout its duration?

References

1. Denning, S. How to Make the Whole Organization "Agile". *Strategy Leadersh.* **2016**, *44*, 10–17. [CrossRef]
2. Hohl, P.; Klünder, J.; van Bennekum, A.; Lockard, R.; Gifford, J.; Münch, J.; Stupperich, M.; Schneider, K. Back to the Future: Origins and Directions of the "Agile Manifesto"—Views of the Originators. *J. Softw. Eng. Res. Dev.* **2018**, *6*, 15. [CrossRef]
3. Kannan, V.; Basit, M.A.; Bajaj, P.; Carrington, A.R.; Donahue, I.B.; Flahaven, E.L.; Medford, R.; Melaku, T.; Moran, B.A.; Saldana, L.E.; et al. User Stories as Lightweight Requirements for Agile Clinical Decision Support Development. *J. Am. Med. Inform. Assoc.* **2019**, *26*, 1344–1354. [CrossRef] [PubMed]
4. Misra, S.; Kumar, V.; Kumar, U.; Fantazy, K.; Akhter, M. Agile Software Development Practices: Evolution, Principles, and Criticisms. *Int. J. Qual. Reliab. Manag.* **2012**, *29*, 972–980. [CrossRef]
5. Pikkarainen, M.; Salo, O.; Still, J. Deploying Agile Practices in Organizations: A Case Study. In *Software Process Improvement, Proceedings of the 12th European Conference, EuroSPI 2005, Budapest, Hungary, 9–11 November 2005*; Richardson, I., Abrahamsson, P., Messnarz, R., Eds.; Springer: Berlin/Heidelberg, Germany, 2005; pp. 16–27.
6. Why New Hardware Businesses Fail and How to Make Sure Yours Doesn't. Available online: https://predictabledesigns.com/13-reasons-why-hardware-startups-fail-and-how-to-make-sure-yours-doesnt/ (accessed on 12 April 2022).

7. Huang, P.M.; Darrin, A.G.; Knuth, A.A. Agile Hardware and Software System Engineering for Innovation. In Proceedings of the 2012 IEEE Aerospace Conference, Big Sky, MT, USA, 3–10 March 2012; pp. 1–10.
8. Joaquin, M.; Alexander, A.; Matthias, B.; Jens, H.; Kristin, P. Enhancing the Additive Manufacturing Process for Spare Parts by Applying Agile Hardware Development Principles. In Proceedings of the 2019 IEEE 10th International Conference on Mechanical and Intelligent Manufacturing Technologies (ICMIMT), Cape Town, South Africa, 15–17 February 2019; pp. 109–116.
9. The Tele and Radio Research Institute—PPTF. Available online: https://pptf.pl/en/pptf-members/the-tele-and-radio-research-institute/ (accessed on 14 May 2022).
10. We Are a Software Producer Using Artificial Intelligence—Meritus Sp. z o.o. Available online: https://meritus.pl/en/ (accessed on 16 May 2022).
11. Serova, E.; Kalmykov, O. On the Issue of Implementation of Agile and Strategy as a Practice Mixed-Method in Strategic Planning. In *Eurasian Business Perspectives, Proceedings of the 35nd Eurasia Business and Economics Society Conference, Rome, Italy, 7–9 April 2021*; Bilgin, M.H., Danis, H., Demir, E., Vale, S., Eds.; Springer: Cham, Switzerland, 2021; pp. 127–139.
12. Mkoba, E.; Marnewick, C. Conceptual Framework for Auditing Agile Projects. *IEEE Access* **2020**, *8*, 126460–126476. [CrossRef]
13. Jiménez, V.; Afonso, P.; Fernandes, G. Using Agile Project Management in the Design and Implementation of Activity-Based Costing Systems. *Sustainability* **2020**, *12*, 10352. [CrossRef]
14. Masood, Z.; Farooq, S. The Benefits and Key Challenges of Agile Project Management under Recent Research Opportunities. *Int. Res. J. Manag. Sci.* **2017**, *5*, 20–28.
15. Cram, W.A.; Marabelli, M. Have Your Cake and Eat It Too? Simultaneously Pursuing the Knowledge-Sharing Benefits of Agile and Traditional Development Approaches. *Inf. Manag.* **2018**, *55*, 322–339. [CrossRef]
16. Shameem, M.; Kumar, C.; Chandra, B.; Khan, A.A. Systematic Review of Success Factors for Scaling Agile Methods in Global Software Development Environment: A Client-Vendor Perspective. In Proceedings of the 2017 24th Asia-Pacific Software Engineering Conference Workshops (APSECW), Nanjing, China, 4–8 December 2017; pp. 17–24.
17. Benefits of Agile Project Management | Kanbanize. Available online: https://kanbanize.com/agile/project-management/benefits-of-agile (accessed on 23 April 2022).
18. Agile Alliance. Agile Alliance | 2015. Available online: https://www.agilealliance.org/ (accessed on 2 May 2022).
19. Paulk, M.C. Agile Methodologies and Process Discipline. *J. Contrib.* **2002**. [CrossRef]
20. de la Barra, C.L.; Crawford, B.; Soto, R.; Misra, S.; Monfroy, E. Agile Software Development: It Is about Knowledge Management and Creativity. In *Computational Science and Its Applications—ICCSA 2013, Proceedings of the 13th International Conference, ICCSA 2013, Ho Chi Minh City, Vietnam, 24–27 June 2013*; Murgante, B., Misra, S., Carlini, M., Torre, C.M., Nguyen, H.-Q., Taniar, D., Apduhan, B.O., Gervasi, O., Eds.; Springer: Berlin/Heidelberg, Germany, 2013; pp. 98–113.
21. Manifesto for Agile Software Development. Available online: https://agilemanifesto.org/ (accessed on 23 April 2022).
22. 12 Principles behind the Agile Manifesto | Agile Alliance. Agile Alliance | 2015. Available online: https://www.agilealliance.org/agile101/12-principles-behind-the-agile-manifesto/ (accessed on 1 May 2022).
23. Flewelling, P. *The The Agile Developer's Handbook: Get More Value from Your Software Development: Get the Best out of the Agile Methodology*; Packt Publishing Ltd.: Birmingham, UK, 2018; ISBN 978-1-78728-073-1.
24. Schwaber, K.; Sutherland, J. The Scrum Guide. In *The Definitive Guide to Scrum: The Rules of the Game*; Creative Commons: Mountain View, CA, USA, 2020.
25. Ahmad, M.O.; Markkula, J.; Oivo, M. Kanban in Software Development: A Systematic Literature Review. In Proceedings of the 2013 39th Euromicro Conference on Software Engineering and Advanced Applications, Santander, Spain, 4–6 September 2013; pp. 9–16.
26. Beck, K.; Hendrickson, M.; Fowler, M. *Planning Extreme Programming*; Addison-Wesley Professional: Boston, MA, USA, 2001; ISBN 978-0-201-71091-5.
27. 15th Annual State of Agile Report | Digital.Ai. Available online: https://digital.ai/resource-center/analyst-reports/state-of-agile-report (accessed on 23 April 2022).
28. Khan, A.A.; Shameem, M.; Kumar, R.R.; Hussain, S.; Yan, X. Fuzzy AHP Based Prioritization and Taxonomy of Software Process Improvement Success Factors in Global Software Development. *Appl. Soft Comput.* **2019**, *83*, 105648. [CrossRef]
29. Hallett, C. Best Practices in Industry and CSE Senior Design. Diploma Thesis, University of Nebraska-Lincoln, Lincoln, NE, USA, 19 April 2021.
30. What Is a Daily Scrum? Available online: https://www.scrum.org/resources/what-is-a-daily-scrum (accessed on 16 May 2022).
31. Lagerberg, L.; Skude, T. The Impact of Agile Principles and Practices on Large-Scale Software Development Projects: A Multiple-Case Study of Two Software Development Projects at Ericsson. In Proceedings of the 2013 ACM/IEEE International Symposium on Empirical Software Engineering and Measurement, Baltimore, MD, USA, 10–11 October 2013.
32. What Is an Agile Iteration? | Wrike Agile Guide. Available online: https://www.wrike.com/agile-guide/faq/what-is-agile-iteration/ (accessed on 16 May 2022).
33. What Is a Sprint? Available online: https://airfocus.com/glossary/what-is-a-sprint/ (accessed on 16 May 2022).

34. What Is an Iteration? *Agile Alliance | 2015*. Available online: https://www.agilealliance.org/glossary/iteration/#q=~{}(infinite~{}false~{}filters~{}(postType~{}(~{}\T1\textquoterightpage~{}\T1\textquoterightpost~{}\T1\textquoterightaa_book~{}\T1\textquoterightaa_event_session~{}\T1\textquoterightaa_experience_report~{}\T1\textquoterightaa_glossary~{}\T1\textquoterightaa_research_paper~{}\T1\textquoterightaa_video)~{}tags~{}(~{}\T1\textquoterightiteration))~{}searchTerm~{}\T1\textquoteright~{}sort~{}false~{}sortDirection~{}\T1\textquoterightasc~{}page~{}1) (accessed on 3 May 2022).
35. Dalton, J. Kanban Board. In *Great Big Agile: An OS for Agile Leaders*; Dalton, J., Ed.; Apress: Berkeley, CA, USA, 2019; pp. 187–188, ISBN 978-1-4842-4206-3.
36. Digital, M. Agile Concepts: The Scrum Task Board. Available online: https://manifesto.co.uk/agile-concepts-scrum-task-board/ (accessed on 16 May 2022).
37. Wirdemann, R. *Scrum Mit User Stories*; Carl Hanser Verlag: Munich, Germany, 2009; ISBN 978-3-446-41656-7.
38. Sreenivasan, S.; Kothandaraman, K. Improving Processes by Aligning Capability Maturity Model Integration and the Scaled Agile Framework®. *Glob. Bus. Organ. Excell.* **2019**, *38*, 42–51. [CrossRef]
39. Putta, A.; Paasivaara, M.; Lassenius, C. Adopting Scaled Agile Framework (SAFe) a Multivocal Literature Review. In Proceedings of the 19th International Conference on Agile Software Development: Companion, New York, NY, USA, 21–25 May 2018; pp. 1–4.
40. Alqudah, M.; Razali, R. A Review of Scaling Agile Methods in Large Software Development. *Int. J. Adv. Sci. Eng. Inf. Technol.* **2016**, *6*, 828–837. [CrossRef]
41. Core Values. Scaled Agile Framework. Available online: https://www.scaledagileframework.com/safe-core-values/#:~{}:text=The%20four%20Core%20Values%20of,participates%20in%20a%20SAFe%20portfolio (accessed on 4 May 2022).
42. Applying SAFe to Hardware Development. "Scaled Agile Framework". Available online: https://www.scaledagileframework.com/applying-safe-to-hardware-development/ (accessed on 4 May 2022).
43. Justice, J.; Sammicheli, P. *Scrum for Hardware*; Independently Published: Chicago, IL, USA, 2018; ISBN 978-1-983373-31-2.
44. Cooper, R.G. Perspective: The Stage-gate® Idea-to-launch Process—Update, What's New, and Nexgen Systems. *J. Prod. Innov. Manag.* **2008**, *25*, 213–232. [CrossRef]
45. Cooper, R.G.; Sommer, A.F. The Agile–Stage-gate Hybrid Model: A Promising New Approach and a New Research Opportunity. *J. Prod. Innov. Manag.* **2016**, *33*, 513–526. [CrossRef]
46. MAHD About Us: MAHD: The Definitive Agile for Hardware Development Resource. Available online: https://agileforhardware.org/introduction-to-the-mahd-framework/ (accessed on 14 April 2022).
47. Agile for Hardware Home > MAHD: The Definitive Agile for Hardware Development Resource. Available online: https://agileforhardware.org/ (accessed on 20 July 2022).
48. An Introduction to the MAHD Framework > MAHD: The Definitive Agile for Hardware Development Resource. Available online: https://agileforhardware.org/complete-agile-for-hardware-framework/ (accessed on 14 April 2022).
49. Lima, G.L.B.; Ferreira, G.A.L.; Saotome, O.; Da Cunha, A.M.; Dias, L.A.V. Hardware Development: Agile and Co-Design. In Proceedings of the 2015 12th International Conference on Information Technology—New Generations, Las Vegas, NV, USA, 13–15 April 2015; pp. 784–787.
50. Laanti, M. Piloting Lean-Agile Hardware Development. In *Proceedings of the Scientific Workshop Proceedings of XP2016*; Association for Computing Machinery: New York, NY, USA, 2016; pp. 1–6.
51. Atzberger, A.; Paetzold, K. Current Challenges of Agile Hardware Development: What Are Still the Pain Points Nowadays? *Proc. Des. Soc. Int. Conf. Eng. Des.* **2019**, *1*, 2209–2218. [CrossRef]
52. Atzberger, A.; Gerling, C.; Schrof, J.; Schmidt, T.S.; Weiss, S.; Paetzold, K. Evolution of the Hype around Agile Hardware Development. In Proceedings of the 2019 IEEE International Conference on Engineering, Technology and Innovation (ICE/ITMC), Valbonne Sophia-Antipolis, France, 17–19 June 2019; pp. 1–8.
53. Does Spacex Use Agile? Available online: https://www.EclipseAviation.com (accessed on 4 April 2022).
54. Tesla Agile Development: Product Management at Its Best | 280 Group. Available online: https://280group.com/product-management-blog/tesla-agile-development/ (accessed on 21 May 2022).
55. General Motors Accelerates Transformation. Available online: https://media.gm.com/media/us/en/gm/home.detail.html/content/Pages/news/us/en/2018/nov/1126-gm.html (accessed on 21 May 2022).
56. Agile Transformation Spotlight: The Success Story of Scrum & Toyota. Available online: https://www.scrum.org/resources/agile-transformation-spotlight-success-story-scrum-toyota (accessed on 21 May 2022).
57. The Tele and Radio Research Institute. Available online: https://itr.lukasiewicz.gov.pl/en/ (accessed on 13 April 2022).
58. About Us. Available online: https://lukasiewicz.gov.pl/en/about-us/ (accessed on 21 May 2022).
59. Hozdic, E. Smart Factory for Industry 4.0: A Review. *IJMMT* **2015**, *7*, 28–35.
60. Poulis, K.; Poulis, E.; Plakoyiannaki, E. The Role of Context in Case Study Selection: An International Business Perspective. *Int. Bus. Rev.* **2013**, *22*, 304–314. [CrossRef]
61. Yang, S.; MR, A.R.; Kaminski, J.; Pepin, H. Opportunities for Industry 4.0 to Support Remanufacturing. *Appl. Sci.* **2018**, *8*, 1177. [CrossRef]
62. Seno, L.; Zunino, C. A Simulation Approach to a Real-Time Ethernet Protocol: EtherCAT. In Proceedings of the 2008 IEEE International Conference on Emerging Technologies and Factory Automation, IEEE, Hamburg, Germany, 15–18 September 2008; pp. 440–443.

63. Abhish, K.; Rakesh, V. Real-Time Analysis of a Multi-Client Multi-Server Architecture for Networked Control Systems. *J. Eng. Appl. Sci.* **2015**, *10*, 7522–7526.
64. Mackey, D.A. Content Analysis. In *The Encyclopedia of Criminology and Criminal Justice*; John Wiley & Sons, Ltd.: Hoboken, NJ, USA, 2014; pp. 1–5. ISBN 978-1-118-51738-3.
65. Krippendorff, K. *Content Analysis: An Introduction to Its Methodology*; SAGE Publications: Thousand Oaks, CA, USA, 2018; ISBN 978-1-5063-9567-8.
66. Stemler, S. An Overview of Content Analysis. *Pract. Assess. Res. Eval.* **2019**, *7*, 17. [CrossRef]
67. Boyce, C.; Neale, P. *Conducting in-Depth Interviews: A Guide for Designing and Conducting in-Depth Interviews for Evaluation Input*; Pathfinder International: Watertown, MA, USA, 2006.
68. Turner, D.W., III. Qualitative Interview Design: A Practical Guide for Novice Investigators. *Qual. Rep.* **2010**, *15*, 754–760. [CrossRef]
69. Mousavi, M.S.; Fararouei, M.; Afsar-Kazerooni, P.; Nasirian, M.; Ghaem, H. Evaluation of Conducting Phone Interviews on Sexual Behavior: An Iranian Population-Based Study. *Shiraz E-Med. J.* **2022**, *23*. [CrossRef]
70. Sharifi, H.; Zhang, Z. Agile Manufacturing in Practice—Application of a Methodology. *Int. J. Oper. Prod. Manag.* **2001**, *21*, 772–794. [CrossRef]

applied sciences

MDPI

Article

Antivirus Evasion Methods in Modern Operating Systems

Dominik Samociuk

Institute of Informatics, Silesian University of Technology, 44-100 Gliwice, Poland; dsamociuk@polsl.pl

Abstract: In order to safeguard one's privacy while accessing the internet, it is crucial to have an antivirus program installed on the device. Despite their usefulness in protecting against malware, these programs are not foolproof. Cybercriminals have access to numerous techniques and tools for circumventing antivirus software, which can greatly aid them in their illicit activities. The objective of this research was to examine the most prevalent methods and tools for bypassing antivirus software and to demonstrate how readily accessible and simple they are to use. The aim of this paper is to raise awareness among readers about the associated risks and to assist internet users in protecting themselves from potential threats. The findings of the research indicate that the efficacy of evasion tools is positively correlated with their age and popularity. Tests have shown that, with the latest updates, contemporary antivirus software is capable of resisting virtually all of the tested methods generated using default settings. However, the most significant aspect of this paper is the section presenting experiments with basic but powerful modifications to established evasion mechanisms, which have been found to deceive modern, up-to-date antivirus software.

Keywords: network security; antivirus evasion; cybersecurity

Citation: Samociuk, D. Antivirus Evasion Methods in Modern Operating Systems. *Appl. Sci.* **2023**, *13*, 5083. https://doi.org/10.3390/app13085083

Academic Editors: Zbigniew Lubniewski, Tadeus Uhl and Przemysław Falkowski-Gilski

Received: 9 March 2023
Revised: 17 April 2023
Accepted: 17 April 2023
Published: 19 April 2023

1. Introduction

In the present day, the internet is widely accessible to people of all ages and backgrounds, allowing them to use it for communication, entertainment, and information acquisition. However, hackers also have access to the internet and can use malicious software to compromise user privacy and important files.

To protect against such attacks, users often install antivirus software that aims to reduce the risk of device infection. Downloading files from unknown sources, even well-known programs like web browsers, can lead to infection. Therefore, it is essential to download content from trusted and verified sources. If a suspicious file is downloaded, the antivirus software will typically display an appropriate message and move the infected file to quarantine, where it can be cleaned or deleted. User trust in antivirus software is essential, and regular updates of the signature database and the tool itself are important.

Hackers use various bypassing and evasion techniques to gain access to devices. Antivirus software developers continuously develop new security tactics and update signature databases to combat emerging threats. However, the increasing number of new viruses implies that current protection technologies may not always be sufficient [1].

This article discusses popular bypassing techniques and tools that should be detectable by most antivirus programs but also shows that complex attack chains that combine evasion techniques can bypass modern and commonly-used antiviruses. The paper is structured into nine sections: introduction, related work, security mechanisms, evasion techniques, testing environment, results of research on bypassing antiviruses, discussion, conclusions, and further research directions.

2. Related Work

The issue of circumventing antivirus security measures starts with understanding the behavior of malware samples [2]. A thorough analysis of malware is crucial for enhancing the operation of antivirus engines, ultimately improving user security and privacy. Techniques to bypass antivirus security measures have been developed since the inception of antivirus software. The perpetual competition between hackers and antivirus developers results in more advanced attacks.

In 2018, Kalogranis [3] tested four tools, namely AVET, peCloack.py, Shellter, and Veil-Evasion, used to bypass antivirus software. Kalogranis' results demonstrated that AVET and Veil-Evasion were efficient at bypassing antivirus software by 60%, while peCloack.py and Shellter were efficient at 40%.

In a subsequent paper [4], the authors compared the effectiveness of antivirus software bypassing tools on the Windows operating system with Kalogranis' work, extending the research. The authors repeated the tests on the tools used by Kalogranis, added a new antivirus bypass tool called TheFatRat, and used a payload generated with Metasploit to extend the research. Veil-Evasion and Shellter failed to bypass security. TheFatRat bypassed one, PeCloak.py four, while Avet bypassed five out of six antivirus software programs used.

Panagopoulos [5] conducted a study on bypassing antivirus software, utilizing malware generation tools such as TheFatRat, Phantom-Evasion, Hercules, and Veil-Evasion in the tests. Phantom-Evasion was the most effective tool, achieving around 65% efficiency, followed by Hercules with around 47%, TheFatRat with 22%, and Veil-Evasion with the lowest efficiency within the range of limits of 10%.

Another work [6] analyzed Bitdefender as one of the best antivirus solutions, and the developers decided to perform all tests only on this antivirus program. The malware generated was Remote Access Trojan (RAT), which was made available to the victim machine using the Apache server. Nine different antivirus bypass tools were tested, and the authors considered not only whether RAT would be detected by antivirus devices but also whether the antivirus would block the triggered Meterpreter session activated by RAT. The effectiveness of these tools was presented as a percentage of the number of methods used for a given tool.

The use of malware and other malicious software is a persistent threat to computer systems, and as a result, antivirus software has become an essential tool for protecting against these threats. However, as antivirus software becomes more advanced, so do the techniques used to bypass it. We explored research related to state of security of antivirus tools. There is also current research on anti-antivirus techniques.

In [7,8], authors presented a technique of obfuscation that involves modifying the code of a piece of malware to make it more difficult to detect. The modifications may include changing variable names, adding extraneous code, and using encryption to hide the true function of the code. Researchers have developed various tools for detecting obfuscated code, but the effectiveness of these tools depends on the level of obfuscation used.

In [9], the researchers presented trends in antivirus evasion techniques, while [10] proposed new details for fileless malware, a type of malware that does not rely on a file to infect a system. Instead, it resides in the system's memory, making it more difficult to detect. This technique has become more popular in recent years as antivirus software has become better at detecting traditional file-based malware.

In [11,12], the authors focused on polymorphic malware, which is a type of malware that can change its code on the fly to evade detection. This technique involves creating multiple variants of the malware, each with a different code signature. When the malware is executed, it selects one of the variants at random and executes it. This makes it more difficult for antivirus software to detect the malware, as each variant has a unique code signature.

Finally, Ref. [13] covered standard signature evasion, while [14] researched detection of malware run as a virtual machine.

From the literature review, it can be concluded that, while no antivirus is infallible, antivirus software bypass tools have their advantages and disadvantages. As observed, there is a significant difference in the effectiveness of certain antivirus software bypassing tools, which can be attributed to various factors such as research measures, test execution dates, the difference in masked malware, its version, the version of the tested antivirus software, and even the pool of tested antivirus solutions. Antivirus tools and tools designed to bypass them engage in an intense war, where one party exploiting the other's disadvantage can lead to a significant difference in results.

As shown, individual bypassing of antiviruses has been studied in the past for old versions of antiviruses. However, to the best of author's knowledge, to date, there has been no comprehensive research conducted on the combination of multiple antivirus bypass techniques. Although individual techniques have been studied and developed, the analysis of their effectiveness in combination has yet to be explored. Given the constantly evolving nature of malware and antivirus software, it is essential to investigate how various techniques can be combined to bypass multiple layers of protection. This research can contribute to improving the understanding of the vulnerabilities in antivirus software and developing more robust and effective security measures.

3. Antivirus Detection Mechanisms

The human immune system, like any system, is prone to errors, and can be compromised, leading to illnesses. Similarly, electronic devices can be compromised by malicious software, and they require protective measures. Antivirus software is essential for devices and should be given priority in terms of installed software. Antivirus software scans files and compares their contents with a blacklist database of known malicious codes. If it detects any potential threats, it removes or quarantines them for cleaning. Antivirus software functions similarly to a doctor by analyzing potential threats and determining how to resolve them. Understanding the methods used by antivirus software can help us better understand attackers' motives when creating new malware. This sections provides an overview of modern antivirus software techniques used to identify malware and protect electronic devices from potential threats.

3.1. Signatures

Signature-based malware detection is a traditional method that has been used for a long time. A signature is a unique pattern or code template of a virus that can be used to identify it [15]. The signature is a short fragment that does not reveal any useful information for virus writers to use. Antivirus software can compare the signature of a scanned file with all the signatures from its blacklist of suspicious files, also known as the signature database. If the signature being compared exists in the database, the scanned file is marked as malicious. This method is effective for threats that are already known, making it a popular threat search technique. However, signature-based detection has some drawbacks [16]. It is powerless against new threats and susceptible to modifications that mask the virus. Scanning a new or modified threat may prove unreliable until the signature database is updated with new or modified signatures. Additionally, storing all signatures leads to an exponential increase in their number, resulting in lower performance and longer search times. However, if the virus software change does not modify the signature directly, the signature can still be useful in identifying the malware.

3.2. Behavioral Detection

Behavioral recognition is a malware detection technique that is based on analyzing the behavior of programs when launched. It was presented by Fred Cohen [17] as a way of defining illegal services and distinguishing them from legal ones within a system. Unlike signature-based detection, which relies on unique patterns, the behavior detector collects programs with similar behavior, which can help in identify different threats based on one behavior signature [18]. This method is particularly useful for dealing with new threats and modifications, as it does not rely on a signature database. However, one disadvantage of behavioral recognition is that it may incorrectly classify safe programs as dangerous if they use device resources in a manner similar to known viruses, resulting in a false positive. Zahra Bazrafshan, Hashem Hashemi, Seyed Mehdi Hazrati Fard, and Ali Hamzeh also highlighted this advantage in their publication [19].

3.3. Heuristic Detection

Dmitry Gryaznov in [20] defined heuristics as a set of rules applied to a program to determine if it contains a virus or not. Seyed Mehdi Hazrati Fard, Ali Hamzeh, Zahra Bazrafshan, and Hashem Hashemi described heuristic-based detection as utilizing machine learning and data mining techniques to learn a program's intentions [18]. The learning model detects suspicious features found in new and modified viruses. Heuristic analysis is classified as static or dynamic. Static analysis compares the possible behavior of the decompiled code with the heuristics database, and dynamic analysis runs a program in an isolated environment to observe its operation [21]. The combination of static and dynamic heuristics allows for the assessment of new and modified threats without exposing the client's device. However, as Gryaznov and Fred Cohen noted, it is impossible to create an algorithm that can definitively distinguish between a virus and an uninfected program with 100% accuracy. Heuristic-based detection also has its limitations. Static analysis may return a false positive, as it can find similarity between safe files and the heuristics database. Dynamic analysis may fail to detect a virus that requires certain conditions to be met before being launched. To address these limitations, combining heuristics with signature-based detection can eliminate these drawbacks.

3.4. Sandboxing

In advanced threat detection methods, such as heuristic-based detection, Sandbox serves as a system that emulates the behavior of a host computer. It functions as a virtual machine where potentially harmful files are executed to monitor their actions. Sandbox offers a secure environment that isolates the executed programs from the client's device, thereby reducing the chances of device infection to negligible levels.

4. Antivirus Evasion Mechanisms

Although modern antivirus software is generally effective in protecting against most current threats, it is not immune to vulnerabilities in its code that can be exploited by hackers. In this section, we will discuss several popular techniques that can be used by hackers to exploit these vulnerabilities.

4.1. Signatures

A vulnerability in antivirus software can lead to the exploitation of the program's flaws by hackers. An example of such a vulnerability is the discovery that poorly-designed signatures in an antivirus program can be used as an attack tool against the user. Researchers have developed a method to obtain the exact virus signatures of a specific antivirus software. The method involves three steps: (1) detecting critical bytes in the malware using feedback from the antivirus, (2) aligning corresponding bytes from samples with the same pattern and combining them into one sequence, and (3) converting the linked sequences into the appropriate format to obtain the original signature. The obtained signatures can be used to implant them in secure data, resulting in the quarantining or deletion of such data, which can lead to significant problems. This vulnerability and its exploitation are detailed in [22].

Currently, there are numerous scientific works utilizing the detection mechanism of signatures for the aforementioned attacks. An exemplary framework was presented in [23], where the authors attacked operating systems by forcing the antivirus to perform specific actions harmful to the user.

4.2. Obfuscation

Obfuscation is a well-known technique employed by attackers to modify the code and signature of a virus to make it difficult for antivirus software to interpret the malicious code. This technique involves adding redundant code, modifying the case of a script, and reordering commands to create confusion and evade detection. Although obfuscation does not affect the malware's functionality, it may reduce its effectiveness [24]. However, relying solely on obfuscation to deceive antivirus software may not be sufficient, as modern antivirus software utilizes heuristic and behavior-based approaches to detect potential threats. To bypass antivirus detection, attackers often use obfuscation in combination with other techniques, such as polymorphism or metamorphism.

Another research study using a systematic approach for antivirus evasion using malware obfuscation was presented in [25]. The authors focused on comparing different mechanisms of obfuscation for bypassing virus detection on different platforms. Obfuscation is a wide term, so new kinds of methods for code obfuscation are presented in the literature, such as ROP gadgets used for antivirus evasion presented in [26].

4.3. Encryption

Encryption is a process that splits a virus into two components, namely the virus body and the decryption loop. The virus body is encrypted using a suitable method, such as XOR, while the decryption loop is a brief program responsible for encrypting and decrypting the virus body. The virus must be decrypted using the decryption loop before it can function. Similarly, an antivirus needs to obtain the decryption portion of an infected file to read its encrypted content, and only then can it compare the signatures [27].

Unlike code obfuscation, which aims to obscure code, many tools use encryption based on cryptographic principles to limit the possibility of static analysis of malware code. This includes both ready-made frameworks such as PEzoNG presented in [28], as well as implementations of well-known encryption algorithms within ransomware code. An analysis of the latter solution, using Cryptolock as an example, was conducted in [29].

4.4. Morphism

Morphism, specifically polymorphic viruses, is a complex technique that aims to make analysis more difficult by creating an unlimited number of different decryptors. Unlike encryption, no part of the virus is encrypted. Polymorphic viruses use numerous obfuscation techniques to change the appearance of the decryption code from copy to copy, making it harder for antiviruses to recognize the threat. In metamorphism, the entire program code changes instead of just the encryptor's code as in polymorphism. Each copy is built differently, with varying sizes or sequences of code, but its behavior remains unchanged. To detect a virus using metamorphism, advanced behavioral and heuristic detection engines are required [27].

Tools for detecting code morphism strive to keep up with the development of frameworks for generating encrypted code, such as the tool presented in [30] for changing shellcode with each subsequent run.

4.5. Process Injection

Process injection is a technique utilized to camouflage a malicious process by running it in the memory space of another process. Additionally, the injected process can inherit permissions of the host process, potentially providing a hacker with more opportunities. The difficulty in detecting this technique lies in the fact that malware can be injected into a program that operates similar processes. Malware can effectively deceive behavior-based analyses and operate surreptitiously for a prolonged period before being detected. The classic DLL injection is one of the most popular process injection methods. This involves writing the path to the malicious DLL library into the address space of another process, followed by executing a remote thread that calls the malicious library during the injection process [31,32].

Process injection is often used with additional techniques mentioned in previous subsections. Research on how to detect process injection, followed by the process injection mechanism itself, is presented in [33], where the authors tried to detect fileless malware with process injection in various forms. There are many other frameworks where process injection is used to create multi-process malware execution in combination with other malware evasion mechanims, such as the above-mentioned ROP programming in the ROPE framework in [34].

5. Research Environment and Toolset

Hackers utilize various tools to evade detection by antivirus engines. Some of these tools do not require a high level of expertise in cybersecurity. Inexperienced attackers can simply download the tool and follow the instructions provided by its creators to bypass weaker antiviruses. As technology improves and antivirus software is updated, some of these techniques become less effective over time. The current discussion focuses on describing the tools used in this research to provide better understanding of the tests that were performed.

5.1. Basic Tools

Kali Linux [35] is a Linux distribution that is free of charge and is based on Debian. It was launched in 2013 and is designed for advanced penetration testing and security auditing purposes. Kali provides a plethora of tools for tasks such as penetration testing, security research, reverse engineering, and computer forensics [36]. The operating system contains over 600 penetration tools, and it is important to mention that it is open-source software, which enables users to modify the software provided according to their requirements.

Metasploit is a widely-used framework by penetration testers and hackers that provides a vast array of tools for testing system vulnerabilities and breaking security [37]. With approximately 600 payloads, 2200 exploits, and 45 encoders, Metasploit is a powerful tool that is built into Kali Linux by default, making installation unnecessary. Once the user logs in to Kali Linux and enters "msfconsole" in the terminal, the tool is ready for use, and the user can listen for connections on the port generated by the payload. More details about the tool and its functionalities will be discussed in a subsequent section.

Msfvenom [38] is a payload generation tool provided by the Metasploit framework. It was utilized to generate payloads that were used in subsequent tests. The program can be invoked through the terminal by entering the Msfvenom command with the required arguments. The tool provides a range of options, including the generation of payloads, encoding, selecting the output file format, and adjusting the number of load coding iterations.

Shikata Ga Nai (SNG) [39] is a polymorphic XOR encoder with additive feedback that is available in the Metasploit framework. This tool has been around for some time but is still in use today. It employs code obfuscation techniques such as altering the order of code instructions, randomly changing registers, and adding junk code to evade signature recognition every time the code is compiled. Additionally, the additive feedback allows the output data to be incorrect in subsequent iteration stages when incorrect input data are received. This is due to the algorithm performing XOR operations on subsequent instructions, using a randomly-generated key, and then adding the current instruction to the key. Code decoding involves performing all steps in reverse order [40]. SNG can be used by adding an extra argument to the previously described example of payload generation, where the argument preceded by the -e flag selects the SNG encoder.

Example usage is as follows:

```
msfvenom -p windows/meterpreter/reverse_tcp LHOST=192.168.224.134 LPORT=8080
-e x86/shikata_ga_nai -f exe > defaultMsfvenom.exe
```

5.2. Antivirus Evasion Frameworks

Hyperion [41] is a real-time encoder that was developed in 2012. The program was initially designed for the Windows operating system. To install it on Kali Linux, it is necessary to install the Wine package [42], which enables support for Windows operations in a Unix environment. Once the Wine package is installed, Hyperion can be downloaded, extracted, and converted into an executable file. Thereafter, the steps involved in using Hyperion are straightforward. The command below can be used:

```
wine hyperion.exe payloadToEncrypt.exe encryptedPayload.exe
```

In the terminal, the use of the Wine package to handle the Hyperion executable is indicated by providing two consecutive arguments. The first argument is the executable file to be encrypted, and the second argument is the output file of the already-encoded executable. The operation of Hyperion can be divided into two parts, namely the encryptor and the container. The encryptor takes the executable file to be encrypted, computes its checksum, and adds it to the file in memory. It then generates a random key that is used to encrypt the input file using the AES-128 algorithm, along with the previously calculated checksum. The container acts as a decryptor, copying and decrypting the encrypted input file and running it. As it lacks the decryption key, it performs a brute-force key space check based on a checksum to determine if the key is valid. Although this may seem like a disadvantage, it actually protects against static and dynamic analysis by antivirus engines. Author Christian Ammann also mentions that the encrypted counterpart has an unknown signature and cannot be analyzed by heuristics, thereby protecting the binary code against reverse engineering or the replacement of encryption procedures with packers to reduce the size of the executable file [43]. However, since the document describing Hyperion is almost 10 years old, the claim about the unknown signature may no longer hold true.

TheFatRat [44] is a potent exploitation tool developed by Edo Maland, used for generating payloads that can run on various systems such as Linux, Windows, Mac, or Android [45]. It uses various evasion mechanisms such as obfuscation and cryptography techniques to hide the payload from antivirus scanners. This includes methods like encrypting the payload, encoding it in various formats, and using anti-debugging techniques. TheFatRat can modify the payload to bypass signature-based detection mechanisms used by antivirus software. This includes changing the file's signature or removing it altogether. Dynamic payload generation: The tool can generate payloads dynamically, meaning that each payload is unique and can not be detected by antivirus software that relies on pre-existing signatures. Polymorphic code: TheFatRat can generate polymorphic code, which means that the code changes each time it is executed, making it more difficult for antivirus software to detect. To install TheFatRat on a Kali Linux machine, the following commands are entered in the terminal:

```
git clone https://github.com/Screetsec/TheFatRat.git
cd TheFatRat
chmod +x setup.sh && ./setup.sh
```

After downloading TheFatRat repository and granting permission to execute the file, the installation process begins. Following this, the program update is carried out using the following commands in sequence:

```
cd TheFatRat
./update && chmod +x setup.sh && ./setup.sh
```

To initiate TheFatRat tool, the user needs to execute a command in the terminal with administrator privileges. The options and tools can be accessed by entering the corresponding numbers assigned to the exploits. To exploit PwnWinds, which is a backdoor program offering the same payload as previously-mentioned tools, option 6 is selected. TheFatRat employs the Msfvenom tool to create payloads. After selecting the desired option, the user is prompted to enter the IP and port for the attack, as well as the name of the output file. Depending on the option selected, the program may also require the user to choose the appropriate type of payload.

Shellter [46] is a software tool that facilitates the dynamic injection of a payload into a secure executable file. Payloads can be either created in Shellter or generated with other tools such as Msfvenom. However, it should be noted that Shellter is currently limited to supporting the injection of code only into 32-bit applications [47]. The installation of Shellter requires the previously-described Wine [42] package to be installed again. The Shellter tool can be installed by running the install command in the terminal.

After installation, the program can be initiated by running the Shellter command, which opens the program window. The user must then choose whether to use automatic or manual mode. For this study, the focus was on the automatic mode to demonstrate the ease of use of the tools employed. Next, the user selects the file to which the previously-generated payload will be injected. The program then prompts the user to enable "stealth mode", which ensures the proper operation of the infected program. Since the attacker typically wants the attack to go unnoticed, the user selects the 'Y' option. Following this, the program presents the option to use pre-existing payloads available in Shellter or to use the previously-generated payload. If the user chooses to use their payload, they can indicate it at this point. However, if a payload from the pre-existing list, such as the reverse_tcp meterpreter, is selected, the user must then indicate the host's IP and port.

Veil-Evasion [48] is a Python-based framework widely used to create payloads, which can effectively bypass antivirus protections [49]. It was designed to operate on Kali Linux, making installation a straightforward process by entering an install command in the terminal.

Once installed, the program can be started by typing the command "veil" in the console, which launches the program window displaying two tools. The tool of interest in this context is the Veil-Evasion option, which provides over 41 payloads, divided into 10 different programming languages, including C, PowerShell, Python, Ruby, and AutoIt. The user can list the payloads by using the "list" command and then selecting the desired payload using the "x" command, where "x" represents the assigned number of the payload. After selecting the payload, a window with various options for the generated payload appears, enabling the user to modify the payload to make it more unique and resistant to antivirus scanners. For payloads with a meterpreter console and reverse TCP connection, the required options are LHOST and LPORT, while Veil-Exploit offers many other optional settings. These include the use of an encoder, RAM checks, and the clicktrack option, which only launches the payload after a specific number of mouse clicks by the victim. After modifying the payload as per the attacker's requirements, the "generate" command generates the payload, and the user can specify the name of the output file and the desired extension.

6. Experimental Procedures and Results for Antivirus Bypass Mechanisms

In order to analyze the effectiveness of antivirus bypass mechanisms, tests were conducted on various antivirus software programs as well as tools developed to combat them. The focus of this section is to present the procedures for and results of these tests.

Attack flow is presented in detail in Figure 1. The first condition that a masking tool must meet to be effective is to bypass static scanning. This means that the software delivered to the victim's machine must be able to bypass the initial static scan to check whether the file is infected or not before the infected file is launched. If the software is detected at this stage, it is pointless to proceed, as the attacker's primary goal is to break through the antivirus defenses and enter the victim's system unnoticed. The second condition for an effective masking tool is to establish a connection to the meterpreter console while the malware is running on the victim's machine.

The research was repeated with the latest versions of both attack tools and antivirus software. In addition, the generated malware was also scanned using the antiscan.me online scanner to visualize the results for even more antiviruses. Usage of multiple antiviruses minimizes the risk of bias of chosen AVs in VM tests. From the obtained results, it can be concluded that, in fact, the best-performing antiviruses were used as stated by [50]. However, scans on virtual machines were given priority during the research. Both the antivirus and malware masking software were downloaded on the same day, meaning that they were the latest versions at the time of testing. However, during re-examination, only the antivirus software was updated to verify whether the results would differ after the antivirus update and whether the antivirus would recognize the old threat over time or not.

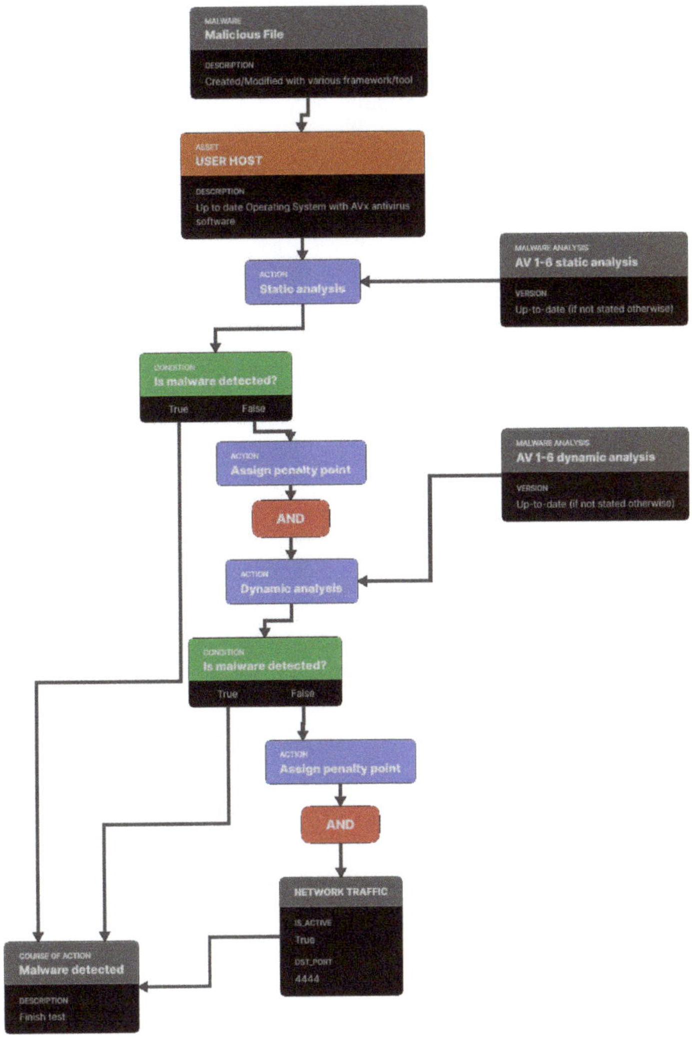

Figure 1. Attack flow and awarding of penalty points during the course of experiments.

Overall, these experimental procedures and results provide valuable insights into the effectiveness of antivirus bypass mechanisms and the importance of staying up to date with the latest versions of antivirus software.

The payload obtained with the Msfvenom tool on Kali Linux was generated as a default file and used as a reference for subsequent tests without any modifications. The payload was then delivered to each of the virtual machines to assess the ability of modern antivirus software to detect it. It was found that all 6 tested software programs immediately detected the potential threat, making it an ideal object for testing other tools. In contrast, an online scan of the payload revealed that only 20 out of 26 available antivirus engines were able to detect the threat, suggesting that the remaining 6 engines may have poor protection and may not be widely used. The same generated payload was used in all test cases; however, each framework modified it in different combinations of evasion mechanisms.

In this way, we compared combinations of antiviruses evasion mechanisms and not the sophistication of the payload itself.

The decision to test free antivirus software was based on a study of nearly 2000 respondents who indicated that users prefer the basic version over the paid one [51]. The choice of antivirus software for testing was informed by a 2021 report from AV-Comparatives [50], which presented the best-performing antivirus programs according to their tests. To maintain confidentiality and prevent misuse by potential hackers, the author of this work uses the acronyms AV1-AV6 to refer to the specific antivirus programs tested. Whole topology, dependencies and connections are presented in Figure 2. All VMs were created on single computer and with a shared hub-like connection to ensure that there were no other security or networking mechanisms blocking malware, such as a firewall or misconfigured routing.

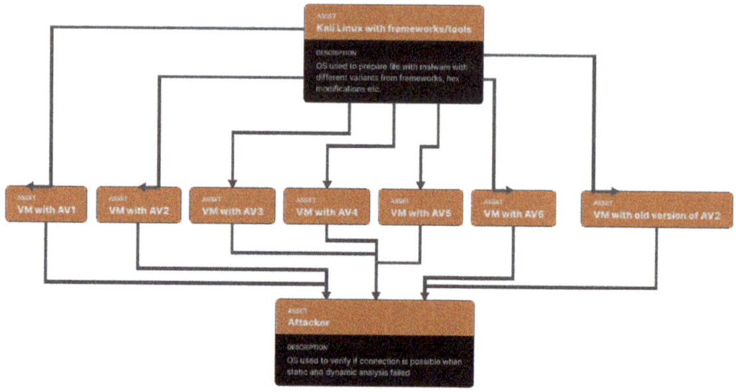

Figure 2. Dependencies between hosts in the created testing topology.

To begin the research, the latest version of the popular malware bypass tool, Hyperion, was downloaded and installed. The research consisted of three distinct parts. First, the payload generated previously was encrypted using Hyperion. The second part involved running the encrypted file in a hex editor to modify specific characters, changing commands to their direct equivalents rather than changing execution flow, in order to generate different hashes/signatures of the payload file. Finally, a new payload was generated using Msfvenom, this time utilizing the Shikata Ga Nai encoder on the payload. The already-encoded file was then subjected to Hyperion's encryption. In the end, all three infected files underwent antivirus scanning and online scanning to assess their effectiveness. Hyperion, being one of the oldest and most widely-used malware bypass tools, was chosen as the starting point for the research.

We used Y/N notation to represent whether the attack was detected (Y) or not (N). As presented in Table 1, the experimentation of expanding the basic payload with encryption or modifying the Hyperion code with a hex editor did not yield significant results for either static scan (S) or for run-time/active state (A). This is likely due to the fact that Hyperion, being an older software created in 2012, is relatively outdated compared to Msfvenom; therefore, similar results were expected. While obfuscation techniques were once effective, contemporary antivirus software has developed heuristic or behavioral detection capabilities that can identify even minor changes in the signature appearance. Notably, one antivirus engine was deceived during online scanning more than with the basic payload. This suggests that, while these basic mechanisms may not be effective against the best software on the market, they may have minimal impact on smaller software producers. In conclusion, it can be stated with confidence that none of the aforementioned tools are currently effective, thus leading to the abandonment of further experimentation with Hyperion.

Table 1. Hyperion results.

Payload	Online Scan	AV Results											
		AV1		AV2		AV3		AV4		AV5		AV6	
		S	A	S	A	S	A	S	A	S	A	S	A
Hyperion	19/26	Y		Y		Y		Y		Y		Y	
+SNG	19/26	Y		Y		Y		Y		Y		Y	
+Hex	19/26	Y		Y		Y		Y		Y		Y	

TheFatRat has been available for a considerable period of time and has a relatively large community, unlike Hyperion. The source code is accessible on GitHub and is regularly updated, which enhances its prospects of evading antivirus software. For the purpose of the study, the research focuses on the solutions that use the same payload as the one initially generated. TheFatRat uses Msfvenom to generate the reverse TCP payload, preserving the idea of the research. After installation and execution of TheFatRat in the console, the sixth option, PwnWinds, was selected from the menu to generate payloads for testing. The first was a file generated in the .bat extension based on PowerShell code, and the second was identical except for the selected port (all previous payloads were functioning on port 8080, whereas this one was set to port 4444). The option that generates a file with an .exe extension based on C# and PowerShell was then selected, followed by a file with an .exe extension based on the C language. To proceed, it was necessary to convert the .bat file to .exe using BatToExe Converter. After generating all the necessary files, the scanning process began.

The results presented in Table 2 are noteworthy. From the table, it is apparent that AV6 performed poorly in the comparison. It was unable to handle payloads generated using C and PowerShell languages. Interestingly, 8 months after the test, AV6 performance remained the same. However, it should be noted that only the static scan level was considered, and it was not possible to establish a meterpreter connection. Furthermore, the payload generated in pure C was also able to evade detection by AV6. A crucial observation is that a payload generated in the .bat extension using PowerShell was more likely to be detected when listening on port 8080 than on port 4444, which is the default port for attacks by various trojans. This finding may suggest that port 8080 was frequently used for attacks in the past, as it is the port commonly used by web browsers. As a result, processes launched by infected files may not have attracted much attention, and, consequently, antivirus developers may have been less attentive to this port in the past. However, half of the tested antiviruses did not detect the static load on port 4444 at the level of static scans.

Due to the longevity of research and the successful evasion of antiviruses, the authors decided to redo all tests after eight months, with updated version of antiviruses and their databases. The installed antivirus updates rendered all six tested antiviruses immune to the old threat. This indicates that antivirus updates are necessary and effective, as evidenced by the results of the online scan. From the results, it can be inferred that the antiviruses tested eight months ago were less effective than the newer ones, except for the generated .bat file that returned worse results in the second study, possibly due to a scan error or a problem with the software engine vendor. In conclusion, the only effective tool appears to be the payload generated in TheFatRat, based on the C language and PowerShell, but only when confronted with the AV6 antivirus.

In another study, we utilized Shellter software to inject a generated payload into a potentially safe program. The process began with downloading an appropriate program that would serve as the host for the payload injection. From a security point of view, the process should be not malicious. Standard processes without changes should not be flagged by antiviruses. The reason for utilizing Shellter was to change not-malicious process into malicious ones, hopefully also bypassing antiviruses. ColorPix was identified as a suitable program for this purpose, given its functionality as a color retrieval tool that does not require installation. However, we should note that any file will render, in most modern cases, the same results, as the process is not tested but the structure is, which has changed from the antivirus point of view. In other words, the payload should be detected, not the transportation mechanism—in this case, usage of ColorPix. Using Shellter, the basic payload was successfully injected into the downloaded program. To evaluate the effectiveness of the injected payload, we subjected it to static scans by six different antivirus software programs. Surprisingly, as shown in Table 3, the injected payload was able to evade detection by five of the six tested antivirus software programs, with only AV5 exhibiting immunity to the injected payload. These results suggest that Shellter software can be an effective tool for evading antivirus software detection, at least at the static scan level.

Table 2. TheFatRat results.

Payload	Online Scan	AV Results											
		AV1		AV2		AV3		AV4		AV5		AV6	
		S	A	S	A	S	A	S	A	S	A	S	A
Results obtained in January 2022													
TheFatRat 62	13/26	Y	Y	Y	Y	Y	Y	Y	Y	Y	Y	**N**	**N**
ThaFatRat 66	10/26	Y	Y	Y	Y	Y	Y	Y	Y	Y	Y	**N**	Y
ThaFatRat 61-B	9/26	Y	Y	Y	Y	**N**	Y	Y	Y	**N**	Y	**N**	Y
TheFatRat 61	15/26	Y	Y	Y	Y	Y	Y	Y	Y	Y	Y	Y	Y
Results obtained in September 2022													
TheFatRat 62	16/26	Y	Y	Y	Y	Y	Y	Y	Y	Y	Y	**N**	**N**
TheFatRat 66	14/26	Y	Y	Y	Y	Y	Y	Y	Y	Y	Y	Y	Y
TheFatRat 61-B	13/26	Y	Y	Y	Y	Y	Y	Y	Y	Y	Y	Y	Y
TheFatRat 61	14/26	Y	Y	Y	Y	Y	Y	Y	Y	Y	Y	Y	Y

Note: 62 - C+PowerShell; 66 - C; 61 - bat+PowerShell; B - Port 4444.

The experiment was repeated using the same payload configuration but with a .raw extension. The results were even more alarming, with half of the tested antiviruses cheated at the static scanning level, and the other half completely fooled, allowing Shellter to establish a connection to the meterpreter console and run directly on the victim's machine. Subsequent testing after an eight-month interval, with updated software and databases, revealed that the antivirus software had improved in detecting the payload. Only AV4 at the static scan level and AV3 at the static scan level and launch level were deceived by the payload based on the .raw extension. Shellter has the potential to deceive signature-based scanning by injecting payloads into countless programs, with each program displaying a different injected payload. Additionally, it pays little attention to the processes it employs, meaning it is able to deceive most static scanners, and potentially can deceive behavioral or heuristic scans. The study demonstrates that Shellter is an effective tool, initially fooling half of the antivirus software tested, but only one after a few months. This underscores the importance of installing appropriate antivirus updates.

Veil-Evasion is a framework similar to TheFatRat software that provides numerous solutions for bypassing antivirus security. Veil-Evasion has a significant following on GitHub, but the framework has not been updated since 2020, which may suggest a lack of active development. To test Veil-Evasion's effectiveness, three variants of generated payloads using Msfvenom were selected for study. The Veil-Evasion study was divided into three parts, including testing the basic payload generated in Python, generating the payload based on PowerShell and Bash, and manually editing the previously-generated payload with a hex editor. The results, shown in Table 4, indicate that each tested antivirus was effective in threat detection. Differences were observed in online scans, where the Python-based payload was more frequently detected by antivirus scanners.

The other two payloads had similar results, but an interesting situation emerged in re-testing after eight months with up-to-date antiviruses. As expected, more scanners detected the threat, but during this period, the payload generated with PowerShell and Bash was caught by one more scanner than the one that was manually edited in the hex. This finding suggests that some antivirus software did not receive an update, or it may be a distortion from the online scan site.

Table 3. Shellter results.

Payload	Online Scan	AV Results											
		AV1		AV2		AV3		AV4		AV5		AV6	
		S	A	S	A	S	A	S	A	S	A	S	A
Results obtained in January 2022													
Shellter PP	5/26	N	Y	N	Y	N	Y	N	Y	Y	Y	N	Y
Shellter PPR	2/26	N	N	N	N	N	N	N	Y	N	Y	N	Y
Results obtained in September 2022													
Shellter PP	15/26	Y	Y	Y	Y	Y	Y	Y	Y	Y	Y	Y	Y
Shellter PPR	3/26	Y	Y	Y	Y	N	N	N	Y	Y	Y	Y	Y

Note: PP—Primary payload, PPR—Primary payload with .raw extension.

Table 4. Veil-Evasion results.

Payload	Online Scan	AV Results											
		AV1		AV2		AV3		AV4		AV5		AV6	
		S	A	S	A	S	A	S	A	S	A	S	A
Results obtained in January 2022													
Python	13/26	Y	Y	Y	Y	Y	Y	Y	Y	Y	Y	Y	Y
Powershell + Bash	10/26	Y	Y	Y	Y	Y	Y	Y	Y	Y	Y	Y	Y
PS + Bash + Hex	10/26	Y	Y	Y	Y	Y	Y	Y	Y	Y	Y	Y	Y
Results obtained in September 2022													
Python	14/26	Y	Y	Y	Y	Y	Y	Y	Y	Y	Y	Y	Y
Powershell + Bash	13/26	Y	Y	Y	Y	Y	Y	Y	Y	Y	Y	Y	Y
PS + Bash + Hex	12/26	Y	Y	Y	Y	Y	Y	Y	Y	Y	Y	Y	Y

In order to be comparable to previous research and to allow for other researchers to reconstruct the authors' tests, the criteria for assessing the effectiveness of the masking tool and antivirus software are similar to those described in previous research. Assigning penalty points for failure to detect malware in static and dynamic scan is a state-of-the-art methodology and thus was used in the research. Instead of using quality measurements or other kinds of rankings, penalty points are direct values that can be compared with past research and can be used by other researchers or in future work. If the antivirus does not detect the malware during the static scan level, it receives one penalty point. If the antivirus does not detect the malware during the static and dynamic scan levels, it receives two penalty points. Similarly, if the payload is not detected during the static scan level, it receives one point, but if it completely bypasses antivirus security and connects to the meterpreter session, it receives two points. The more points the antivirus software receives, the less effective it is. Alternatively, the more points the masking technique obtains, the more dangerous it is.

Table 5 displays the effectiveness of the tested tools, with Shellter ranking as the most dangerous tool in the first two positions: first with a raw load and second with a basic load. TheFatRat with the payload generated using the C language and PowerShell remains in third place, while the other tested mechanisms appear to pose only a slight threat. It should be noted that the tested tools offer many more possibilities and modifications, but this study focuses on the simplest and most accessible solutions available on the web. This highlights the fact that even with limited knowledge, a potential attacker has all the tools necessary to threaten an unsecured user.

Table 5. Summary results for evasion tools and their combinations.

	Static	Runtime (active)	Summary
Basic payload	0	0	0
Hyperion	0	0	0
Hyperion + SNG	0	0	0
Hyperion + SNG + Hex	0	0	0
TheFatRat 62	0	1	**2**
TheFatRat 66	1	0	**1**
TheFatRat 61-B	3	0	**3**
TheFatRat 61	0	0	0
Shellter PP	5	0	**5**
Shellter PPR	3	3	**9**
V-E Python	0	0	0
V-E PS + Bash	0	0	0
V-E PS + Bash + Hex	0	0	0

To complete the study, older versions of the AV2 antivirus software were tested. Versions 2008 and 2016 were obtained from the website oldversions.com and installed on virtual machines. After installation, infected files were sent to the virtual machines, and the malware scanning process was initiated. The results, as shown in Table 6, indicate that the older antivirus software was unable to detect the threat, even in the case of the basic payload. This is expected, as Msfvenom was developed much later. However, AV2 from 2016 was able to detect only basic threats, which is consistent with the fact that all payloads that were not identified either appeared after 2016 or had multiple updates after that year.

Table 6. Old versions of AV2 results.

	AV2—Version from 2008		AV2—Version from 2016	
	S	A	S	A
Basic payload	N	N	Y	Y
Hyperion	N	N	Y	Y
Hyperion + SNG	N	N	Y	Y
Hyperion + SNG + Hex	N	N	Y	Y
TheFatRat 62	N	N	N	N
TheFatRat 66	N	N	N	N
TheFatRat 61-B	N	N	N	N
TheFatRat 61	N	N	N	N
Shellter PP	N	N	N	N
Shellter PPR	N	N	N	N
V-E Python	N	N	N	N
V-E PS + Bash	N	N	N	N
V-E PS + Bash + Hex	N	N	N	N

7. Discussion

This research has demonstrated that there are readily-available software tools that can be used to mask malicious files, effectively bypassing antivirus security regardless of the level of advancement of the attacker. The effectiveness of these tools depends on several factors, including the popularity of the tool, time, and how the tool works. When a given mechanism is frequently used, identical versions of malware may have been generated before, making the file useless if an antivirus engine detects it. The effectiveness of the software is also influenced by time, as antivirus software often introduces updates related to malware signatures, along with successive patches to fill defense gaps. This forces malware developers to update their programs with newer and newer solutions. The most important factor in the effectiveness of the cloaking mechanism is how it works. Changes in the hex editor or the use of an encoder to obfuscate the code are insufficient in the face of advanced heuristic searches combined with signature-based searches. Among the programs tested, Shellter was found to be the most concerning, as it can create a stable connection of meterpreter sessions on half of the tested antiviruses. The effectiveness of Shellter is also influenced by the injected payload, as the basic payload could bypass most antiviruses, but only on a static level, whereas the payload with the .raw extension could bypass all of them, half of which could be bypassed completely.

8. Conclusions

The study has demonstrated that, regardless of one's knowledge of cybersecurity, every individual is capable of infecting another person's device with varying degrees of success to steal or destroy valuable data. The tests conducted in this study employed a payload that establishes a connection between the victim's machine and the attacker's machine through a meterpreter session. This section of the study concentrates on several threats and commands that a hacker could utilize to launch an attack on someone else's machine. The objective of presenting these threats is to raise awareness of the extent of the threat posed by weak security and how much damage a hacker can inflict on a victim's machine.

9. Future Work

The study of bypassing antivirus mechanisms is an ever-evolving field, with new threats emerging daily and with antivirus software continuously attempting to thwart them. Further research is necessary to improve internet security by exploring more advanced techniques for bypassing antivirus software. The techniques examined in this study were either default ones or minimally modified, and there are many untested and lesser-known tools available on GitHub that could exploit previously-unknown vulnerabilities. Investigating how antivirus masking techniques function on file types other than payloads may also yield different results depending on the malware's triggered processes. While this study focused on Windows, attacks on other systems such as MacOS or Android may produce completely different outcomes. It is important to remember that, in this research, the performance of current computers is powerful and should not be a problem for modern antiviruses. However, in personal and/or Internet of Things devices, the relationship among security, usability, and performance should also be taken into consideration in subsequent research. As smartphones have become ubiquitous and are used for a variety of sensitive activities such as taking pictures, making calls, and conducting financial transactions, it is crucial to maintain the highest level of protection against potential hacker attacks.

Funding: This research received no external funding.

Institutional Review Board Statement: Not applicable

Informed Consent Statement: Not applicable

Data Availability Statement: Test results are not shared due to ethical restrication/misuse.

Acknowledgments: This research was funded by project SUBB-MN/BKM 2022, grant number 02/120/BKM22/0019.

Conflicts of Interest: The author declares no conflict of interest.

References

1. AV-Test Malware. 2022. Available online: https://www.av-test.org/en/statistics/malware/ (accessed on 15 April 2023).
2. Sharif, M.I.; Lanzi, A.; Giffin, J.T.; Lee, W. Impeding Malware Analysis Using Conditional Code Obfuscation. In Proceedings of the 4th Virus Bulletin International Conference, San Diego, CA, USA, 10 February–13 February 2008.
3. Kalogranis, C. AntiVirus Software Evasion: An Evaluation of the AV Evasion Tools. Ph.D. Thesis, University of Piraeus Department of Digital Systems, Piraeus, Greece, 2018.
4. Aminu, S.A.; Sufyanu, Z.; Sani, T.; Idris, A. Evaluating the effectiveness of antivirus evasion tools against windows platform. *J. Sci.* **2019**, *4*, 112–119.
5. Panagopoulos, I. Antivirus Evasion Methods. Ph.D. Thesis, University of Piraeus Department of Digital Systems, Piraeus, Greece, 2020.
6. Garba, F.A.; Yarima, F.U.; Kunya, K.I.; Abdullahi, F.U.; Bello, A.A.; Abba, A.; Musa, A.L. Evaluating Antivirus Evasion Tools Against Bitdefender Antivirus. In Proceedings of the International Conference on FINTECH Opportunities and Challenges, Karachi, Pakistan, 18 October–19 October 2021.
7. Collberg, C.; Thomborson, C.; Low, D. A taxonomy of obfuscating transformations. *ACM Comput. Surv.* **1997**, *29*, 1–69.
8. Chen, Y.; Liu, Y.; Chen, Y. A survey on android malware obfuscation. *J. Comput. Sci. Technol.* **2016**, *31*, 901–921.
9. F-Secure. State of Cyber Security 2018. 2018. Available online: https://www.f-secure.com/documents/996508/1030743/F-Secure_State_of_Cyber_Security_2018.pdf (accessed on 15 April 2023).
10. Mandiant. M-Trends 2018: A View from the Front Lines. 2018. Available online: https://www.fireeye.com/content/dam/fireeye-www/services/pdfs/mandiant-2018-mtrends.pdf (accessed on 15 April 2023).
11. Christodorescu, M.; Jha, S.; Seshia, S.A. Static analysis of executables to detect malicious patterns. In Proceedings of the 12th Annual Network and Distributed System Security Symposium (NDSS'05), San Diego, CA, USA, 27 February–3 March 2005; pp. 1–14.
12. Moser, A.; Kruegel, C.; Kirda, E. Limits of static analysis for malware detection. In Proceedings of the 23rd Annual Computer Security Applications Conference (ACSAC'07), Miami Beach, FL, USA, 10–14 December 2007; pp. 421–430.
13. Ligh, M.; Adair, S.; Hartstein, B.; Richard, M. *Malware Analyst's Cookbook and DVD: Tools and Techniques for Fighting Malicious Code*; John Wiley & Sons: Hoboken, NJ, USA, 2014.
14. Bishop, M.; Gates, C. Combating virtual machine detection and fingerprinting. *IEEE Secur. Priv.* **2009**, *7*, 56–60.
15. Kephart, J.O. Automatic Extraction of Computer Virus Signatures. In Proceedings of the 4th Virus Bulletin International Conference, Abingdon, UK, 1994; pp. 179–194.
16. Zemlyanaya, D.; Boldyrikhin, N.; Svizhenko, A.; Yukhnov, B. Virus signature detection algorithm. *Intell. Inf. Technol. Math. Model.* **2021**, *2131*, 022086. [CrossRef]
17. Cohen, F. Computer Viruses: Theory and Experiments. *Comput. Secur.* **1987**, *6*, 22–35. [CrossRef]
18. Bazrafshan, Z.; Hashemi, H.; Fard, S.M.H.; Hamzeh, A. A Survey on Heuristic Malware Detection Techniques. In Proceedings of the Conference on information and Knowledge Technology, Shiraz, Iran, 28–30 May 2013; pp. 113–120.
19. Cohen, F. Computer Viruses. Ph.D. Thesis, University of South California, Los Angeles, CA, USA, 1986.
20. Gryaznov, D. Scanners of The Year 2000: Heuristics. In Proceedings of the Fifth International Virus Bulletin Conference, Boston, MA, USA, September 1999; pp. 225–234.
21. Kaspersky. usa.kaspersky.com. Available online: https://usa.kaspersky.com/resource-center/definitions/heuristic-analysis (accessed on 15 April 2023).
22. Wressnegger, C.; Freeman, K.; Yamaguchi, F.; Rieck, K. From Malware Signatures to Anti-Virus Assisted Attacks. *arXiv* **2016**, arXiv:1610.06022.
23. Wressnegger, C.; Freeman, K.; Yamaguchi, F.; Rieck, K. Automatically inferring malware signatures for anti-virus assisted attacks. In Proceedings of the 2017 ACM on Asia Conference on Computer and Communications Security, Abu Dhabi, United Arab Emirates, 2–6 April 2017; pp. 587–598.
24. Balakrishnan, A.; Schulze, C. Code Obfuscation Literature Survey. *CS701 Constr. Compil.* **2005**, *19*, 31. Available online: https://pages.cs.wisc.edu/~arinib/writeup.pdf (accessed on 15 April 2023).
25. Kaushik, K.; Sandhu, H.S.; Gupta, N.K.; Sharma, N.; Tanwar, R. A systematic approach for evading antiviruses using malware obfuscation. In *Emergent Converging Technologies and Biomedical Systems: Select Proceedings of ETBS 2021*; Springer: Singapore, 2021; pp. 29–37.
26. Ntantogian, C.; Poulios, G.; Karopoulos, G.; Xenakis, C. Transforming malicious code to ROP gadgets for antivirus evasion. *IET Inf. Secur.* **2019**, *13*, 570–578. [CrossRef]
27. Rad, B.B.; Masrom, M.; Ibrahim, S. Camouflage in Malware: From Encryption to Metamorphism. *Int. J. Comput. Sci. Netw. Secur.* **2012**, *12*, 74–83.

28. Bernardinetti, G.; Di Cristofaro, D.; Bianchi, G. PEzoNG: Advanced Packer For Automated Evasion On Windows. *J. Comput. Virol. Hacking Tech.* **2022**, *18*, 315–331. [CrossRef]
29. Scaife, N.; Carter, H.; Traynor, P.; Butler, K.R. Cryptolock (and drop it): Stopping ransomware attacks on user data. In Proceedings of the 2016 IEEE 36th International Conference on Distributed Computing Systems (ICDCS), Nara, Japan, 27–30 June 2016; pp. 303–312.
30. Nicula, S.S. Generating antivirus evasive executables using code mutation. *Inform. Econ.* **2018**, *22*, 81–86. [CrossRef]
31. Redcanary.com. Process Injection. 2022. Available online: https://redcanary.com/threatdetection-report/techniques/process-injection/ (accessed on 15 April 2023).
32. Ashkan Hosseini. Ten Process Injection Techniques. 2017. Available online: https://www.elastic.co/blog/ten-process-injection-techniques-technical-surveycommon-and-trending-process (accessed on 15 April 2023).
33. Sanjay, B.N.; Rakshith, D.C.; Akash, R.B.; Hegde, V.V. An approach to detect fileless malware and defend its evasive mechanisms. In Proceedings of the 2018 3rd International Conference on Computational Systems and Information Technology for Sustainable Solutions (CSITSS), Bengaluru, India, 20–22 December 2018; pp. 234–239.
34. D'Elia, D.C.; Invidia, L.; Querzoni, L. Rope: Covert multi-process malware execution with return-oriented programming. In Proceedings of the Computer Security–ESORICS 2021: 26th European Symposium on Research in Computer Security, Darmstadt, Germany, 4–8 October 2021; Proceedings, Part I 26; Springer International Publishing: Berlin, Germany 2021; pp. 197–217.
35. Kali Linux Team. Kali Docs. 2022. Available online: https://www.kali.org/docs/ (accessed on 15 April 2023).
36. G0tmi1k. What is Kali Linux. 2022. Available online: https://www.kali.org/docs/introduction/what-is-kali-linux/ (accessed on 15 April 2023).
37. Rapid7. Metasploit. 2022. Available online: https://metasploit.com/ (accessed on 15 April 2023).
38. Rapid7. Msfvenom. 2021. Available online: https://github.com/rapid7/metasploit-framework/blob/master/msfvenom (accessed on 15 April 2023).
39. Rapid7. Shikata Ga Nai Repo. 2017. Available online: https://github.com/rapid7/metasploitframework/blob/master/modules/encoders/x86/shikata_ga_nai.rb (accessed on 15 April 2023).
40. Miller, S.; Reese, E.; Carr, N. Shikata Ga Nai Encoder Still Going Strong. 2019. Available online: https://www.mandiant.com/resources/shikata-ga-nai-encoderstill-going-strong (accessed on 15 April 2023).
41. Belial. Hyperion. 2020. Available online: https://nullsecurity.net/tools/binary.html (accessed on 15 April 2023).
42. Wine Team. WineHQ. 2022. Available online: https://www.winehq.org/ (accessed on 15 April 2023).
43. Christian Ammann. Hyperion: Implementation of a PE-Crypter. 2012. Available online: https://www.exploit-db.com/docs/english/18849-hyperion-implementation-ofa-pe-crypter.pdf (accessed on 15 April 2023).
44. Edo Maland. TheFatRat. 2021. Available online: https://github.com/screetsec/TheFatRat (accessed on 15 April 2023).
45. BlackHat. TheFatRat. 2020. Available online: https://www.blackhatethicalhacking.com/tools/thefatrat/ (accessed on 15 April 2023).
46. kyREcon. Shellter. 2022. Available online: https://www.shellterproject.com/introducing-shellter/ (accessed on 15 April 2023).
47. kyREcon. Shellter Readme. 2022. Available online: https://www.shellterproject.com/Downloads/Shellter/Readme.txt (accessed on 15 April 2023).
48. VeilEvasion Team. Veil. 2020. Available online: https://github.com/Veil-Framework/Veil (accessed on 15 April 2023).
49. Srinivas. Antivirus Evasion Tools. 2021. Available online: https://resources.infosecinstitute.com/topic/antivirus-evasion-tools/ (accessed on 15 April 2023).
50. AV-Comparatives. Summary Report 2021. 2021. Available online: https://www.av-comparatives.org/tests/summary-report-2021/ (accessed on 15 April 2023).
51. Statista Research Department. Which Version of Antivirus Program Do You Use on Personal Devices? 2021. Available online: https://www.statista.com/statistics/1252587/free-vs-paid-antivirus-usage-by-brand-russia/ (accessed on 15 April 2023).

MDPI
St. Alban-Anlage 66
4052 Basel
Switzerland
www.mdpi.com

Applied Sciences Editorial Office
E-mail: applsci@mdpi.com
www.mdpi.com/journal/applsci

www.ingramcontent.com/pod-product-compliance
Lightning Source LLC
Chambersburg PA
CBHW061834210926
51850CB00006B/888